THE PHILOSOPHY AND PRACTICE OF WILDLIFE MANAGEMENT

THIRD EDITION

THE PHILOSOPHY AND PRACTICE OF WILDLIFE MANAGEMENT

THIRD EDITION

by

Frederick F. Gilbert, PhD
Lakehead University
Ontario

and

Donald G. Dodds, PhD
Cervid Consulting
Nova Scotia

KRIEGER PUBLISHING COMPANY
MALABAR, FLORIDA
2001

Original Edition 1987
Second Edition 1992
Third Edition 2001

Printed and Published by
**KRIEGER PUBLISHING COMPANY
KRIEGER DRIVE
MALABAR, FLORIDA 32950**

Copyright © 1987, 1992, 2001 by Krieger Publishing Company

All rights reserved. No part of this book may be reproduced in any form or by any means, electronic or mechanical, including information storage and retrieval systems without permission in writing from the publisher.
No liability is asssumed with respect to the use of the information contained herein.
Printed in the United States of America.

FROM A DECLARATION OF PRINCIPLES JOINTLY ADOPTED BY A COMMITTEE OF THE AMERICAN BAR ASSOCIATION AND A COMMITTEE OF PUBLISHERS:
This publication is designed to provide accurate and authoritative information in regard to the subject matter covered. It is sold with the understanding that the publisher is not engaged in rendering legal, accounting, or other professional service. If legal advice or other expert assistance is required, the services of a competent professional person should be sought.

Library of Congress Cataloging-In-Publication Data

Gilbert, Frederick F., 1941-
    The philosophy and practice of wildlife management / Frederick F. Gilbert and Donald G. Dodds.—3rd ed.
      p. cm.
    Includes bibliographical references and index.
    ISBN 1-57524-051-3
    1. Wildlife management.   I. Dodds, Donald G.   II. Title.

SK355 .G45 2000
639.9—dc21
                                                                  00-041216

Contents

Preface to the Third Edition xi

**INTRODUCTION—HEARING OUR VOICES—
WHAT DO THEY SAY?** 1

Thoughts Concerning Wildlife Professionals and Natural Systems 1
Our Experience 3
Progress 4
Responsibilities 5
References 7

**1. THEN AND NOW—WILDLIFE MANAGEMENT,
JURISDICTIONAL RESPONSIBILITIES,
LEGISLATION, AND ADMINISTRATION** 9

Perspectives from History 9
The Evolution of Modern Wildlife Management—United States 14
The Evolution of Modern Wildlife Management—Canada 19
Jurisdictional Responsibility in the United States and Canada 21
Legislation: The United States and Canada 24
The Land and Wildlife: Ownership, Rights, and Restrictions 26
Administration and Policy: The United States and Canada 28
Nongovernment Wildlife Organizations in
   the United States and Canada 36
Bibliography 37
Recommended Readings 39

**2. MAN AND WILDLIFE—CULTURE, CONFLICTS, AND
VALUES** 41

Religion, Man, and Wild Animals 44

v

	Conflict: Man Versus the Animals	46
	Questions of Responsibility	47
	Attitudes and Values	51
	Culture, Conflicts, and Values—An Example	61
	Bibliography	63
	Recommended Readings	64
3.	**NATIVE AMERICAN ACCESS TO WILDLIFE**	**67**
	Before the Europeans	67
	After European Contact	69
	Government Policies in the United States and Canada	70
	Some Examples of the Native American–Wildlife Complex	72
	Into the Future	77
	Note	78
	Bibliography	78
	Recommended Readings	80
4.	**SOME BIOLOGICAL BASES FOR AND APPROACHES TO MANAGEMENT**	**83**
	Computer Models	83
	Nutrition and Energetics	86
	Behavior	91
	Populations	98
	Resource Partitioning	100
	Bibliography	100
	Additional Readings	110
5.	**PARASITES, DISEASE, AND WILDLIFE**	**111**
	Rabies	112
	Canine Distemper	114
	Hemorrhagic Disease	115
	Bluetongue	115
	Arboviruses	116
	Rinderpest	116
	Canine Parvovirus	117
	Pasteurellosis	118
	Tularemia	119
	Brucellosis	120
	Anthrax	120
	Yersinosis or Pseudotuberculosis	122

	Chronic Wasting Disease	123
	Ectoparasites	123
	Lyme Disease	125
	Elaeophorosis	125
	Parelaphostrongylosis	126
	Lungworm-Pneumonia Complex	129
	Strongyloidosis	132
	Summary	132
	Bibliography	135
	Additional Readings	152
6.	**MANAGEMENT SYSTEMS**	**153**
	Some Management Principles	155
	Regulatory Management	158
	Bibliography	163
7.	**HABITAT MANAGEMENT**	**165**
	Habitat Evaluation and Land-Use Planning	165
	Land-Use Planning Controls	173
	Purchase of Property Rights	174
	Incentive Programs	175
	Declining Wildlife Habitat—Old-Growth Coniferous Forest	176
	Habitat Management—Can It Be Achieved?	183
	The Lakeshore Capacity Study	186
	Other Approaches	187
	Bibliography	189
	Recommended Readings	194
8.	**SPECIES MANAGEMENT**	**197**
	Ungulates—Caribou (*Rangifer tarandus*)	197
	Bibliography	202
	Marine Mammals—Harp Seal (*Pagophilus groenlandicus*)	204
	Bibliography	206
	Furbearers—Beaver (*Castor canadensis*)	206
	Bibliography	210
	Waterfowl—Black Duck (*Anas rubripes*)	210
	Bibliography	213
	Upland Game Birds—Ruffed Grouse (*Bonasa umbellus*)	215
	Bibliography	217

Raptors—Peregrine Falcon (*Falco peregrinus*)   217
    Bibliography   220
Black-footed Ferret—(*Mustela nigripes*)   220
    Bibliography   223

9. SOME SPECIALIZED AREAS OF MANAGEMENT   225

Protected Areas   225
    Bibliography   231
Exotic Species   232
    Bibliography   236
Shooting Preserves and Put-and-Take   237
    Bibliography   239
Migratory Animals   239
    Bibliography   242
Urban Wildlife   242
    Bibliography   245
Depredations   246
    Bibliography   250
Humane Trapping   251
    Bibliography   256

10. ENDANGERED SPECIES—SOME MANAGEMENT STRATEGIES   257

The Convention on International Trade in Endangered Species   264
Strategies for Wildlife Managers   267
Protection Through Area Preservation: Parks, Preserves, and Buffer Zones   268
Protection Through Use: Utilization and the Marketplace   270
Protection of Captives: Species Survival Plans   274
Reintroductions   277
Bibliography   279
Recommended Readings   282

11. ENVIRONMENTAL IMPACT ASSESSMENT (EIA)—THE NEW DIMENSION   283

The EIA in the United States—Background   284
The EIA in Canada—Background   287
Some Approaches to the EIA   290
The Process—From the Beginnings   297
The Assessment Proposal   299
The Environmental Impact Statement   302
Assessment Problems   304

Different Viewpoints	306
The Administrator's Perspective	306
The Proponent's Perspective	306
The Consultant's Perspective	307
The Research Scientist's Perspective	307
Areas of Concern to the Wildlife Biologist	308
The Future of the EIA in North America	311
The EIA Elsewhere	312
Bibliography	314
Recommended Readings	316
12. WILDLIFE INTERNATIONAL AID POLICIES, PROBLEMS, AND MANAGEMENT	**319**
Wildlife Policy and Management: Some Concerns in Africa	324
Wildlife Management: The Status in Central Europe and Elsewhere	328
A Problem to Ponder	334
Bibliography	337
Recommended Readings	339
Epilogue	341
Bibliography	343
Appendix: Important Canadian and United States Federal Legislation Affecting Wildlife Resources and Management	345
Index	349

Preface to the Third Edition

This book is not meant to be exhaustive in any one area but endeavors to expose the preprofessional and professional wildlife biologist to concepts necessary for effective wildlife management. It is meant to be thought provoking rather than to provide a rigid model. We assume that students who read this book have already been adequately exposed to such subject areas as animal physiology, ecology, wildlife nutrition, and wildlife biology. We are attempting to apply this basic knowledge to management of wildlife species. The constraints, be they social, economic, political, or biological, need to be addressed in individual cases. An effective manager balances these objectives and communicates them to different interest groups. Where compromise is possible, it should be employed. Where special interests must be met, (e.g., endangered species), the manager must eloquently explain why other interests cannot be accommodated. A wildlife manager is a spokesperson for wildlife, a custodian of a variety of public interests, and a key element in the decision-making process regarding land use. To meet effectively this demanding role requires basic knowledge of wildlife species and their biological requirements and a consummate skill in people management. We hope this book provides sufficient insight into the appropriate mechanisms for achieving these aims. Both of us share a background of public, private, and institutional involvement. We have worked for state, provincial, and federal agencies, have consulted for private developers, park services, and foreign governments, and have been heavily involved in the academic milieu. We hope this experiential background coupled with our knowledge of wildlife has allowed us to impart some useful tips to aspiring and practicing wildlife managers.

We are grateful to the many people without whom this project would never have come to fruition. We acknowledge their contributions and in the case of the third edition especially thank Dennis Colson and Bill Foreyt for their critical input.

Finally, there is something to be said for coauthoring a book. With all the other demands on our time it is unlikely that either of us, alone, would have been able to complete the task of any of the editions of this book. The sense of obligation hung heavy at times, and the end result benefited immensely from the constant critical review of two writers. We each lost some of our sensitivity to protecting our "sacred prose" and also discovered unknown idiosyncrasies in our writing styles. We hope that you, the reader, find the effort to have been worthwhile.

Two final comments are necessary. Wherever the masculine gender has been used throughout the text it is merely as a convention of the English language, with no pejorative connotation intended. Citations have been minimized to increase readability and the reader is directed to the bibliographies within each chapter for all the literature used as reference material. Most of these same comments were made at the beginning of the first edition of this book and they are still valid. It is exciting to have the opportunity to fine tune our work and update it one final time. What is almost overwhelming is the bulk of new material which our profession continues to generate and the magnitude of the environmental issues. It is difficult to remain optimistic about the fate of our planet and the species *Homo sapiens* when reviewed through the myriad problems we face, most of which we have created. We hope that other wildlife managers may find some succour, some commiseration, and perhaps even some enlightenment from our words. If so, the task of rewriting the book will have been worth the efforts.

The book has chapters leading the reader from the historical bases of wildlife management and the evolution of wildlife legislation and administration in North America through some of the more controversial management problems at the turn of the century. Chapter 1 explores the special relationships that exist between the human species and wild creatures from prehistoric to modern times. Although the emphasis is on North American culture and the various user societies which have existed there, parallels can be found in other geographical areas of the planet. Chapter 2 details the development of wildlife management as a special discipline, showing the relationship to other environmental concerns and the underlying legislative and administrative structures in Canada and the United States. The issues surrounding the rights of native Americans as they relate to wildlife is the basis of Chapter 3. Chapter 4 examines the necessary biological understanding that underpins effective wildlife management. It highlights rather than details the intricate and comprehensive knowledge that must ultimately be made available at the species and community levels. Chapter 5 closely examines the issues of disease and parasitism as they affect wildlife populations. Regulatory management systems are looked at briefly in Chapter 6. Chapter 7 reviews habitat management primarily from an assessment perspective for integration into the land-use planning process. Several species examples and specialized areas of management are highlighted in Chapters 8 and 9. Endangered species management is reviewed in Chapter 10, and Chapter 11 looks at environmental impact assessment as it impinges on the professional wildlife manager. Chapter 12 is an attempt to expose the reader to the international arena and the socioeconomic realities of wildlife management in the underdeveloped countries of Africa and some of the sophisticated management approaches in Europe. Finally, the Epilogue is our attempt to predict the future role of the wildlife manager and the changes to which he will be exposed or will help bring about.

We have enjoyed the collaborative experience of writing this book and the sharing of ideas garnered during our separate professional careers—careers which have had many surprising parallels and have touched each other academically and professionally several times in the past four decades. Our experiential base has depended on our interactions with so many other professionals throughout North America and the rest of the world that we are really products of these reciprocal influences, our past and present management responsibilities, and our attempts to transmit this information to students in a university environment.

INTRODUCTION—Hearing Our Voices—What Do They Say?

Thoughts Concerning Wildlife Professionals and Natural Systems

"Here, then, is our idea of man's relationship to Nature: Man is within Nature as an active, functional being. But his natural propensity is to control Nature through transformative activities whose goals cannot themselves be ascertained in Nature. Man is both within Nature, and without it. His relationship to Nature is therefore distinctively dualistic."

—Thomas Colwell, Jr.

In this quotation from Thomas Colwell, Jr. (1970) it is implicit that goals, in regard to nature, are the goals of man. Evolution is not, we know, directed toward predetermined ends. Nature exists and living things change with time according to the dictates of natural selection, sometimes interrupted by cataclysmic occurrences and sometimes by a rather rapid adaptation to necessity within the realm of a single species. It is, rather, our own propensity to alter natural systems and humankind's goals, or lack thereof, with which we are concerned.

There are many philosophical works on man and nature that have influenced generations past and present. It may be helpful for us to consider the insights into man and nature some of them have provided, for their concerns may be as relevant today as when they were written.

Humans probably always have sought to understand their relationship with natural systems. Though early humans were surely concerned primarily with successful predation upon other creatures, wonderment must have existed at times concerning the birth and death of living things and of natural events like earthquakes, lightning, rain, wind, and fire. We know that as man evolved, his earliest depictions of life did indeed portray many such thoughts and concerns.

It is natural for us to ponder the sphere of reality in nature. Reason requires this to be so. Thus, in the realm of nature and man, Spinoza (cited in Wild, 1930) held that "we should seek the knowledge of the union existing between the mind and the whole of nature" and though he was a pantheist, believing that God is All, in fact known as a man intoxicated with God, his belief that what is morally good for humanity can be found in a study of nature is a belief often held today by many humans from pantheists to atheists. It was, in fact, a feeling oft expressed with great emotion by many people of various atheistic, agnostic and theistic views in the 1960s who rebelled against all they believed was destroying our natural systems.

For the utilitarian John Stuart Mill (1950), however, nature was to be used for human good. All human action he said, consists in altering, and all useful action in improving the spontaneous course of nature, and he also noted that man should be perpetually striving to amend the course of nature "and bringing that part of it over which we can exercise control, more nearly into conformity with a high standard of justice and goodness." Mill's philosophy became one of the foundation blocks for the research into, and, eventually, control over, natural phenomena which has so improved the human existence in the nineteenth and twentieth centuries. How much that improvement occurred at the expense of our natural systems was and continues to be a concern of those seeking solace in nature.

Almost 250 years before Mill, Francis Bacon (cited in Burtt,1939) had also evidenced a utilitarian approach in his reasoning about nature. Among his *Aphorisms Concerning the Interpretation of Nature and the Kingdom of Man*, Bacon stated what we may feel to be the obvious when he noted that, man, being the servant and interpreter of nature, can do and understand so much and so much only as he has observed in fact or in thought of the course of nature; beyond this he neither knows anything nor can do anything. But this was followed with a clear statement suggesting intervention and manipulation when he wrote, "Neither the naked hand nor the understanding left to itself can effect much. It is by instruments and helps that work is done, which are as much wanted for the understanding as for the hand. And as the instruments of the hand either give motion or guide it, so the instruments of the mind supply either suggestions for the understanding or cautions."

Many who consider the complexities of man and nature today believe that Spinoza was right and that much good for man may be found in nature. Thus solutions to practical environmental problems might exist within some of the changing value systems that have been proposed by ecologists since the ecological revolution of the late 1960s and early 1970s. But perhaps because the conflict of man being both within and outside of nature does, in a sense, truly exist, the changes in human values so many of us thought were real in 1970, in regard to our environments, were apparently not real at all. Or, if real, they were real for a minority of people only. The majority of the developed world's citizenry were not ready then to be jerked into an age of environmental mythology. And perhaps this was for the best, for seeds of change in public attitudes were planted then and now the more viable seedlings are beginning to bear fruit. We are making some progress in developing the institutions and structures within, and between, states to deal with nature and with the assaults upon our natural systems that will help both wild things and humankind realize the maximum good. But we have, of course, only begun and in our profession of wildlife management questions can justly be posed. As carriers of real environmental torches in our every day work, have we done enough? What is our role today? Can we not do more? Must we not do more?

Our Experience

In our profession the duality of man, both within and outside of nature, has certainly been recognized, if not understood, for some time. And, perhaps no wildlife professional has ever echoed this complex within us more effectively than Aldo Leopold. In *Thinking Like a Mountain,* Leopold (1949) relived this contradiction for millions of readers when he wrote these passages that have become so familiar to so many.

> "A deep chesty bawl echoes from rimrock to rimrock, rolls down the mountain, and fades into the far blackness of the night. It is an outburst of wild defiant sorrow, and of contempt for all the adversities of the world.
> Every living thing (and perhaps many a dead one as well) pays heed to that call. To the deer it is a reminder of the way of all flesh, to the pine a forecast of midnight scuffles and of blood upon the snow, to the coyote a promise of gleanings to come, to the cowman a threat of red ink at the bank, to the hunter a challenge of fang against bullet. Yet behind these obvious and immediate hopes and fears there lies a deeper meaning, known only to the mountain itself. Only the mountain has lived long enough to listen objectively to the howl of a wolf.
> In those days we had never heard of passing up a chance to kill a wolf. In a second we were pumping lead into the pack, but with more excitement than accuracy: how to aim a steep downhill shot is always confusing. When our rifles were empty, the old wolf was down, and a pup was dragging a leg into impassable slide-rocks.
> We reached the old wolf in time to watch a fierce green fire dying in her eyes. I realized then, and have known ever since, that there was something new to me in those eyes—something known only to her and to the mountain. I was young then, and full of trigger-itch; I thought that because fewer wolves meant more deer, that no wolves would mean a hunter's paradise. But after seeing the green fire die, I sensed that neither the wolf nor the mountain agreed with such a view."

And although Leopold understood our contradiction in 1949 and probably had since the 1920s, as did a few other wildlife workers, we have, as a profession, failed over these many years to effect the changes in our approaches to land and water our natural systems cry out for. So, like the dying wolf, our marshes and estuaries and old-growth forests continue to slip into oblivion and the dependent seals or whales or primates, the plankton and tropical forests, die their unnatural deaths.

But there is hope. The wildlife biologist and ecologist know something of nature and wild things. We understand something of how natural systems function and something of the requirements of species and populations. And we are individually committed to our work. So, while we may admit to contradictions within us in our understandings of nature and humankind, we also know that our professional approaches to management of habitats, wild populations, and human users of wild resources should seek to benefit all in a balanced manner. We know that we will not, by our actions, eliminate portions of delicate systems or endangered populations and that our flexible management hand must be light at times where necessary, yet intensive when otherwise required. We leave it to others to ponder man's relationship with the systems we study and manipulate. Ours is a dedication to the lands, the waters, the wild things, and to the humans who benefit. We are, indeed, within nature, a part of the systems we occupy. We are also managers of those same systems. We are dedicated to both sameness and change. Sameness that will guarantee the lives of natural systems and change that alters other systems for the betterment of man. But we know, too, that man cannot exist in a world bereft of wild creatures and their homes—these natural systems we deal with.

In our efforts to right the wrongs done to our lands, waters, and wildlife, the manager is the key professional among us. Wildlife managers are at the fulcrum between the research biologist and the policy maker where a tilt in the balance may either hinder or help. It is the manager's responsibility to determine the kinds of information needed to help a system or a population prosper and to pass these needs on to those involved in research and to apply results effectively to manage both habitats and animals.

The manager is also the one who understands what changes in direction are required in policy, what to protect and how best to do it, what to grow, what to harvest, what to introduce, what to reintroduce, what to manage in the wild, and what to leave alone. He is the one to advise policy makers who then must try and effect the means and seek to provide the material goods through appropriate legislation, structures, and policy.

Progress

Today, management of our wildlife must often be accommodated within parameters set by the management of other resources. This is particularly true in regard to our forests. As the manager's knowledge of minimum habitat size for life forms increases, so will we be better able to integrate our forest and wildlife management across the landscape. Managers should be working toward integration in both silvicultural and harvest planning, projected decades down the road to ensure adequate continued existence of systems in perpetuity while required wood is continuously removed. Wildlife management must also continue to adapt to changing agricultural methods and changing land use patterns in managing indigenous species wherever possible, and where appropriate, managing populations of the most suitable exotics. It is our professional managers, as well, who should be in the vanguard reclaiming surface mine areas, land, and water, for recolonizing wildlife communities. And where development continues to expand, whether it be housing, industry, or roads, it is the professional wildlife manager who should be there advising to help mitigate impacts on populations, habitats, and, when possible, to enhance conditions for wildlife. We should always be ready to try and make certain that wastelands are no longer spawned by careless developers or insensitive officials. Almost every urban area can serve as a habitat for some wild creatures and almost every development, whether it be an apartment complex, a dam, a mine, or a smelter can be made less of an assault on natural systems by an appropriate siting, by adequate waste disposal, and by harmonic design and construction, while adjacent landscapes can often be improved for certain species of wildlife. Wildlife management today and tomorrow must be coordinated, side by side, with our economic developments.

In considering such coordination, governments around the world are developing guidelines and strategies basic to planning for sustainable systems management and the resulting maintenance of both species and genetic diversity. The stimulus for this action comes from the World Conservation Strategy of 1980, the Brundtland Report (*Our Common Future*) of 1987, *Caring for the Earth, A Strategy for Sustainable Living,* published in 1991, and the 1992 United Nations Conference on the Environment and Development. The structures contributing to policy development are roundtables, advisory councils, government agencies, and think tanks comprised of stakeholders from native persons, naturalists, and environmentalists to professional resource persons, business, and industry. These are beginnings, perhaps suggesting that our dualistic relationship, within and outside of nature, with its attendant conflicts of seeking goodness as we continue to conquer and suppress, are beginning to be recognized and considered seriously.

Responsibilities

Wildlife biologists, by the nature of their profession, must assume a large share of the responsibility to help alleviate the effects of a mountainside scarification or the draining of a marshland system. The businessman and the politician are less likely to feel the touch of nature upon their shoulders for it may be their lot to understand only the domination and control of nature, seeking good largely in resultant profits or power. So, we have a certain struggle ahead. It will not be easily won.

In facing up to our challenge, we have a responsibility not to indulge in environmental or wildlife radicalism. Ours is an honest struggle, based on facts as we know them from our understanding of the diversity and complexity of the systems we struggle to manage. Our path is one of continuity, perhaps pragmatic, but seldom compromising. Our tolerances for compromise exist only within those ecological parameters that allow for the continued integrity of natural systems. And if we continue to speak honestly and rationally in resource matters, our voices will be positive influences on governments and publics the world over.

Ours is also a collective responsibility to all species and habitats. For us, wild species and populations must be paramount. Ours is a struggle for the many, for local and regional populations, the habitats and the species, and the systems upon which life depends.

Sometime before the year 1850, William Wordsworth sat with quill in hand and penned lines probably more appropriate to us today than they were when he wrote them. Wise words remain wise for any generation.

> "The world is too much with us: late and soon,
> Getting and spending, we lay waste our powers:
> Little we see in Nature that is ours;
> We have given our hearts away, a sordid boon!
> The Sea that bares her bosom to the moon;
> The winds that will be howling at all hours,
> And are up-gathered now like sleeping flowers;
> For this, for everything, we are out of tune."

Perhaps, as we struggle to evolve the institutions, structures, and policies needed to establish the wildlife manager's input into the oversight of our natural systems, we can help humanity harmonize with nature just a little bit better. It may be that someday within the life span of students today, the dualism of humankind—a part of us within and a part of us, seemingly, outside of nature—will be so perfectly understood that our habitats and our wildlife will indeed prosper alongside appropriate human developments for the benefit of people everywhere. To this end we must keep the fierce green light alive in the eyes of wolves. We must remember the mountain.

Adapted from paper: Hearing Our Voices—What Do They Say? Thoughts Concerning Game Biologists and Natural Systems. Trans. XIX IUGB, Trondheim. 1989. (Pub. Trans. 90).

References

Bacon, F. 1939. Aphorisms Concerning the Interpretation of Nature and the Kingdom of Man. In: E. A. Burtt (ed.), *The English Philosophers from Bacon to Mill*. Random House, Inc.
Colwell, T. B., Jr. 1970. Ecology and Philosophy. In: E. Laszlo and J. B. Wilbur (eds.), *Human Values and Natural Science*. Gordon and Beach.
Leopold, A. 1949. Thinking Like a Mountain. From: *A Sand County Almanac: and Sketches Here and There*. Oxford University Press.
Mill, J. S. 1950. Nature. In: J. H. Randall, Jr., J. Bachler, E. U. Shirk (eds.) *Readings in Philosophy*. Barnes and Noble, Inc.
Wild, J. (ed.). 1930. *Spinoza: Selections*. Charles Scribner's Sons.

1 THEN AND NOW—Wildlife Management, Jurisdictional Responsibilities, Legislation, and Administration

Perspectives from History

When Noah released the animals from the ark he was in fact reintroducing birds and mammals. According to Genesis 8:19: "All the animals and birds went out of the boat in groups of their own kind." Since Noah had been told to protect "two pairs of each kind of ritually clean animal, but only one pair of each kind of unclean animal" (Genesis 7:12), we can also assume that wild animals had been recognized and used, by species, according to religious law for some time previously. Regulatory harvesting was the first manipulation of wildlife that can be inferred from the Biblical record. Regulatory management reappears in Leviticus 11:4–6 and again in Deuteronomy 14:4–20. These verses indicate that specific animals were designated fit for human consumption, including deer, wild sheep, and wild goats. Others, including the rabbit and rock badger (probably the hyrax), were declared unfit. Then in Deuteronomy 22:67, the management becomes somewhat more sophisticated when Moses includes a recognition of the need to keep the breeding females alive: " . . . If you happen to find a bird's nest in a tree or on the ground with the mother bird sitting either on the eggs or with her young, you are not to take the mother bird. You may take the young birds, but you must let the mother bird go . . ." Probably some kind of harvest selection has occurred since our Neanderthal ancestors hunted, although the taking of one species, sex, or age rather than another may well have been more a function of evolved animal behavioral patterns than a design established by predatory man. With the arrival of Cro-Magnon man about 35,000 B.C., harvest selection was modified as a component of ritual as cultures evolved and replaced one another. Similar ritual-oriented selection continues today among many cultures exhibiting a primitive existence.

The written historical record of wildlife management beginnings was reviewed by Aldo Leopold in the first chapter of his classic work, *Game Management* (Table 1.1). Chronological examples for the United States until 1900

Table 1.1 — Some Wildlife Management Beginnings[1]

Source	Approximate Dates	Original Concern	Type of Action	Category of Management
Solon	About 600 BC	People	Forbade people to hunt	Regulatory
Marco Polo	Late 13th century	Animals	Restricted hunting and planted grain	Regulatory and habitat
Edward, Duke of York	14th century; recorded 1406–1413	Animals and privileged people	Controlled methods, seasons, sex & ages taken	Regulatory
Henry VII	1485–1509	People	Trespass protection	Regulatory
Henry VIII	1536 ±	Animals and privileged people	Closed seasons and areas	Regulatory and reservation of land
James I	1603–1625	Animals and privileged people	Trespass protection and closed areas	Regulatory and reservation of land
	1631	Animals	Artificial propagation	Stocking
William and Mary	1694	Animals	Prohibited burning of cover	Regulatory and habitat
Mulmesbury	1799	People	Cover control for efficiency of harvesting	Regulatory

[1] In part from Leopold, 1933. Game Management. Scribners, N.Y.

are presented in Table 1.2, and a similar chronological list for Canada appears in Table 1.3. A literature survey turns up many management happenings such as the ones we have chosen to list. Alison (1978), for example, traced Egyptian hunting records to the sixth dynasty (2625–2475 B.C.) and notes records left by Greeks, Etruscans, Persians, and Aztecs. In Egypt, an office of government was established to deal exclusively with waterfowl hunters and the marshes, implying management.

With the exception of Kublai Kahn, who engaged in both regulatory and habitat management; Henry VIII, who established reserves; and James I, who propagated and stocked mallards, the management activities between 1200 and 1800 were primarily regulatory in nature. Often the attempts to restrict harvest or prevent trespass were implemented for the benefit of the privileged classes rather than to enhance the populations or habitats of animals. So it was the general rule in Europe that regulatory measures were initiated by kings or others in high authority. Bubenik (1976) noted that to the end of the Middle Ages, the right to reduce game to possession was the "privilege of sovereign" who entrusted the right to "the highest aristocracy, clergy, and later on to free towns."

In the United States, the early history is almost solely regulatory in nature and severe penalties were handed out to offenders. For example, the Act for the Preservation of Deer, passed at the first convening of the Vermont legislature in February 1779, protected bucks, does, and fawns from 10 January to 10 June, and it was noted that anyone convicted of violating the act "shall forfeit and pay for every such offence, the sum of fifteen pounds." Half of the fine went to the prosecutor and half to the town treasury. If the offender was unable to pay, he was "assigned in service" to the complainer or another person for a "sufficient term." Royal Provincial Law enacted in 1741 in New Hampshire protected deer from 31 December until 31 August and provided for a fine of ten pounds or forty days work for the local government. For a second offense the penalty was increased to fifty days work! Each town in New Hampshire chose two officers known as deer reeves or deer keepers to enforce the law and to prosecute violators. They were also given authority to search premises without warrant.

To control wolves and panthers in 1779, Vermont offered a bounty of "eight pounds paid out of the public treasury; and half so much for every wolf's whelp that sucks, which he shall kill and destroy" (Slade 1823). To prevent fraud the heads were brought to selectmen or constables who cut off the ears. Stealing from another's pit or trap brought a penalty of whipping "on the naked back,—not exceeding ten stripes." Fines of about twenty-five dollars (ten pounds) amounted to several weeks' salary for the working man in 1779. Forty days' work was an approximate equivalent of the fine. A bounty on wolves

Table 1.2—Early Legislation Dates Reflecting the Beginnings of Wildlife Management in the United States to 1900[1]

Source	Date	Concern	Area	Management
Trippensee, 1948	1623	People	Plymouth Colony	Free hunting and fishing—no mgmt.
	1646	Deer	Portsmouth, R.I.	Regulatory (closed season)
Dasmann, 1981[1]	1677	Exports	Connecticut	Regulatory
	1699	Deer	Virginia	Regulatory
Leopold, 1933	1718	Deer	Massachusetts	Regulatory
Dasman, 1981	1730	Deer	Maryland	Regulatory
	1738	Deer	Virginia	Regulatory ("firelight" prohibited)
McClintock, 1888	1741	Deer	New Hampshire	Regulatory (with wardens)
Slade, 1823	1779	Deer	Vermont	Regulatory
	1779	Wolves-panthers	Vermont	Regulatory
Phillips, 1928	1789	Exotics	New Jersey	Stocking (ring-necked pheasant)
Leopold, 1933	1790	Exotics	New Jersey	Stocking (Hungarian partridge)
	1850	Game	Massachusetts	Regulatory (protection)
	1850		New Hampshire	Regulatory (warden system)
	1850	Nongame	New Jersey	Regulatory
	1864	Game	New York	Regulatory (license required)
Dasmann, 1981	1872	Wildlife	Wyoming	Preserve—National Park (Yellowstone)
Leopold, 1933	1875	Game and nongame	Arkansas	Regulatory (market hunting prohibited)
	1878	Game	Iowa	Regulatory (bag limits)
Dasmann, 1981	1878	Game	California	State game agencies
			New Hampshire	State games agencies
Trefethen, 1961	1885	Animals	United States	Bureau of Biological Survey
Severinghaus, 1974	1887	Deer	New York	Parks for propagation
Dasmann, 1981	1900	Game	United States	Regulatory (Lacey Act, prevented interstate trade of game taken illegally)[2]

[1]In part, from Graham. E. H. 1947. The Land and Wildlife. Oxford. N.Y.

[2]One of the most important pieces of legislation for wildlife in the United States. It elevated the Bureau of Biological Survey to be a strong base for the Fish & Wildlife Service and prevented importation of foreign species. This act also curtailed market hunting.

Table 1.3 — Early Legislation Dates Reflecting the Beginning of Wildlife Management in Canada to 1900

Source	Date	Concern	Area	Management
Dagg, 1974	1793	Wolves	Ontario-Upper Canada	Predator control (bounty)
Clarke, 1976	1821	Deer	Ontario-Upper Canada	Regulatory
Dagg, 1974	1821	Game	Ontario-Upper Canada	Regulatory
Dodds, 1982	1839	Wolf	Newfoundland	Predator control (bounty)
Benson and Dodds, 1980	1843	Moose	Nova Scotia	Regulatory (methods of harvest)
Dagg, 1974	1856	Game	Ontario-Upper Canada	Regulatory
Benson and Dodds, 1980	1862	Caribou	Nova Scotia	Regulatory
Dodds, 1960	1864	Snowshoe hare	Newfoundland	Stocking (introduction)
Dagg, 1974	1865	Exotic	Quebec-Lower Canada	Stocking (house sparrow)
Dagg, 1974	1867	Resources (wildlife)	Canada	Provincial jurisdiction
Benson and Dodds, 1980	1874	Deer	Nova Scotia	Regulatory (closed season)
Pimlott, 1953	1878	Moose	Newfoundland	Stocking (introduction)
Nova Scotia (Statutes of)	1884	Game species	Nova Scotia	Regulatory (license required)
Benson and Dodds, 1980	1894	Deer	Nova Scotia	Stocking (reintroduction)
Dagg, 1974	1895	Game	Ontario	Regulatory-Game and Fisheries Act
Dagg, 1974	1895	Game species	Ontario (Pelee)	Stocking
Nova Scotia (Statutes of)	1896	Game	Nova Scotia	Regulatory-Game and Fisheries Act
Piers, 1898	1897	Exotic	Nova Scotia	Stocking (ring-necked pheasants, introduction)
Piers, 1898	1898	Starp-tailed grouse	Nova Scotia	Stocking (introduction)

was established in Ontario in 1793 and even though the animal was becoming increasingly scarce, a bounty was placed on wolves in 1839 in the Colony of Newfoundland. Deer were first protected, from 10 January to 1 July, in Ontario in 1821, and regulatory management was extended to the whole of Upper Canada in 1856. Penalties were also severe in Lower Canada and late in the nineteenth century the Consolidated Statutes of Nova Scotia, passed on the nineteenth day of April 1884, provided for a fine of five dollars minimum and ten dollars maximum for taking a single grouse, partridge, woodcock, snipe, or teal out of season. The loss of one or two weeks' pay for killing one bird might have been cause for public concern except that precious few people were apprehended. Doubtless the harsh penalties in both Canada and the United States in years past attest to the importance placed upon wildlife resources by a people who prior to emigration had been forced to poach animals to which they were denied title. In North America, the settlers were originally zealous in their desire to guarantee free access to wildlife, but as stocks declined, the sportsman and naturalist became equally zealous in protecting their precious wildlife resource from abuse by the greedy and selfish among them.

The Evolution of Modern Wildlife Management—United States

The records show that wildlife management is an ancient practice, but they also tell us something of a people's relationship to animals as well as something about the upper classes of society that protected what was for many years a private access to wild animals. North American patterns of management developed first from earlier experiences in Europe and later from the very different North American experience and reaction against European traditions. After settlement in the New World, the public's role in wildlife management was gradually defined and expressed over a long period. For almost two hundred years, the freedom to hunt, including commercial hunting, combined with such devastating effects of settlement as destruction of vast areas of the continent's eastern hardwood forests spelled disaster for much of the New World's wildlife. As the nineteenth century drew to a close,

> "the black cloud of extermination hung low over the game ranges of North America. The great herds of bison, that a scant generation earlier, had offered the most thrilling spectacle in nature had vanished from the plains. The dainty pronghorn persisted in dwindling numbers in secluded prairies that were taken up by ranchers and settlers as rapidly as they could be located" (Trefethen, 1961).

According to Leopold, the dominant preoccupation in the minds of those dealing with wildlife in America until about 1905 was to perpetuate hunting. Managers generally believed that restriction could "string out the remnants of the virgin supply." Leopold also noted that game laws reflecting this feeling were "essentially a device for dividing up a dwindling treasure which nature, rather than man, had produced." Game protection failed to halt the accelerating decline of many species, and necessary changes in approach came about ever so slowly over a period of many years. There was no single action that appeared to alter government thinking or public policy, but a reasonable starting point is Theodore Roosevelt's melding of forest and wildlife and the evolution of his doctrine of conservation. In 1893, Roosevelt noted that "the preservation of forest and game go hand in hand. He who works for either works for both" (Trefethen 1961). Whether precisely accurate or not, we have in that statement the beginning of a systems approach that grew and began to flower in Roosevelt's presidency. Leopold notes that with men such as Roosevelt and Gifford Pinchot leading, "conservation" became the label of a national issue overnight. In Roosevelt's mind conservation was indelibly tied to "wise use" and he thought that the renewable organic resources of wildlife, forests, ranges, and waterpower "might last forever if they were harvested scientifically, and not faster than they reproduced" (Leopold, 1933). Roosevelt's doctrine is divided into four parts and is still relevant today:

1. A recognition of outdoor resources as integral systems
2. A recognition of conservation through wise use as a public responsibility
3. The recognition of private resource ownership as a public trust
4. A recognition of science as a means of discharging the responsibility of resource management

And so the ground work was accomplished for conservation in the United States. In the area of preservation, the first bird sanctuary (Pelican Island, Florida) was established in 1903 and by 1905 three more such sanctuaries were proclaimed. Other wildlife and big game refuges followed in the United States in 1905 and 1908. The first thirty years of this century were still incubation years. The ideas that wildlife was a product of the land and its use could be perpetuated, although present, were not as yet specifically addressed by conservation leaders. Increased protective legislation, the establishment of more refuges, and the development of natural history studies in schools continued. In 1930, the realization of the marriage of land and wildlife was finally articulated in a national platform at the seventeenth American Game Conference. The American Game Policy was produced by a committee of fourteen, chaired by Aldo Leopold. It clearly stated the basic requirements of wildlife and its management by recognizing:

1. Protection, food, and cover requirements.
2. Inducements for landowners.
3. A classification of game into farm, forest and range, and wilderness.
4. The need for facts, skills, funding, and public-sportsman cooperation.

The program included the extensions of public ownership and management, the recognition of the landowner as the custodian of public game, the need to experiment, the need to find facts, the need to recognize the nonshooting protectionist and scientist, and the need to train managers. In addition to recognition, the policy detailed these needs and spelled out both public education programs and funding requirements. Here indeed was a major breakthrough, or at least it should have been. Perhaps if governments and sportsmen alike had followed the direction of the American Game Policy today's massive social and land use conflicts regarding wildlife in the United States might have been partly avoided. But a policy does little by itself. There were certainly men of good will in 1930, it's true, but with an economic depression spreading throughout North America there was little money to implement policy. The United States and most of the rest of the industrialized world were in the trough of the most catastrophic economic decline of the century. Then in 1932, Franklin Delano Roosevelt, the second "conservation President" with the Roosevelt name, intervened and by 1933 money had been promised to support the conservation work programs of Roosevelt's massive National Recovery Administration, and most particularly, for wildlife. With strong leadership from men like Leopold, J. N. "Ding" Darling, and Dr. Ira Gabrielson, the thirties became the great decade of positive change for wildlife in the United States. The Migratory Bird Treaty had been signed between the United States and Canada in 1916 and the Fur Seal Treaty with Canada, Japan, and Russia in 1911, but the thirties saw the passage of the Federal Aid in Wildlife Restoration Act (1937), a federal-state cooperative approach to wildlife research funded from taxes on arms and ammunition. Known generally as the Pittman-Robertson or PR Act, this bill gave impetus to wildlife research and management which has never been equalled anywhere in the world. In February 1936, the first North American Wildlife Conference, replacing the American Game Conference, was called by Franklin Roosevelt. This gathering brought representatives of competing land-use agencies together more effectively than ever before and provided forums for each group to air its views and discuss programs. The National Wildlife Federation constitution was drafted at that conference through the efforts of Darling, and although not a formal part of the proceedings, The Wildlife Society was also born at the 1936 conference. Arthur A. Allen, one of North America's great ornithologists; R. T. (Terry) King, a forest-wildlife professor at the University of Minnesota and later Syracuse; W. L. McAtee of the United States Bureau of Biological Survey; and Miles Pirnie of the Kellogg Bird Sanctuary in Michigan were primarily responsible for getting The Wildlife Society started. This joining of professionals was an important step in the growth of wildlife management. There were four objectives noted at the time of the Society's formation. Two of them have been responsible, to a considerable degree, for elevating the wildlife profession in North America to a level of scientific competence recognized the world over. The four objectives were:

1. To develop wildlife management on a biological basis.
2. To establish a medium for publication.
3. To maintain professional standards.
4. To protect the interest of its members.

It is the first two that have been so successful not only in producing an excellent body of North American wildlife research and management literature, but at the same time, helping to establish biologically oriented wildlife management practices throughout much of the continent.

The 1930s produced more for wildlife. The Wildlife Management Institute (1946), which had its origins in 1912 as the American Game Protection and Game Propagation Association, got its start as an important lobbying and educational body with the 1936 conference. At that time it was named the American Wildlife Institute, after a 1930 change in title from the American Protective Association. The Institute flourished from the start. Whereas The Wildlife Society would obtain its funding from membership dues and the National Wildlife Federation from both private funding and dues, the Institute would always lean heavily on the sporting goods manufacturing industry for its financing. The objectives of the Institute were to support research, to educate the public and professionals, and act as a lobbying organization to influence the public's political conscience. Through its research input at the Delta Waterfowl Research Station in Manitoba, and for a time at the University of New Brunswick in Canada, and through its support role in the development of cooperative wildlife research units at land grant institutions throughout the United States, the Institute has had a marked effect on both professional research and education. In addition it has funded research directly, in Canada as well as in the United States. The Institute provided support for moose research in Newfoundland soon after that colony confederated in 1949. It also assisted Canada's newest province in laying the groundwork for both its Wildlife Act and its original wildlife policy. The Institute has published a number of important books, but its educational efforts have been used best in its constant positive influence for wildlife on elected political representatives at the federal level. Its Washington policy statements and newsletter, which are concerned with important conservation issues, reach thousands of conservation-minded citizens each month.

None of the wildlife organizations that came to life in the thirties would have been successful without leadership. One such leader was Ding Darling, the great cartoonist-administrator, who originally piloted the National Wildlife

Federation. Another was Dr. Ira N. Gabrielson, who guided the Bureau of Biological Survey and later the Wildlife Management Institute for many years. Still others were the first Wildlife Society presidents (Rudolph Bennet, Arthur A. Allen, and Aldo Leopold), who were responsible for an eruption of wildlife interests that has carried through to the present.

The Society of American Foresters (1900) was organized thirty-six years ahead of The Wildlife Society, but their respective federal agencies for "trees and animals" began at about the same time. Gifford Pinchot became chief of the Division of Forestry in 1898 and the division became the U. S. Forest Service (USFS) in 1905. The Biological Survey (1885) in the Department of Agriculture became the Bureau of Biological Survey in 1905, and at the end of the thirties (1940) it became the U. S. Fish and Wildlife Service (USFWS) in the Department of the Interior. Later the USFWS became the Bureau of Sport Fisheries and Wildlife, but in 1974 an act of Congress renamed the bureau and it became the USFWS once again. The research activities of the USFWS were transferred to the National Biological Survey, also within the Department of the Interior, in 1993. The National Biological Survey was amalgamated with the U.S. Geological Survey in 1996 to form the Biological Resource Division. Other advances in the 1930s decade included the establishment of the Soil Conservation Service in 1935, which provided a mechanism to counter wetland drainage and elimination of natural cover, and the Migratory Bird Hunting Permit (Duck Stamp) Act of 1934 which, following the Migratory Bird Conservation (Refuge) Act of 1929, provided a steady source of funding for refuge acquisition. It was a final federal touch required to provide for effective migratory bird management.

The conservation movement that began at the turn of the century progressed rapidly in the thirties driven not only by government legislation, agencies, and conferences, but also by sportsmen's organizations, sporting magazines, and the results of education and training. There were, and are, many hunting, fishing, and naturalists' groups, local and national, but the most prestigious, influential, and wealthy organization until about 1940 was the Boone and Crockett Club founded on the initiative of Theodore Roosevelt in 1888. The club had the distinction that all of its members were wealthy and therefore influential. The club's influence in wildlife policy matters at federal and state levels, its movements in support of game refuges, and its early support in other areas of wildlife management were detailed in 1961 by James B. Trefethen in his *Crusade for Wildlife*.

Perhaps three people stand out in Trefethen's narrative; Theodore Roosevelt, Major Charles Sheldon, and George Bird Grinnell. Charles Sheldon was a conservation crusader, naturalist, hunter, and author leading the fight for big game wilderness preservation. Grinnell was the editor of Forest and Stream

Weekly which later became Forest and Stream and merged with Rod and Gun. He was also primarily responsible for the founding of the National Association of Audubon Societies in 1902. Along with many other important professional zoologists, curators, and naturalists such as Clinton Hart Merriam, Carl Akeley, and Edmund Heller, he also served on the governing board of Forest and Stream until the early twenties. T. Gilbert Pearson, another Audubon Society leader, and Ernest Thompson Seton, the naturalist-author, later served on the magazine's advisory board.

For wildlife education and training, the years 1919, 1929, and 1932 were landmarks. The Cornell School of Game Farming and Roosevelt Wildlife Experiment Station of the New York State College of Forestry at Syracuse began in 1919. Aldo Leopold began lecturing on wildlife management at the University of Wisconsin in 1929, and in 1932 Ding Darling was responsible for the first cooperative wildlife program between Iowa State College and the Iowa Fish and Game Commission. We should also make note of the year 1933, for that was the year Leopold's masterful book *Game Management* first appeared.

The Evolution of Modern Wildlife Management—Canada

Traditional use of wild animals for food and shelter is as old in Canada as it is elsewhere in North America, but unlike in the United States, where with the exception of Alaska, such use has declined drastically, it has steadfastly remained a practice in much of northern Canada. Beaver helped open the western lands to settlement as trappers sought them in what might seem to us now almost inaccessible areas. Because of the beaver's fur value in Europe, beaver populations were gradually reduced from the east to the west coast, and by the early nineteenth century the beaver was almost gone from most of what is now Canada. But the species came back; first with the inadvertent help of man as land-use practices changed from pure exploitation to human settlement, agriculture, and forestry, and later with the purposeful intervention of man through management. In Newfoundland, closed seasons were initiated in 1923, and transplantations from small remaining populations began in 1935. Protection and reintroduction from remnant populations occurred in most provinces until beaver were once again found throughout acceptable range.

The saga of the buffalo in Canada is not a pleasant story. Between the 1820s and the 1860s the Metis of the Canadian prairies killed some 200,000 buffalo a year, and others were taken by white American (U.S.) and Native American hide hunters. Aboriginals who had once depended upon buffalo were starving for lack of meat as early as 1850, and by 1889 only 256 individual buffalo were known to exist in Canada. So the same "black cloud of extinction" that hung over much of the United States also covered a part of Canada by the turn of the century. Ernest Swift, writing about the United States in 1958, spoke for all of North America when he said:

> "there would have been no conquest of (North) America if there had been no wildlife. There would have been no Indians to meet the Spanish, or John Smith, or Champlain, if the Continent had not been well stocked with game. There would have been no fabulous journeys of exploration, no Radisson, no Hudson's Bay Company, no Lewis and Clark. . . . Game was elemental to survival, a basic part of the ecology; it still has a mystic and profound grip on the lives and emotions of millions of (North) Americans. The killing off of the wildlife and its general disappearance through other causes, has probably aroused more national attention and produced more concerted effort to the general tenets of conservation than any other issue."

At the federal level, wildlife management began in Canada with the Commission of Conservation, which was constituted by Sir Wilfred Laurier under the Conservation Act (an act of Parliament) in 1909 (amended in 1910 and 1913). Following the lead of Teddy Roosevelt and the eighth principle emanating from his 1908 Conservation Conference of Governors which called for such commissions, Laurier and his government constituted the Canadian commission with federal membership from the Ministries of Interior, Agriculture, and Mines. Provincial members, ex officio, were from each natural resource agency, while the governor-general-in-council appointed twenty university members and the chairman. The chairman, from the commission's inception in 1909 until its demise, was Sir Clifford Sifton, a one time minister of interior, great parliamentarian, avid conservationist, and controversial politician. In 1916, an interdepartmental Advisory Board on Wildlife Protection, also with members appointed by the governor-general-in-council, was established to work closely with the commission. With C. Gordon Hewitt serving as secretary, the advisory board drafted legislation under the Migratory Bird Treaty and revised the Northwest Game Act of 1906, both in 1917. It cohosted several national conferences with the commission, including the National Conference on Game and Wildlife Conservation of February 1919, when Hewitt gave a stirring plea for a "nationwide effort in wildlife conservation". Following changes in responsibility for administering new legislation, and no doubt missing the influence of Hewitt, who died in 1920, the commission was dissolved in 1922.

Gordon Hewitt is the conservation giant who stands out as Canada's primary wildlife crusader during conservation's formative years from 1909 until his untimely death in 1920. He was the one primarily responsible for securing the Migratory Bird Treaty, drafting the Migratory Bird Convention Act Regulations, and preparing the (revised) Northwest Game Act. He was an entomologist by profession and today is recognized annually by the Canadian Society of Zoologists, but perhaps his greatest accomplishments were in the area of wildlife policy. He served as consulting zoologist to the government and the Commission of Conservation as well as serving as secretary to the Advisory Board on Wildlife Protection. Hewitt also guided the Entomological Service as it evolved into an important branch of the Department of Agriculture. His book, the *Conservation of Wildlife in Canada,* and his representation on behalf of Canada at the major international wildlife conferences, where he served so ably, were contributions that helped launch Canada's wildlife management movement. Beginning with the ratification of the Migratory Bird Treaty in 1917 and continuing until 1947, the federal wildlife legislation in Canada was administered by the Parks Branch of the Department of Interior. In 1918 Hoyes Lloyd, a chemist and amateur ornithologist, was hired by the Branch as an ornithologist with a mandate to administer the Migratory Bird Regulations. Lloyd soon was named supervisor of wildlife protection and given authority to administer the Northwest Game Act. After his retirement Lloyd served as chairman of the International Commission for Bird Preservation, Pan-American Section.

Since 1932, the Royal Canadian Mounted Police have maintained a primary responsibility for enforcement of the Migratory Bird Convention Act although provincial officials are also empowered to enforce these federal regulations. The more senior Canadian federal and provincial wildlife authorities meet annually to enhance federal-provincial wildlife cooperation. This annual meeting is known today as the Federal-Provincial Wildlife Conference, which began under a different heading back in 1922. Finally, the Canadian Wildlife Service (CWS), the country's primary wildlife agency, was established in 1947 as a division of the National Parks Branch. Today the CWS continues within the Environmental Conservation Service of the Department of Canadian Heritage.

Jurisdictional Responsibility in the United States and Canada

In the United States, the federal government maintains five general areas of responsibility over wildlife resources. The government is fully responsible for the resource on all federal lands including refuges, national parks, and about 191 million acres of national forests. Federal jurisdictional responsibility is also present relative to the government's tax levying authority (e.g., the duck stamp) and its power to make treaties. Offshore, the federal government is primarily responsible for marine mammals and fish. Finally, federal authority extends to the control of the interstate transport of wildlife. More specifically, the federal government exercises primary authority over migratory birds, endangered species, marine animal resources, interstate and international traffic in wildlife, and all animals on federal land. State jurisdictional responsibility is maintained over all sedentary or resident animals, both "game" and "nongame," over reptiles, fishes, and amphibians. The states also exercise tax levying authority through the issuing of various licenses and permits.

In Canada, federal jurisdiction extends to migratory birds, all animals on federal lands, including refuges and national parks, and mammals migrating across provincial-territorial boundaries when in the territories. Federal responsibility in the Yukon and Northwest Territories has lessened in recent years while cooperative management programs between the territorial governments and indigenous peoples have been initiated following major land claim settlements. Since Canada's Constitution was patriated in April 1982, and the British North America Act of 1867 no longer serves as the sole basis on which to determine jurisdictional authority, increased federal-provincial cooperation is resulting in effective agreements to benefit the wildlife resource. The federal government maintains tax levying power (e.g., the Migratory Bird Permit) as well as treaty authority, as in the United States. Marine mammals have been considered a federal responsibility in the past and remain so today. The provinces have jurisdiction over resident game birds and mammals and over migratory mammals when inside provincial boundaries. Responsibility for fish is confusing at present. The federal government has primary responsibility for anadromous and fresh water fish in all coastal provinces except Quebec; however cooperative agreements may provide the other provinces with greater responsibility. Additionally, provincial acts relating to water, pollution, or environmental matters may affect responsibility through the control of fish habitat. Inland provinces presently exhibit authority over both their fish and fish habitat, even though the original responsibility was federal under Section 91 of the British North America Act. British Columbia recently concluded an agreement with Ottawa that exemplifies this with respect to salmon species.

Of the many federal agencies involved in wildlife management in the United States, the USFWS is probably the best known. The predecessors of this agency originated in the Department of Agriculture, with economic concerns and predator or depredator control problems as early justification for their existence. Today the Fish and Wildlife Service is in the Department of the Interior under the assistant secretary for fish and wildlife. This agency has initiated national and international wildlife legislation and is the federal body that oversees administration and enforcement of most federal wildlife acts and their regulations. The Service is also concerned with the management of migratory birds. Some 450 federal refuges encompassing about 90 million acres come under the aegis of the Fish and Wildlife Service and it is this agency that is most closely involved in international cooperation. Federal wildlife research now is administered from three regional Biological Research Division offices. There are 16 science centers and 94 field research stations throughout the United States. The Biological Research Division also serves as the federal agency cooperating with state agencies, universities and the Wildlife Management Institute in the operation of 58 Cooperative Research Units. The Units meet research needs as well as educate graduate students for the wildlife profession. The National Resource Conservation Service replaced the Soil Conservation Service of the Department of Agriculture in 1996. It is somewhat different than most of the U. S. federal agencies in its approach to wildlife management in that its biologists may work directly with landowners. Biologists are involved in habitat management programs on land areas managed within a soil conservation district for multiple use. Subsidy programs can allow landowners, singly or organized, to take advantage of the knowledge and skills of SCS biologists to enhance field and forest edges, establish marshes, ponds, hedgerows, or windbreaks to benefit wildlife populations, from songbirds to rabbits. This agency also administers the Conservation Reserve Program.

The Forest Service is in the Department of Agriculture as well, and its biologists conduct research and experimental management in both forest-wildlife complexes and forest-grazing-wildlife complexes. The National Park Service is in the Department of the Interior. Its wildlife biologists work only in the parks within the framework of National Park policy. Other agencies in the federal government involved in wildlife work are the Tennessee Valley Authority in the southeastern United States and the U.S. Army Corps of Engineers, where development mitigation efforts are of considerable importance. There are many agencies whose land-use practices and policies affect wildlife in addition to the ones mentioned. The Bureau of Land Management, which administers about 271 million acres of public land, the Reclamation Service, the Armed Forces, and the Bureau of Indian Affairs are examples of such agencies. Other agencies may be involved somewhat indirectly; the Treasury Department through its Coast Guard, which may enforce portions of the Lacey Act relating to the importation of exotics, and the Customs Service, which may be involved with the importation of exotics and endangered species.

In Canada, the Canadian Wildlife Service (CWS) is most closely involved with wildlife matters on a federal level. Branch personnel are involved in administration and enforcement relative to all federal wildlife legislation, including the Migratory Birds Convention Act and the Convention on International Trade in Endangered Species of wild fauna and flora. The Branch conducts research on terrestrial, fresh water, coastal, and pelagic wildlife species and associated systems throughout Canada. It is also involved in socioeconomic evaluations, wildlife toxicology, environmental assessment, and biodiversity

24 • *The Philosophy and Practice of Wildlife Management*

research. Additional responsibilities include the administration and management of some eighty-two sanctuaries and thirty-nine wildlife areas. The CWS cooperates with Parks Canada in research and management within National Parks and Historic Sites and with the Canadian Forest Service in sustainable forests research.

All states and provinces have designated branches of government to administer wildlife legislation within their respective jurisdictions, to cooperate with other agencies at state, provincial, and federal levels, to conduct research and management, and to carry out interpretive educational and/or public relations endeavors.

Legislation: The United States and Canada

Federal wildlife acts in the United States allow for the development of wildlife regulatory management, wildlife research, and habitat management where federal responsibility exists (Table 1.4). Beginning in 1900 with the Lacey Act, federal legislation complemented the various state acts and their regulations to provide a substantial body of law for the protection and management of all species. In Canada, federal wildlife legislation includes the Territorial Ordinances, the Migratory Bird Convention Act of 1917 (revised), the Migratory Bird Hunting Permit of 1966, the Canada Wildlife Act of 1973, the Export, Import and Interprovincial Transport of Wildlife Act which has replaced the Game Export Act and a Species at Risk Act which is going through the legislative process. Federal legislation also complements specific provincial acts or portions of acts to provide a basic body of statute law for wildlife protection and management in Canada.

In addition to the Migratory Bird Treaty, both Canada and the United States have been principals in other international legislation of importance. The Fur Seal Treaty of 1911, which included Russia and Japan as participants, is credited with saving the seals from extinction and perpetuating a healthy and economically viable population. The first whaling treaty in 1937 included Norway and Germany and was an important initial step in whale preservation and resource management. The whaling and fur seal treaties have been revised and continue to apply today. Canada and the United States were participants in the 1975 Convention on International Trade in Endangered Species, which sought to protect threatened species of wildlife from extinction wherever they may exist. The United States and Canada have legislation specifically designed to protect rare and endangered species such as the bald eagle and musk ox. Some legislation is general (such as the U. S. Endangered Species Preservation Act of 1966, 1969, 1973), while other specific state and provincial regulations are in effect to protect regionally threatened wildlife.

Table 1.4 — Some Important Federal Wildlife Legislation — the United States

Title	Date	Intent
Lacey Act	1900	Regulated market hunting, controlled import of exotics, controlled interstate transport of illegally taken game, controlled trespass on refuges
Migratory Bird Treaty Act	1918	Protection of migratory birds, either complete or through regulation
Migratory Bird Conservation Act	1929	Provided for establishment of refuges
Migratory Bird Hunting Stamp Act (the Duck Stamp Act)	1934	Provided for federal migratory bird license; proceeds to finance acquisition of refuges and production areas
Fish and Wildlife Coordination Act	1934	Authorized conservation measures in federal water projects and required consultation with USFWS and states concerning any water project
Convention. USA and Mexico for the protection of migratory birds and game mammals	1936	Protection of certain migratory birds; cooperation and control concerning both mammals and birds along the border
Pittman-Robertson Federal Aid in Wildlife Restoration Act	1937	Provided for an excise tax on sporting arms and ammunition to finance research on federal-state cooperative basis
Convention. USA and Mexico, for the protection of migratory birds and game mammals	1940	Committed U.S. to protection and husbandry including refuges; implementation in 1973 (Endangered Species Act)
Fish and Wildlife Act	1956	National wildlife policy and general considerations
Endangered Species Conservation Act	1973	Protected species threatened with extinction or in imminent danger of becoming extinct
Fish and Wildlife Conservation Act	1980	Authorized development of wildlife conservation plans and actions to benefit wildlife, especially non-game types
North American Wetland Conservation Act	1989	Authorized funding for wetland restoration and conservation linked to supporting the North American Waterfowl Management Plan

The Land and Wildlife: Ownership, Rights, and Restrictions

One of the most important areas of concern affecting wildlife management in North America is the public wildlife ownership-land tenure issue. It was William the Conqueror who imposed the feudal land system in 1066. Because he personally owned all the real property in England, he was able to establish a pyramidal system of control with himself at the top. At each descending level in the hierarchy, there was an exchange of a form of service for the privilege to use the land. The service-to-use ratio increased with each step downward in the following hierarchy:

King
Knight of Service
Grand Sergeanty
Petty Sergeanty
Frankalmein
Villeins, hewers and drawers

Total work and allegiance were demanded from those at the bottom in exchange for space to accommodate a bleak and bare existence.

As the church was also a strong force at the time and ecclesiastics deigned to maintain certain land use rights, in 1290 conveyances to the church were forbidden without the king's permission; so politics entered the trespass scene in England. In 1660, "a fixed annual sum" (rent) was required from the land holders (the Statute of Tenures). In Canada today, a holdover from the early tenure of 1660 is the "Doctrine of Escheat," confirmed by statute in some jurisdictions, (e.g., Escheats Act R.S.N.S. 1957, c.91.) which prevents title to land lying in abeyance. This principle also can be seen in one of the cardinal rules of conveyancing: the Rule Against Perpetuities states that "title to land must vest within the lives in being plus twenty-one years." That is, if a person dies intestate, leaving no heirs, the land returns to the Crown after twenty-one years.

Today land is not owned absolutely. The owner has tenure over the land he has registered and described but there are some restrictions on the use the owner may make of the land, wherever it is located in Canada or the United States. Landowners only have rights to exclusive possession and use not prohibited by statute. In some jurisdictions, for instance, zoning legislation may prevent the land being used for purposes other than those stated. Forest management or agricultural production may be restricted relative to harvest or crop production. Access may be guaranteed to certain classes of persons for specific purposes such as to reach water areas where fishing is allowed. The extent to which a landowner is controlled in his use of land varies greatly between states and between provinces. In Nova Scotia, for instance, there are some twenty-four provincial acts affecting one's title and sixteen or more which may restrict use.

According to common law, the ownership of domestic animals is absolute, but the ownership of wild animals is more complicated. Prior to the beginning of the concept of land title, man had a kind of property right in wild creatures and there were no restrictions on pursuit. With the right of property (land tenure) pursuit became more difficult. Under Roman law, common ownership was vested in the people subject to the right of the landowner to forbid the killing of game on his property.

At one time in England, the title to all wild animals was considered to be vested in the sovereign, and this right was dealt with by Henry III in 1225 in the Magna Carta and the Charter of the Forest. The position of ownership at common law since 1225 is that ownership of wild animals is in the state in its collective sovereign capacity as a representative of all its citizens. It has long been the common law of England that landowners have a qualified property in wild animals and that this qualified property becomes absolute upon the death of the animal. There is an inherent conflict between the hunter's right to pursue game which is held in trust for the people and the landowner's right to protect himself and his property from trespassers, particularly those carrying firearms. Another position from which to view the same problem is that game which belongs to all, but to no one person, is found on lands belonging to a few, and those few can prohibit access of all people to the game.

Absolute property is not confined to animals killed by the owner but extends to trespassers as well, for a trespasser may have no property rights in the animal he kills. Thus at common law, the right of killing was annexed to the soil although the landowner did not own the animals while they were living.

Common law did not afford the landowner protection, however, for killing an animal on another's land was not a punishable offense, and the offender could only be sued for nominal damages, which was not a deterrent to trespass, nor is it today. In North America, we reverted to statute law to strengthen the position of both landowners and trespassers. In many jurisdictions, the landowner is protected by legislation and the regulations require no advertising concerning boundaries of ownership. Other legislation may protect the landowner under certain specified conditions of land use rather than ownership, and in some jurisdictions, there may be no protection except by civil suit for common trespass, which is really no protection. Legal decisions are often based upon precedent, which means that decisions determined in a particular case based upon statute law are a determinant in future cases of a similar nature. Such case law develops over time so that those jurisdictions with the earliest legislation also should be the ones with the richest case law. The United States has accumulated a greater amount of case law based on precedents involving wildlife and other conservation concerns than Canada has, and doubtless more than Mexico but perhaps less than England.

In Canada today, the title to free-living wild animals is vested in the Crown in the right of the province where the animals exist. Similarly, title to wild animals in the United States is held by the public (the people) represented by their respective state governments. In jurisdictions throughout the United States and Canada, the statutory possession of living wild animals and the control of pursuit or hunting of wild animals is detailed in one or more legislative acts. If allowed at all, permission is usually required to keep wild animals in captivity. Endeavors such as fur farming, game farming, and game ranching usually are controlled by legislation often through agricultural agencies. The hunting, trapping, possessing, and pursuing or harassing of wildlife normally are controlled by regulations under a wildlife act.

Administration and Policy: The United States and Canada

With every new act and the regulations thereunder more decisions must be made and more paperwork is required. This is the stuff of administration: the determination of manpower schedules, finalizing budgets, approving reports, preparing and submitting reports, developing policy, and handling correspondence. Administering a wildlife agency differs from the administration of another resource agency primarily in its subject matter. The headings of budgets are wildlife oriented and letters are concerned with wildlife-people problems and the interpretation of policy. It is paper work, telephone work, meetings, and conferences. But a good administrator is also a good manager, and the department or branch head, or director must be both a manager and an efficient administrator. The management part of the job has to do with people. The reports he approves are written by people. The budgets he finalizes are drafted originally by staff, and the manpower schedules he posts effectively determine someones's daily and yearly work programs. It is necessary to know something about wildlife to aspire to a wildlife administrative post, but it is even more important to get along with others who are both senior and junior to you. The administrator-manager should know the field and command respect while also being understanding and insightful.

It is the wildlife administrator's job to know procedure within the wildlife agency and between that agency and others. The administrator must also understand the processes by which legislation is approved and the limits of authority in the public service versus the political domain. At the end of the authority lines as well as at the end of the procedure and protocol lines in all North American government agencies, there is an elected representative of the people. These lines of authority are typically represented by flow charts. Five examples are shown (Figures 1.1–1.5). Each chart has been simplified to show the structure and major headings only. Usually the amount of paper work and the degree of authority and

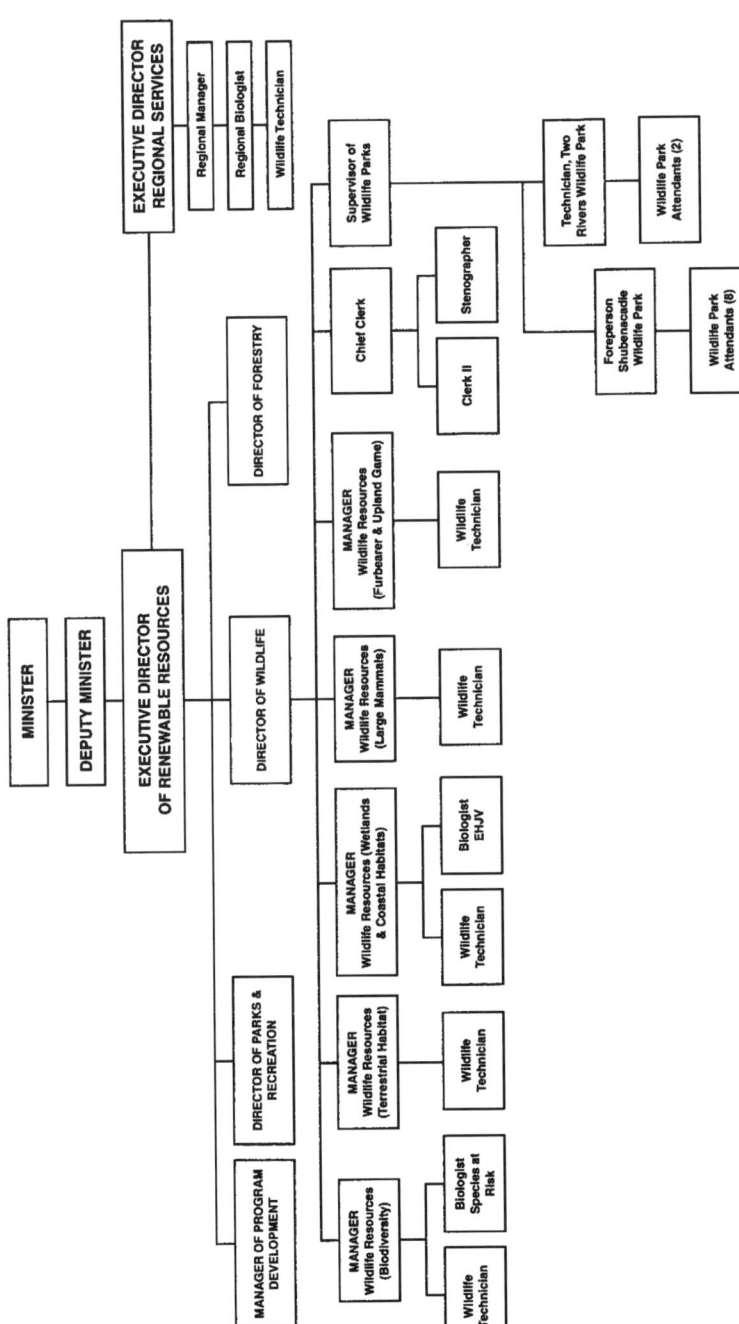

Figure 1.1—Organization chart for the province of Nova Scotia's wildlife department.

influence tend to increase as one moves toward the apex of a flow chart. In the provincial example, which does not necessarily typify all provinces, there is no direct line from planning biologists to the biologists carrying out the field programs (Figure 1.1). In such cases, an informal shortcut may be employed with letters or memos of information directed to the senior administrators involved; it often depends upon the size of the organization and understandings that have developed. In the state examples given (Figures 1.2–1.3) the responsibilities are also noted. These are the kinds of general responsibilities found in equivalent positions elsewhere throughout North America. The next example is a Canadian federal regional one (Figure 1.4) and the last is the Region 6 U.S. Fish and Wildlife Service (Figure 1.5). Each example should help to give you an idea of the complexities of modern wildlife bureaucracies, especially if you realize that the charts are simplified. Because bureaucracies tend to grow and change these examples only are valid now as structural formats of resource agencies.

In the Canadian parliamentary system of government, the senior civil servant of a general resource agency, including wildlife, is the deputy minister. The deputy minister and the assistant deputy minister give final approval to regulatory changes, revisions of existing statutes, or other policy deemed necessary. The deputy minister recommends revisions to the minister, who is an elected representative appointed to the portfolio by the premier (provincial) or prime minister (federal). Before the revisions are approved by the minister, they receive a thorough legal review by the attorney general's department (provincial) or the solicitor general's department (federal). A bill or regulatory revisions may also be reviewed by one or more legislative committees or even debated by legislative bodies. In some cases, regulatory changes may be accomplished by order-in-council (governor-general-in-council or governor-in-council), which means that the minister and cabinet can approve without legislative approval. In the Canadian parliamentary system, the order-in-council route usually means there is less political influence involved, for if a regulatory change must go through the legislative assembly, all elected representatives may have a voice, if they choose to, on second and third readings, and each representative could be influenced by people in the constituency. The elected representative in a Canadian province is known as the MLA (Member of the Legislative Assembly) or MPP (Member of Provincial Parliament). Federally, the elected member is the MP or Member of Parliament. Representation is by electoral district and is approximately based on population. A knowledge of provincial and state processes is important for anyone interested in wildlife because most legislation and regulatory changes relating to wildlife take place within the government arenas of provinces and states. National and international legislation are important, but new acts and amendments occur in these areas less frequently.

In a U.S. state, the position equivalent to deputy minister is usually that of secretary of a commission depending upon whether the state operates with a

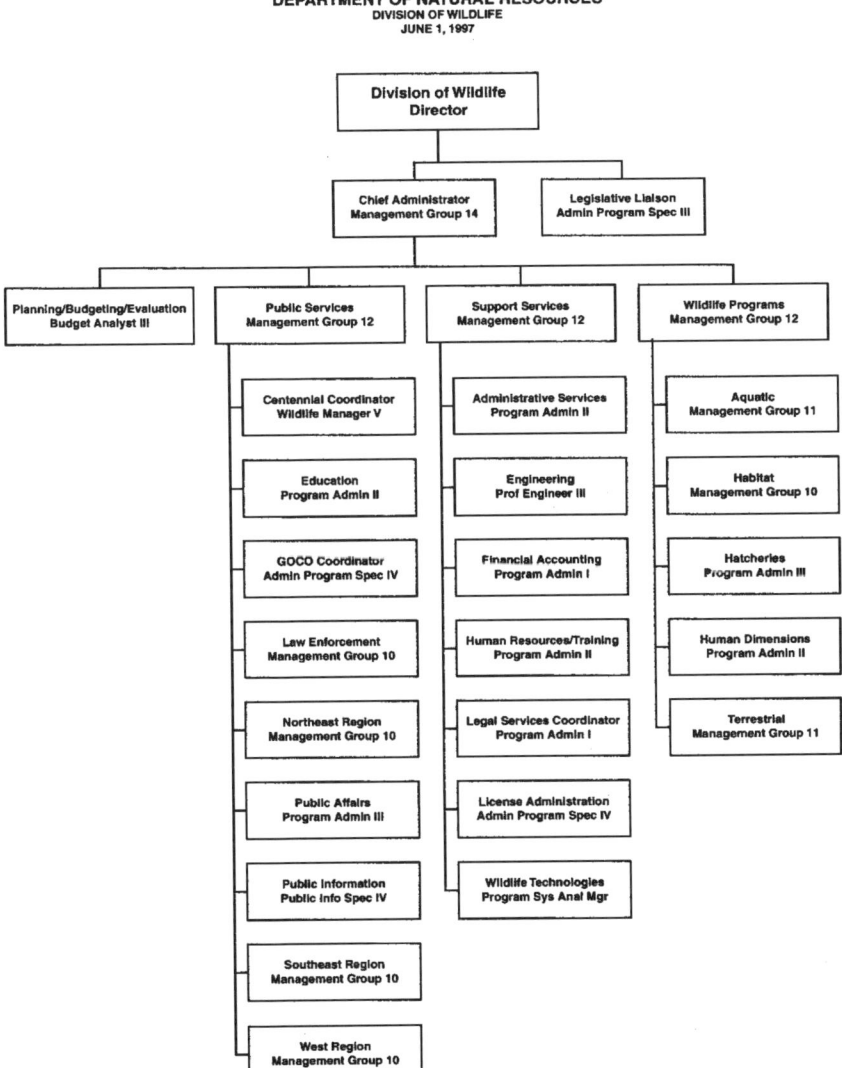

Figure 1.2—Organization chart for the Colorado Division of Wildlife.

commission system or with a line system. In the state of Colorado example provided (Figure 1.2), the senior post is that of chair of the wildlife commission. The representative equivalent to a minister may be either a commissioner or a commission chair, appointed by the state governor, or according to the

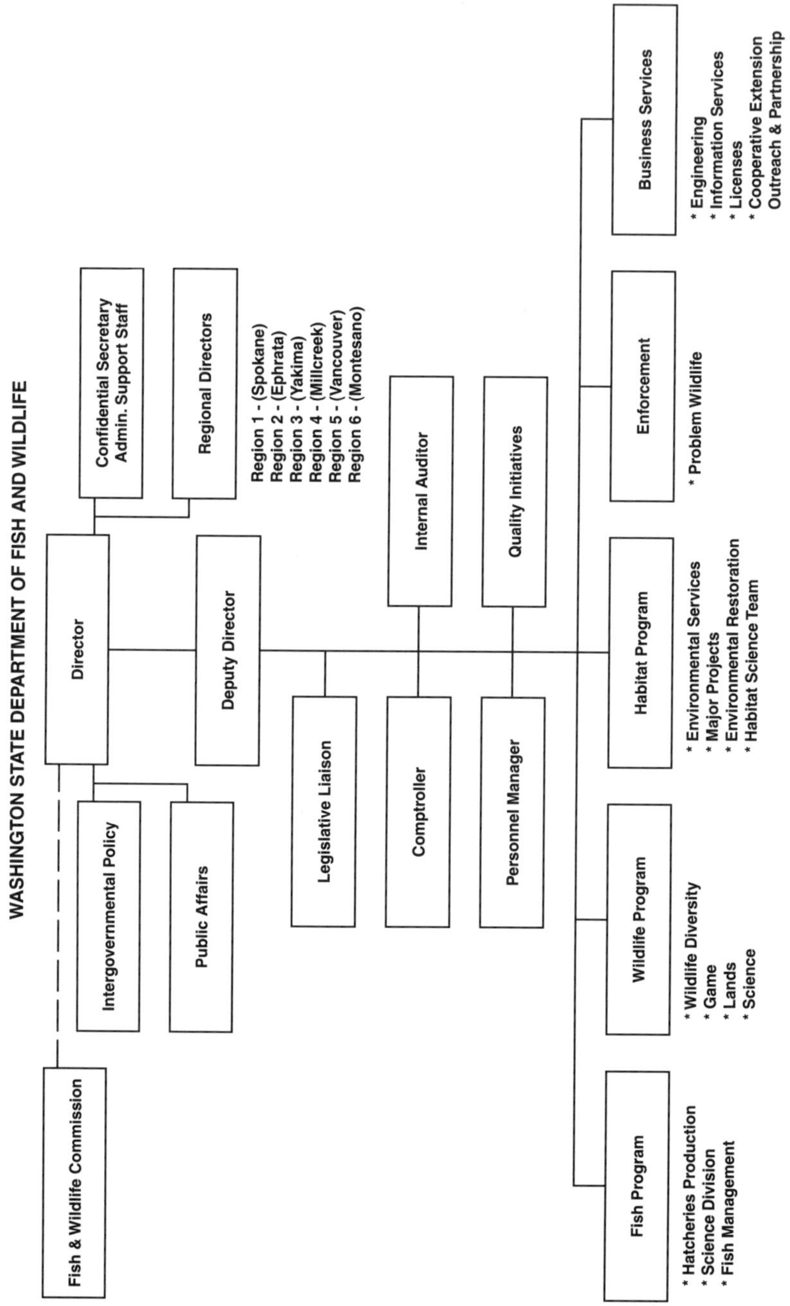

32 • *The Philosophy and Practice of Wildlife Management*

Figure 1.3—Organization chart for the Washington State Department of Fish and Wildlife.

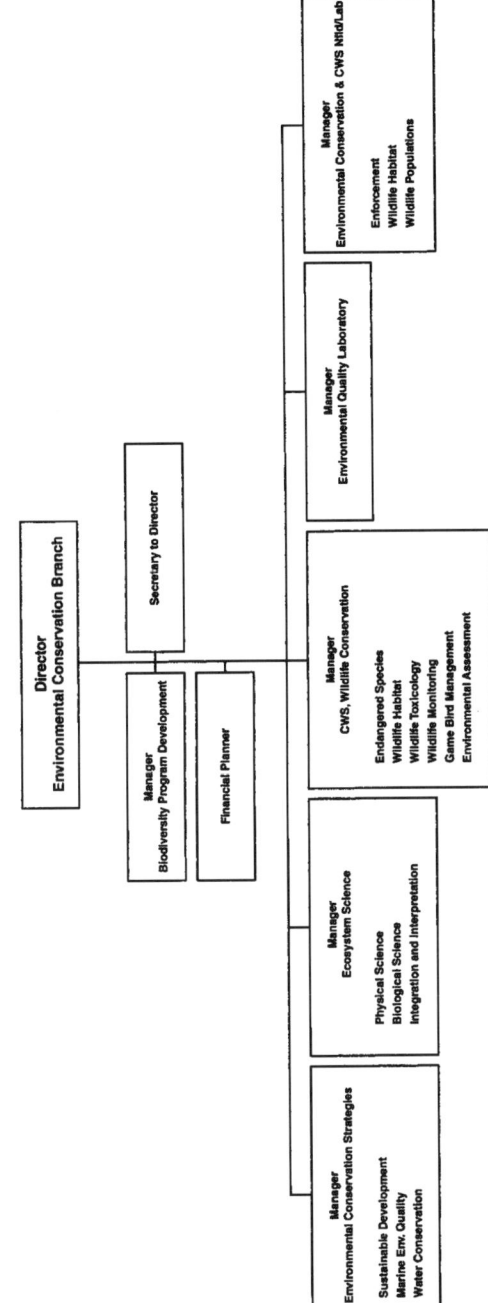

Figure 1.4—Organization chart for the Atlantic Region of the Canadian Environmental Conservation Branch.

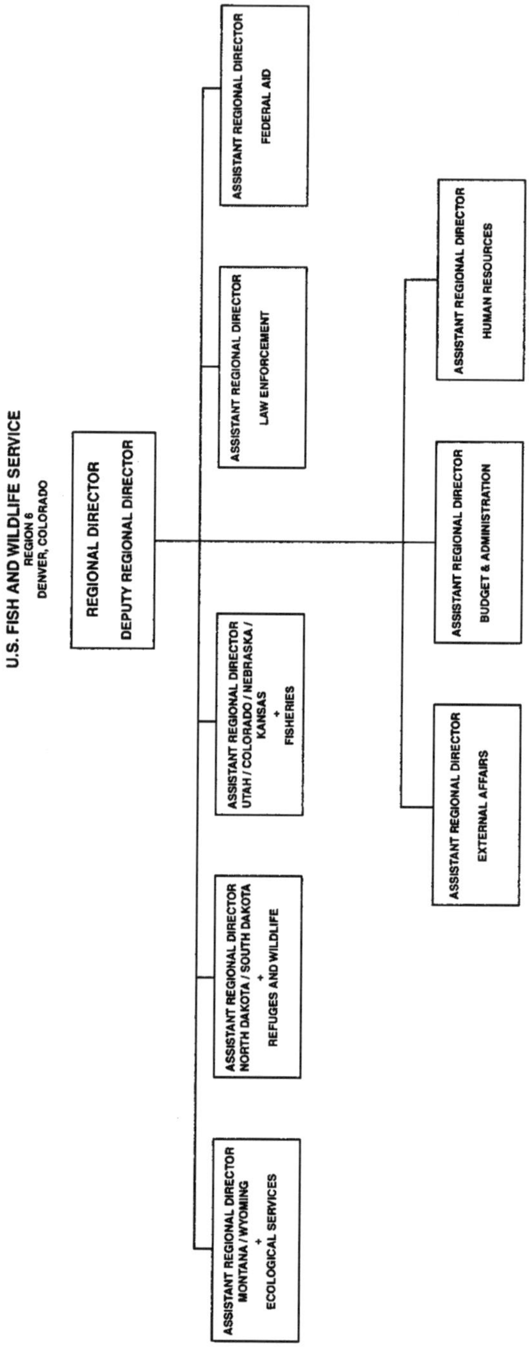

34 • *The Philosophy and Practice of Wildlife Management*

Figure 1.5—Organization chart for Region 6 of the U.S. Fish and Wildlife Service.

system, there may be several commissioners. In the latter instance, some states have formulae that disallow one governor to appoint more than a certain number of commissioners during the term of office. This procedure originally was designed to reduce partisan political influence. The process for a regulatory change is not greatly different in the states from that of the provinces. The people involved may have different titles but the end result is pretty much the same. Many states and provinces have added advisory boards or councils. In the United States the advisory groups most often represent the hunting, fishing, and nonhunting public, universities involved in government-supported research (through Pittman-Robertson), federal cooperative agencies, and sometimes, major private land owners or industrial interests. In Canada, advisory groups may reflect various publics concerned with wildlife and/or provincial geographical representation. Such bodies serve as effective screening groups, advising against unsound policy or poor regulatory changes and improving the better recommendations which emanate from biologists or interest groups.

Poole (1980) has provided a thorough discussion of the U.S. federal legislative process. A basic difference of some importance between the U.S. and Canadian federal systems should also be understood. In the United States, the senior federal legislative body is the Senate, composed of two elected members from each state regardless of size or population. The House of Representatives is comprised of congressmen elected from states, according to their populations. One state may have several congressmen but always only two senators. Both houses comprise Congress and each house must pass the federal wildlife legislation in question. In Canada, the honorable members of the House of Commons are elected on the basis of population and it is the important legislative body. Unlike elected term senators in the United States, the senators in Canada's parliamentary system, until recently, were appointed for life by the governor-general (actually the prime minister) when vacancies occured. Now they serve a term when appointed to age 75 or until death (whichever comes first). The Senate in the Canadian parliamentary system generally serves to improve rather than initiate legislation.

Over periods of several years, Canada and the United States have developed policy statements concerning wildlife. The United States has a "wildlife policy" serving as a guide for its agencies and for the nation. In 1982, Canada completed its "Guidelines for Wildlife Policy in Canada" generally subscribed to by federal and provincial agencies involved with wildlife. In 1990, a "Wildlife Policy for Canada" was approved by the Federal authorities responsible and the provincial ministers. Both countries originally established waterfowl management plans which complement each other on a flyway and a continental basis. Cooperation between the United States and Canada has always been exceptionally good regarding wildlife resources, and this cooperation extends from policy guidelines at senior levels of government down to regional, technical matters of species management considered by biologists. The North American Waterfowl Management Plan (NAWMP) is such an example in which the International Association of Fish and Wildlife Agencies was instrumentally involved. This organization represents all North American jurisdictions and forms an important base for policy development and implementation with respect to wildlife resources. In Canada, several joint ventures have been initiated under the NAWMP which address either regional conditions (e.g., Western Joint Venture) or species concerns (e.g., Snow Goose Joint Venture). Overall funding is essentially 50 percent from the United States and 50 percent from Canada, but, the formula may vary to 75:25 depending upon the ratio of "in kind" to "cash" support in Canada. A typical joint venture project funding approach would find Canada's portion of a project supported one third federally, one third provincially, and one third by a nongovernment organization, such as Ducks Unlimited, Canada. Other nongovernment organizations involved might typically be Wildlife Habitat Canada, the Nature Conservancy or local sportsmens' associations. Projects submitted within joint venture approaches, generally originate with provincial waterfowl biologists, are screened by the Canadian Wildlife Service and then considered by a joint venture committee composed of senior Canadian, United States government and nongovernment organization representatives. Each project must be approved by all joint venture participants. Depending upon the project, a Funding Administrative Agency or "banker" might be Ducks Unlimited or Wildlife Habitat Canada. Because the Wildlife Habitat Canada mandate is broader than Ducks Unlimited's and encompasses cooperative terrestrial habitat projects as well as wetland programs, Ducks Unlimited, at present, is the more active nongovernment organization in projects within joint ventures under NAWMP.

Nongovernment Wildlife Organizations in the United States and Canada

We have previously referred to the importance of citizens' organizations, professional, and otherwise, in the evolution of wildlife conservation in the United States. There are several organizations that continue to be important. Without them the wildlife cause might soon dwindle and die in the United States. In Canada, nongovernment organizations have also become a strong influence in shaping both provincial and federal wildlife policy.

The Wildlife Management Institute is the watchdog organization in Washington. Its *News Bulletin* carries the latest happenings from congressional committees and from the floors of both houses. As a lobbying organization gaining support before the votes are taken, it serves the profession and the wildlife resource throughout the United States. The National Wildlife Federation is also an effective lobbying organization. Its Washington office is on top of the wildlife scene throughout North America and its influence is also felt through its educational materials and its publication *National Wildlife*. The National Audubon Society, originally concerned primarily with birds, has over the years supported many general wildlife initiatives, including the national wildlife refuge system. It remains a powerful support force for wildlife conservation and *Audubon* ranks among the top wildlife publications in the world. Among the other national, regional, state, and local groups which support sound wildlife conservation are the Sierra Club, The Wilderness Society, Ducks Unlimited, The Nature Conservancy, and the National Parks Association. The Conservation Foundation and New York Zoological Society are organizations concerned with wildlife problems in North America and other countries. The Wildlife Society and professional bird, mammal, fish, and herpetological societies all lend weight to conservation causes.

Gilbert et al. (1980) noted that the presence of a "larger, better organized, and more vocal public" helped the United States progress more rapidly than Canada in developing a major wildlife presence and movement. This is true, but the number of environmental, naturalist, and other wildlife related organizations with public support in Canada is increasing. The Canadian Wildlife Federation, which originated with the 1961 Resources for Tomorrow Conference, is a strong voice for conservation causes and boasts the largest membership of all Canadian conservation groups. The Canadian Arctic Resources Committee, Canadian Nature Federation, Canadian Parks and Wilderness Society, Ducks Unlimited Canada and The Nature Conservancy of Canada are playing roles in molding a stronger wildlife conservation movement. Professionally, Canadian wildlife people have been less successful in forming a strong Canada-wide organization; however, regional groups remain viable throughout the country. As an example, in 1959, the Canadian Society of Wildlife and Fishery Biologists was organized and its first general meeting was held in 1960. The name was later changed to the Canadian Society for Environmental Biologists and nonwildlife professionals were welcomed into the organization. In 1971, the Canadian Society of Zoologists added a wildlife section as a specialty group but its membership was composed primarily of academics and in 1990 it was disbanded. Then, The Wildlife Society of Canada was formed in 1981. It, too, has failed to obtain the support needed to provide professionals with a society that would benefit the resource and its professional following, yet student and provincial chapters of The Wildlife Society remain viable.

Bibliography

Alison, R. M. 1978. The earliest records of waterfowl hunting. Wildl. Soc. Bull. 4:196–199.

Bailey, L. H. 1909. *The Nature-study Idea*. The MacMillan Co., New York, NY.

Benson, D. A., and D. G. Dodds. 1977. *Deer of Nova Scotia*. Department of Lands and Forests, Halifax, Nova Scotia.
Bubenik, A. B. 1976. Evolution of wildlife harvesting systems in Europe. Trans. Fed-Prov. Wildl. Conf. 40:97–105.
Clarke, C. H. D. 1976. Evolution of wildlife harvesting systems in Canada. Trans. Fed-Prov. Wildl. Conf. 40:122–139.
Dagg, A. I. 1974. *Canadian Wildlife and Man*. McClelland and Stewart Ltd., Toronto, Ontario.
Dasmann, R. F. 1981. *Wildlife Biology*. Second ed. John Wiley and Sons, New York, NY.
Deuteronomy 14:4–20.
Deuteronomy 22:6–7.
Dodds, D. G. 1983. Terrestrial mammals. In: G. R. South (ed.), *Biogeography and Ecology of the Island of Newfoundland*. pp. 509–550, Junk, The Hague, Holland.
Foster, J. 1978. *Working for Wildlife*. Univ. of Toronto Press, Toronto, Ontario.
Genesis 7:12.
Genesis 8:19.
Gilbert, F. F., E. Bossenmaier, L. Carbyn, D. Hebert, M. Hoefs, J. Huot, G. Mitchell, W. Prescott, A. Simmons, and L. Sudgen. 1980. The Wildlife Society of Canada-A reality in 1981? Wildl. Soc. Bull. 8 (3):179–198.
Gustafson, A. F., C. H. Guise, W. J. Hamilton, Jr., and H. Ries. 1949. *Conservation in the United States*. Comstock Publishing Co., Ithaca, NY.
Hewitt, C. G. 1919. The need of a nation-wide effort in wildlife conservation. In: *National Conference on Game and Wildlife Conservation*. pp. 8–18, Ottawa, Ontario.
Johnson, M. K. 1979. Review of endangered species: policies and legislation. Wildl. Soc. Bull. 7 (2):79–83.
Leopold, A. 1948. *Game Management*. Charles Scribner's Sons, New York, NY.
Leviticus 11:4–6.
Lewis, H. F. 1965. The Canadian Wildlife Service-Its functions and scope. Oryx 2 (3):173–178.
Lewis, H. F. 1967. Wildlife conservation through the century. Rod and Gun in Canada (Jan–Feb).
McClintock, J. N. 1889. *History of New Hampshire*. B. B. Russell, Cornhill, Boston, MA.
Munro, D. A. 1961. Legislative and administrative limitations on wildlife management. In: *Resources for Tomorrow,* Vol 2:867–880. The Oueen's Printer, Ottawa, Ontario.
Nova Scotia. 1884. Consolidated Statutes.

Phillips, J. C. 1928. *Wild Birds Introduced or Transplanted in North America.* U.S. Dept. Agric. Tech. Bull. No. 61, Washington, DC.
Piers, G. 1899. In: *Annual Report of the Nova Scotia Game and Inland Fishery Protection Society for 1898.* Halifax, Nova Scotia.
Pimlott, D. H. 1953. Newfoundland moose. Trans. N. Am. Wildl. Conf. 18:563–581.
Poole, D. A. 1980. The legislative process and wildlife. In: *Wildlife Management Techniques Manual.,* pp. 489–498. The Wildlife Society, Washington, DC.
Severinghaus, C. W. 1974. Return of the deer. The Conservationist (Aug–Sept).
Slade, W., Jr., 1823, *Vermont State Papers. Being an Account of the First and Second Councils of Censors.* J. W. Copeland, Middlebury, CT.
Swift, E. 1958. *The Glory Trail. The Great American Migration and Its Impact on Natural Resources.* The National Wildlife Federation, Washington, DC.
Tober, J. A. 1981. *Who Owns the Wildlife? The Political Economy of Conservation in Nineteenth Century America.* Greenwood Press, Westport, CT.
Trefethen, J. B. 1961. *Crusade for Wildlife.* The Telegraph Press, Harrisburg, PA.
Trippensee, R. E. 1948. *Wildlife Management, Upland Game and General Principles,* Vol. I. McGraw-Hill Book Co. Ltd., New York, NY.
Wildlife Management Institute. 1968. *Organization, Authority and Programs of State Fish and Wildlife Agencies.* Wildl. Manage. Inst., Washington, DC.

Recommended Readings

Audubon Wildlife Reports. Annual or biannual reports that began in 1985 considering major U.S. and international wildlife issues, featuring one U.S. wildlife related government agency in each volume. No longer published.
Foster, J. 1978. *Working for Wildlife. The Beginning of Preservation in Canada.* Univ. of Toronto Press, Toronto, Ontario.
Langenau, E. E., Jr. and C. W. Ostrom. Jr. 1984. Organizational and political factors affecting state wildlife management. Wildl. Soc. Bull. 12:107–116. An interesting hypothetical model of state wildlife administration developed to examine the relationships between the factors of demand, resource base, management structure and public benefits.
Sherwood, M. 1981. *Big Game in Alaska. A History of Wildlife and People.* Yale Univ. Press, New Haven, CT. A generally even presentation of

some of the political, social and jurisdictional complexities in the history of wildlife conservation prior to Alaskan statehood.

Tober, J. A. 1981. *Who Owns the Wildlife? The Political Economy of Conservation in Nineteenth Century America.* Greenwood Press. Westport, CT. An excellent, well-documented study of the evolution of wildlife conservation and wildlife policy in the United States.

Trefethen, J. B. 1961. *Crusade for Wildlife. Highlights in Conservation Progress.* The Telegraph Press, Harrisburg, PA. The history of the Boone and Crockett club, but also a good review of the wildlife conservation movement in the continental United States.

2 MAN AND WILDLIFE—Culture, Conflicts, and Values

One school of thought among anthropologists holds that the American Indian believed himself and the animals he hunted to be brothers derived from one Creator; they were of the same species. Whisker (1981) notes that it was believed that the Creator made man a hunter and all other animals the hunted. Animal mythology was passed from the old to the young, and a deep respect for their animal brothers usually made native Americans careful killers.

> "Nature and its animals were supposed to be the standard of goodness for the Amerindian. All human behavior could be measured against nature and the animals in precise measurements. Animals were generally seen to be courteous, kind, loyal, strong, brave, reverent, clean, alert and loving. Before taking an animal in the hunt one had to be prepared to attest to his own purity of heart. He could not kill that which was more moral than himself. This, of course assumed an intimate knowledge for the problems of moral behavior of the animals he hunted.
>
> Subarctic Indians were obsessed with the responsibilities that man and animal had to one another. The injunction against killing too many animals ranked as the first law the Amerindian must obey. That he may be killed by a man who was his moral equal was the first law for animals. Thus the animal, when seeing that his meat, hide and so on was needed by the true hunter, was required to voluntarily surrender itself to the hunter. A hunter or an animal which did not obey the law violated the rules of nature." (Whisker 1981).

Man was the animal with the higher intelligence and, therefore, greater responsibility. Man communicated with animals through the hunt. Here was a built-in conservation ethic supported by ritual that guaranteed the continued existence of the wild populations upon which man was partly dependent. According to Martin (1978), this mutually beneficial relationship, which was basic to the Indian religious experience, was quickly broken following the initial European visits to the New World. The Indians misinterpreted the cause of increased sickness and death (resulting from European contact) and sought revenge by punishing the animals they thought caused the illnesses. Whether this "war between the Indians and animals" existed or not, the Indian-animal brother relationship did break down with the advent of trading with the Europeans, and most especially, with the beginnings of the fur trade. Culture changed rapidly when a market economy was introduced into North America, for the animals were soon moved from a primary support position and an importance in ritual to a means of obtaining status through material possessions.

The Micmac Indian of the maritime region of Canada was probably among the first to undergo an attitudinal and cultural change relative to wild animals because European contact was earlier in that area of North America than elsewhere. The Micmac's dependence upon hunting and gathering was total; they were not cultivators and used animals daily for food, clothing, and shelter for themselves and food for their dogs. Many species of fish, shellfish, birds, and mammals were also used in ritual feasts like the Tabagie. Because of this dependency upon wild things, the cultural and dietary changes that occurred at the time of European contact were rapid and shocking. The Indians' dependency soon moved from mother nature to brother humans (French missionaries and traders in the case of Micmacs). As it turned out, these were changes for the worse. Similar changes apparently occurred across much of North America with differences only in degree, depending upon the diversification of the Indians' natural economies and the importance of the fur trade to both Europeans and Indians. Attendant upon this, however, was provision by the white man of the means to overexploit populations of wild animals. The primitive weapons of the Indians may have been as responsible for their "conservation ethic" as were their beliefs.

In Africa, the Bisa, of the Luangwa Valley of Zambia, once ritualized both the hunt and the kill, and the relationships between the hunter, the animal, and the hunter's ancestors were a strong influence on the hunter and his future successes. The ritualistic position of the animal was most important. Certain animals that were both fat and brown in color, such as the buffalo (*Cyncerus cafer*), were particularly sought for food while others, colored black and white, such as the zebra (*Equus* sp.), were seldom taken. This selective killing based upon religious belief with its attendant rituals, partly determined population levels and might well be considered a sort of culturally induced species management. Some aspects of ancient Bisa culture, as they affect dealing with animals, are still present, but they were already eroding long ago when the ivory trade first brought an economic dimension into the picture. Now bound by laws imposed upon them by government and restricted in their movements by national park boundaries, the Bisa are gradually moving into a modern era more dependent upon developed society and less dependent upon nature. Their altered environment sometimes reflects considerations for game management which may well be less beneficial for some wild species. In some instances, these changes have already made violators out of honest people, as regulations may be based more on biological considerations than cultural tradition.

If we continue to go backward in time to observe the relationship between man and animals, we learn that man has probably always hunted.

> "With man's capacity for both abstract thought and speech, the meal provided by the hunt offered a truly splendid opportunity to become social. There is every logical reason to assume that man talked about the past hunting successes, grumbled about the hunt's failures, and planned the next day's hunt. As he spun his tale, fanciful elements would have been introduced. (Pretty much like the hunting yarns you and I tell). Primitive man could easily have formed the basis of religion and myths.
>
> The meals would have provided an excellent opportunity to choose leaders and to establish various forms of rank within the group. The great hunter would logically have been honored at such feasts, and the unsuccessful hunter might, with equal logic, be demoted or disgraced. Certainly, the criteria for promotion and demotion would be clear in a society so tied to the hunt as its single supreme necessity.
>
> There is also considerable evidence that primitive man thought of the hunt as the most logical subject of his art. In addition to the probable magic content of the cave paintings and rock drawings that early man fabricated with probable reference to success in the hunt, he also exhibited his purely esthetic powers in the creation of weaponry more beautiful than would ever be needed merely for practical purposes. Some of the many truly beautiful knives and spearheads may have been badges of authority or magical instruments, but they are also quite enviable works of art in themselves." (Whisker, 1981).

Regarding Cro-Magnon man about 35,000 B.C., Howell (1971) has written:

> "There are more than 50 known pictures of strange-looking sorcerers or shamans-human figures clad in the skins of animals, sometimes depicted with animal heads or horns, often appearing to be engaged in some kind of dance. These may have been attempts, by illustrating it in advance, to guarantee successful stalking by hunters disguised as animals. Or they may have been more highly symbolized projections of the hunter's feeling that a ritual dance by a magician or spellbinder would work more potent magic on the game. Or they may even have been attempts to portray a superhuman figure, the spirit of the hunt or the deity of the animals."

We can only speculate about these tantalizing and long lost rituals, but they have so many parallels in hunting societies of more modern times that there is no doubt at all that Cro-Magnon man was a ritualist too. Any society that lives by hunting spends most of its time thinking about the animals that it hunts, and many elaborate systems of totems and taboos are still known among hunting tribes today, telling them what they must and must not do. These range from propitiating the spirit of the animal, so that it will submit easily, and gracefully to being killed, to attempts to disarm its spirit after death so that it will not come back to haunt or harm the killer. The cultures of Eskimos, American Indians, and many of the primitive tribes of subarctic Siberia, all of them, like Cro-Magnon man, cold-weather followers of big game, were steeped in rituals of this kind.

Another thing that preoccupies hunting societies is the problem of fluctuations in the game supply. Cro-Magnon man apparently dealt with this in his magic system by emphasizing the fertility of many of the beasts he painted. Pairs of animals were often shown together, sometimes in the act of mating. Horses, does, and cows were painted with the swollen bellies of advanced pregnancy. In others, the udders were enlarged, as if to emphasize the rich supply of milk that the mother would be capable of giving to any offspring that might be born. That scarcity of game was periodically a problem with Cro-Magnon man is likely. During the colder episodes of the last glacial period he probably did all right. Mammoths, woolly rhinoceroses, ibex, a cold-adapted shaggy little steppe horse and particularly, reindeer flourished in large numbers in the tundra environment that came with the cold. As it warmed up, from time to time he undoubtedly switched over to the deer, bison, and wild cattle that replaced the cold-loving species. But increasing numbers of men, and the beginnings of a tendency toward a settled life (in winter, at least) hinted at by cave occupancy, may well have led to local depletion of the game in many areas and seriously complicated Cro-Magnon man's ability to make a comfortable living. If so, he must certainly have turned to sympathetic hunting magic to help him out.

Religion, Man, and Wild Animals

It is understandable that archaic animistic, and early spiritualistic societies would have evolved with animal-human relationships that were beneficial to man. Any other conditions would not only have been unreasonable but could well have slowed man's development or even threatened his existence. The major modern religions, however, usually recognize animals as responsibilities of man, either as hunted prey or as domesticated creatures. Ritualization is still present in the Jewish tradition, and in the past, the initial slaughter by designated people was not only symbolic but also a means of guaranteeing that proper care was taken in the handling of animal food to be eaten. Thus ritual killing logically followed the designation of "clean" or "unclean" as a matter of health and discipline. The people who obeyed the religious laws remained healthy and were reminded of the law each time they ate meat. Mosaic law also forbade man to eat animals which had died a natural death as well as those killed by other predators. Presumably the health of the people was also a concern here. Orthodox Jews do not kill today and ritualized slaughter continues. Neither do Orthodox Jews hunt, for hunting could result in killing, although the Old Testament nowhere forbids hunting. On the contrary, there are references to eating wild meat and hunting. Another reason to restrict the number of people having contact with the act of taking an animal's life (done only by the religious leaders) might have related to a desire to maintain a peaceful, nonaggressive nature among the masses. After all, killing is a violent act. A casual view of history and today's newspapers suggests that either those responsible for the law were wrong in their belief that preventing men from killing animals would necessarily create a more peaceful people or that this reduction of violence had nothing whatever to do with ritualized rabbinical killing. Kertzen (1955) had this to say concerning the sanctioned or "kosher" laws:

> "The Old Testament (Leviticus) sets down certain definite dietary restrictions: 1. It is forbidden to eat the meat of certain animals (such as the pig and horse) and certain sea foods (shrimp, lobster, crab, oyster). 2. Meats must be slaughtered according to ritual and must meet specific health standards. 3. Meat products and dairy products may not be eaten together. (The Bible says that meat must not be boiled in milk. This was a pre-Biblical, pagan custom).
>
> Maimonides, a distinguished physician as well as philosopher, said that "kosher" food restrictions were health measures—particularly in the case of pork, which deteriorates rapidly in warm climates. He also saw important moral values in applying restraint to eating habits—for if we practice discrimination in satisfying our appetite, we may be more self-controlled with the other temptations of life. Many of the laws concerning kosher food deal with the method of slaughtering the animal: it must be done without pain to the beast, with the greatest possible speed, and by a God-fearing man. Incidentally, Orthodox Jews are forbidden to hunt.
>
> Jews who follow the dietary laws do not feel a sense of deprivation. They regard kosher practices as a symbol of their heritage, a daily lesson in self-discipline, and a constant reminder that human beings must feel pity for all living things. How many Jews obey the dietary law today? No one can answer authoritatively. A safe guess is that less 20 percent of the Jews in America conform strictly to the laws governing kosher food."

It is also possible that the dietary restrictions of the Old Testament were prompted by parasitic diseases which caused many deaths before the taboos were in place. Trichinosis in pork and "red tide" in shellfish are but two examples. Both the New Testament and the Book of Mormon include references to man's use of wild animals as food. In the New Testament it was fish while in the Book of Mormon it was wild animals or wild beasts. The liberal Christian church today proclaims a respect and a concern for all matters environmental. For many Christians today, the old-time stewardship often dealing only with financial matters is usually replaced by a "stewardship of life" concept which extends to God's providential sovereignty over all things and making man, as the servant of God and Christ, fully responsible for God's work.

The relationship of man to other animals covers a wide spectrum among the eastern religions of the world. Often animals and man are viewed as possessors of nonphysical souls which upon death may be either relocated or reincarnated in another creature, often another species. Souls or spirits are born with the child or with other animals that may have inhabited a human or a nonhuman, in a previous existence. Animals other than man are often central to religious belief, and some species have always been particularly important in mythology and legend.

Today, the importance of religion and culture continues relatively unchanged from what it was a thousand years ago or more in many societies. Tradition still dictates the role wild and domestic animals will play in the lives of humans in these areas. In what is known as the developed or industrialized countries, our relationships to animals other than man have become much more complex. Sometimes, the manner in which a group of people treats a species, or a population, becomes a public issue advancing into economic and political arenas and competing in the media for a share of our time, our money, and our emotions.

Conflict: Man Versus the Animals

Throughout the world, mammalian herbivores and birds are competing with man for cultivated and wild food. For millions of Africans, the loss of stored grains to small mammals and birds may help keep them close to the edge of starvation. For thousands of commercial farmers in North America and thousands of collectivized cultivators in socialist countries, losses of growing and stored crops can mean a lower standard of living, and for some, less food in the stomachs of their families. In Africa, the marauding elephant or a herd of buffalo on the move can destroy the crops of an entire village in minutes. In North America, deer may eliminate much of the next year's fruit crop through winter browsing. These conditions are well known, and the man-animal conflict for the same food items has certainly been around since the first precultivators gathered their wild plants. What man has done to exacerbate the issue is to make conditions easier for the wild animal by growing large acreages of a single crop favored by one or more species of bird or mammal.

He has also created dumps, dirty harbors, and coastal airports where birds congregate for food or rest. Man's agricultural activities have been partly responsible for increases in the numbers of some of his most effective competitors such as the red-winged blackbird, while the gulls have responded to the "riches" of city harbors along our coastlines.

Predators also compete with man and become depredators of sheep and cattle. The African farmer protects his cattle and his dog from the leopard. The North American sheep farmer guards his flock against coyotes. Sometimes man even gives up his life when a lion hunts near an African village or a tiger inhabits a forest area in India. The predators that have most recently been viewed seriously as competitors with man are the marine mammals. Harp seals and other pinnipeds consume thousands of tons of fish that humans might also harvest or which serve as part of an important trophic level in a complex ecosystem; porpoises and dolphins also consume as much. In Newfoundland, the fisherman competes with whales, particularly pilot whales, that may destroy both gear and fish. Wherever the conflicts occur, wild animals are destroyed. Such conflicts lead us to questions of animal consciousness and awareness to animal rights and human perceptions of both the wild and the domesticated.

Questions of Responsibility

For centuries, there have been a few individuals who have been well-known locally, nationally, and sometimes internationally for their understanding of, and relationship to, certain animals. There are a great many popular works to choose from that consider our origin and our behavior often in relation to other animals, particularly certain primates and cetaceans. Man's closeness to chimpanzees, whales, cats, and dogs may be discussed seriously by some who even claim to "talk" telepathically to animals they know, both near and far. Their claims to personal relationships with individual animals sometimes surpass the experiences of St. Francis of Assisi, Ernest Thompson Seton, and Grey Owl combined. But the communication cannot be all one way, for to continue a monologue without response would hardly be rewarding. The mammals, or birds, must respond of course, and it is the matter of this response that brings us to the question of animal awareness. In a studied review of the question involving both humanists and scientists, Griffin (1976) concluded:

> "Language has generally been regarded as a unique attribute of human beings different in kind from animal communication. But on close examination of this view, as it has been expressed by linguists, psychologists, and philosophers, it becomes evident that one of the major criteria on which the distinction has been based is the assumption that animals lack any conscious intent to communicate, whereas men know what they are doing. The available evidence concerning communication behavior in animals suggests that there may be no qualitative dichotomy, but

rather a large quantitative difference in complexity of signals and range of intentions that separates animal communication from human language."

Human thinking has generally been held to be closely linked to language, and some philosophers have argued that the two are inseparable or even identical. To the extent that this assertion is accepted, and insofar as animal communication shares basic properties of human language, the employment of versatile communication systems by animals becomes evidence that they have mental experiences and communicate with conscious intent. The contrary view is supported only by negative evidence, which justifies, at the most an agnostic position.

Opening our eyes to the theoretical possibility that animals have significant mental experiences is only a first step toward the more difficult procedure of investigating their actual nature and importance to the animals concerned. Great caution is necessary until adequate methods have been developed to gather independently verifiable data about the properties and significance of any mental experiences animals may prove to have.

It has long been argued that human mental experiences can only be detected and analyzed through the use of language and introspective reports, and that this avenue is totally lacking in other species. Recent discoveries about the versatility of some animal communication systems suggest that this radical dichotomy may also be unsound. It seems possible, at least in principle, to detect and examine any mental experiences or conscious intentions that animals may have through the experimental use of the animal's capabilities for communication. Such communication channels might be learned, as in recent studies of captive chimpanzees, or it might be possible, through the use of models or by other methods, to take advantage of communication behavior which animals already use.

We may continue then to accept a difference in the nature of mental experiences between man and other creatures but we are not yet able to measure the consciousness levels of other creatures in relation to man, although communications behavior is rapidly increasing our understanding. We must conclude that animals are certainly aware of man, but we do not yet know how aware and although we may accept that nonhuman animals think, such thought may not equate with human processes. As Richard Byrne (in Small 1996) has suggested, "Thinking is a process that can take known information and work it out; that is, simulate or imagine the consequences of change that haven't happened in the real world. I believe it is very hard to detect this . . . in non human animals."

The question of animal rights, too, is basic but quite different. Where the neurophysiologist and behaviorist can measure neurophysiological responses and stimulus-response times in studying consciousness levels, the rights question cannot be held up to scientific testing as easily. Here we are more likely to be within moral, religious and philosophical parameters. The question of animal rights has been reviewed by Whisker (1981), with the earliest writings dating from about 1723 to the present. Opinions lie roughly in the opposing views that: (a) the animals have rights to remain naturally free and free of pain and maltreatment as clearly as humans have rights, and (b) only humans have rights under God but humans, as stewards, have responsibilities for the welfare of animals. In the case of (a) the questions of domestication, genetic change, animal production and slaughter as well as of habitat degradation and loss, among others, must often be rationalized, and in the case of (b), it is understood that although humans have responsibilities for the welfare of animals, the welfare of humans comes first. This may necessitate the use of animals for medical research, beasts of burden, food, or any other use that may benefit some aspect of man's existence. The differences often boil down to the right of an animal to live versus the right of humans to take an animal's life. We may clarify these matters slightly or we may muddy them further by turning briefly to the domestic and then back to the wild animal. Are they the same or are they different?

One of the interesting views on domestication we have read does not come from either the scientific or humanistic literature but from a children's story, *The Little Prince,* by Antoine de Saint-Exupéry. It occurs when a little boy (the prince) alone in a strange world finds himself confronted by a fox, and a conversation ensues about the word *tame.* We readily admit to the anthropomorphism but judge the lesson worth the risks of its use.

> "What does that mean 'tame'?"
> "It is an act often neglected," said the fox, "It means to establish ties."
> " 'To establish ties'?"
> "Just that," said the fox. "To me, you are still nothing more than a little boy who is just like a hundred thousand other little boys. And I have no need for you. And you, on your part, have no need of me. To you, I am nothing more than a fox like a hundred other foxes. But if you tame me, than we shall need each other. To me, you will be unique in all the world. To you, I shall be unique in all the world . . . "

But is this really what domestication is all about in the world today? Sometimes it is and sometimes it is not.

The reasons man has tamed animals have never been altruistic. The animal's welfare was not a question of concern. It was man's welfare that was to be improved using formerly wild animals as the means. Man has used domestic and domesticated animals to serve as beasts of burden, to provide milk, to give him fur and hides for clothing and shelter, to assure a ready supply of meat, and to increase his pleasure. Cats were probably domesticated in Egypt where they were pampered by those with means. A cat god, Bast, existed and was worshipped in its own city, Bubastis. Dogs, from many *Canis* stocks were tamed in several areas around the world. They often served to aid in the hunt. Micmac Indians gave their best hunting dogs to friends as an indication of respect but also ate dogs at feasts. Pigs were tamed in Europe and Asia and became important as food and in religion and ritual. Pig rituals continue today in Borneo, and sacred cattle continue their fascinating but sometimes nearly marginal existence in India. Pet fads are growing in the world today, and a sizable industry exists to provide cat and dog (or other pet) fanciers with food and various, largely unnecessary, paraphernalia for them. Food is by far the largest industry based upon animals man has tamed, whether it is for pets or animals destined for slaughter, and the tentacles of the animal food industry extend to many areas of most modern economies.

There are two categories of domestic animals. The first category includes the pet, the watchdog, and often, the beast of burden that are creatures affecting the lives of one or a few humans at most. A unique relationship exists; the animal is totally dependent upon a human for its continued existence in that it could not (or would not be allowed to) thrive if it were released or forced to fend for itself. Close ties have been established. The second category includes animals that are raised in meat factories to produce poultry, beef, or pork in a specific manner to meet our peculiar tastes and desires. Literally millions of genetically altered animals are cared for scientifically for a specific time period, then slaughtered and processed for human consumption. Variations occur within each category throughout the world, but basically we are concerned with either an intimate one-on-one animal-man relationship in which the animal is meant to live as long as possible, or a mass production of 1000 or more animals to one human relationship in which the animals are meant to die at the time they provide us with either the choicest food, the most profit, or both.

Clutton-Brock (1981) divides animals in captivity into exploited captives where breeding is under the influence of natural selection and man-made animals which are essentially controlled by artificial selection. She considers cats and elephants exploited and dogs and domestic livestock man-made.

Because man first tamed wild animals and then, through selective breeding, reduced many once wild creatures to almost complete dependency, our responsibility to these animals should be very great. The domesticated creature usually can not get along without man and the specific conditions it has been bred to exist in. Dependency is often unilateral, although it is true that domestic animals may fill a definite human, social or psychological need in some instances. The responsibility we have for the domestic animal is often to an individual animal, and whether to individual or group, never extends beyond a confined situation. That means that a population or a species is not a matter of concern here. Management is intensive and it is the individual animal that is most important.

Man on the other hand has not reduced wild creatures to a dependency requiring his daily attention although, to be sure, the wild animals' continued existence depends upon him. The individual wild creature is not usually dependent upon one or a few humans. Our responsibility is not to the individual but to a species, and often to a population of wild animals. As the fox told the Little Prince, this is the difference between the tame and the wild. With tamed or domesticated species, primary responsibility is to the individual or group, and with wild animals, primary responsibility is to the species, or population. The Little Prince would have a specific responsibility to one fox if he tamed that fox. Perhaps, as with the Little Prince, we have a collective responsibility to wild populations and to all species. In dealing with the many wild animal-human conflicts such as those between dolphins, seals, and man all preying upon one food supply, the questions of animal awareness, animal rights, and the differences in our responsibilities between the tame and the wild may all come into consideration. The wildlife manager must try to understand as many of the conflicting scientific, humanistic, and historical factors as possible as well as why different people feel as they do.

Attitudes and Values

Few people in our developed societies will fail to express an opinion on an issue involving wild birds or mammals and many people in underdeveloped countries also will speak out. Sometimes it is obvious that a person's opinions are affected strongly by the degree of dependence on a wild animal population. In other cases, it is obvious that a farmer suffering losses from depredating deer will have opinions on what should be done with the deer! For others, however, the reasons for their emotions about wild animals and issues concerning them are less clear-cut. Wildlife managers should be knowledgeable about attitudes and the elements involved in developing them among the many human population subgroups concerned in any wildlife question. Possibly the only way of gaining this understanding in each instance is to conduct intensive surveys, sampling the various subgroups present. For our purposes, an understanding of how attitudes may differ and what subgroups there are, is perhaps most important.

One of the most detailed studies involving attitudes in North American society was completed for the U.S. Fish and Wildlife Service through the Yale University School of Forestry and Environmental Studies. The attitudes believed to exist among American publics as defined by Kellert (1980a) are as follows:

Naturalistic: Primary interests and affection for wildlife and the outdoors.
Ecologistic: Primary concern for the environment as a system, for inter-relationships between wildlife species and natural habitats.
Humanistic: Primary interest and strong affection for individual animals, principally pets. Regarding wildlife, focus is on large attractive animals with strong anthropomorphic associations.
Moralistic: Primary concern for the right and wrong treatment of animals, with strong opposition to exploitation of and cruelty toward animals.
Scientistic: Primary interest in the physical attributes and biological functioning of animals.
Aesthetic: Primary interest in the artistic and symbolic characteristics of animals.
Utilitarian: Primary concern for the practical and material value of animals.
Dominionistic: Primary satisfaction derived from mastery and control over animals, typically in sporting situations.
Negativistic: Primary orientation on active avoidance of animals due to fear or dislike.
Neutralistic: Primary orientation a passive avoidance of animals due to indifference.

As might be expected, Kellert found that moralistic and utilitarian attitudes conflicted "around the theme of exploitation of animals," while the negativistic and humanistic attitudes tended to conflict "around the theme of affection for animals." A general picture of public attitudes concerning wildlife in the United States as derived from the Kellert studies is provided in Table 2.1.

The kinds of factors influencing or forming attitudes in North American society are extremely diverse. Perhaps this is a reflection of the cultural complexity and the rapidly changing concepts affecting developed societies today. Kellert examined regular demographic parameters such as sex, age, marital status, education, income, and profession; he also checked wildlife-related or animal-related interest groups, some occupations, and whether or not individuals attended church (termed religiousity in the reports). We might simplify this matter somewhat by looking at attitude formation as reflections of what we do or what we are, and who we are.

Attitudes are not always inclusive; not all wildlife may be viewed as sacred, worthy of preserving, or worthy of killing—depending upon the issue. Rats, for instance, are viewed quite differently from deer both by those who would preserve deer and those who would hunt them. It depends upon the animal. All lives of wild creatures are not equal in the eyes of humans except in a few instances, perhaps, such as with the Jains of India, or a certain few individuals who extend a reverence for life to animals, as Schweitzer appeared to have done for a part of his life. Meadow voles are viewed differently from seals, and starlings differently from warblers. Such variability can lead to an overemphasis on a narrow segment of wildlife, as noted by Kellert (1980a), and "overlook more basic considerations of ecological relationships between wildlife and their natural habitats."

Table 2.1 — Attitude Occurrence in American Society[1]

Attitude	Estimated % of American Population Strongly Oriented Towards the Attitude[1]	Common Behavioral Expressions	Most Related Values/Benefits
Naturalistic	10%	Outdoor wildlife related recreation— backcountry use, nature birding and nature hunting.	Outdoor recreation
Ecologistic	7%	Conservation support, activism and membership, ecological study.	Ecological
Humanistic	35%	Pets, wildlife tourism, casual zoo visitation	Consumptive, utilitarian
Moralistic	20%	Animal welfare support/membership, kindness to animals	Ethical, existence
Scientistic	1%	Scientific study/hobbies, collecting	Scientific
Aesthetic	15%	Nature appreciation, art, wildlife tourism	Aesthetic
Utilitarian	20%	Consumption of furs, raising meat, bounties, meat hunting	Consumptive, utilitarian
Dominionistic	3%	Animal spectator sports, trophy hunting	Sporting
Negativistic	2%	Cruelty, overt fear behavior	Little or negative
Neutralistic	35%	Avoidance of animal behavior	Little or negative

[1]From Kellert, 1980, in Shaw and Zube, Eds., 1980.

In addition to this selectivity of human attitudes toward wildlife, views are often a partial reflection of what people do. This is particularly true of those who obtain their living from the land as farmers, woodsmen, and trappers. In each instance, the resource as an economic entity is of primary importance, as might be expected. Farmers and agriculturalists in ancillary services and industrial arenas may view the crop as paramount. If a wildlife species is a pest requiring control, their view of the species is likely to be negative. If a farmer is also a naturalist, a hunter, or both, his views may sometimes be ambivalent. Or if no wildlife species affects the individual detrimentally and that person has a naturalist's or hunter's interest, his view may be positive. On balance, the general view from agriculture is likely to be negative, as are many of the agricultural land use practices which reduce type diversity or eliminate wetland areas. Of course, some agricultural practices have also benefited many wildlife species.

Foresters, too, will normally have a negative view of any species that may damage reforested areas or restrict natural regeneration and stocking rates. In many instances where single species tree management is encouraged on natural mixed forest sites, wildlife species will suffer because of the decline in plant diversity. Harvesting practices can be either detrimental or beneficial to certain species of wildlife, but since economic considerations will generally determine what these practices are, they will often end up being harmful. As in the case of agriculturalists, and indeed with people in any profession, anywhere, there are some who view wildlife ambivalently and some who are sympathetic to it.

Trappers tend to see wildlife as an economic entity to be managed in order to provide maximum numbers of animals for harvest and optimum numbers for habitats. They usually have a positive view, wanting to retain species and maintain populations as high as the habitat will allow. They may also be naturalists and invariably are more familiar with their prey's habits than is either the average hunter or the average preservationist. Successful trappers must be knowledgeable and their view is neither more nor less selfish than that of preservationists or farmers. It is in regard to trapping that the humaneness issue has become most important in affecting the management of wildlife. Nowhere has the humane movement been more effective, for better or for worse as far as species are concerned, than in North America and most especially in Canada. The Federal-Provincial Committee for Humane Trapping report (1981) is an example of people's concern with the individual wild animal rather than the population or the species. It resulted in changes in trapping regulations in most Canadian jurisdictions.

Apart from the primary land-based occupations and professions there are a few generalities worth considering. It is obvious perhaps, that if there is an economic dimension involved, a person's view will tend to favor whatever is seen to benefit the species in order to continue benefitting the pocketbook. Job or profession related familiarity with wildlife may also help formulate a person's view. Usually the less direct contact people have with the land and its ecosystems, including wildlife, the more likely they are to have protective feelings for all animals, especially if their first experiences in natural settings begin later in life.

Another area which may influence people's views about animals is related to what they are: their sex, age, income, education, and whether they own land or not. Where they live may also be important and whether they are first or second generation residents of the city, country, or certain geographic regions (as Kellert studied in relation to Alaska, the South, and the Rocky Mountain states). Sometimes what or who we are may be the most important element in determining our views on wildlife. Immigrants from a culture with a utilitarian concept of animals may well have different attitudes from a fourth generation American or Canadian whose ancestors came from England. Our ethnic origins may affect our opinion, and sometimes so may our personal theology.

Understanding the basis of attitudes among the various publics they deal with can be helpful to wildlife managers who face a great challenge in coping with the conflicting and politicized views expressed, concerning wildlife issues today. As Kellert (1980a) has noted:

> "While the frequency of positive feeling and concern for animal welfare in America today is somewhat pleasing to note, the emotional rather than intellectual basis for this interest, and its greater focus on pets and limited wildlife species, poses some potential problems."

And further that:

> "It will require much patience, empathy and tolerance, and a willingness to be involved with many different kinds of people. The challenge is great, but so are the stakes, and the future well-being of our wildlife resources may depend on the outcome."

The attitudes expressed by Americans in the studies referred to here may be different only in degree from those in other developed countries. In 1981, the European Common Market voted to restrict the sale of manufactured seal commodities in Europe, and this decision was a reflection of the seal hunt controversy in Canada. In the 1990s, the European Union has tried to ban the importation of fur from certain wildlife species captured with leghold traps, reflecting the controversy over trapping in North American society. Perhaps then, attitudes are not greatly different among the industrialized societies, at least concerning issues such as seals, furbearers, and whales. We know, however, that attitudes may be very different between populations of underdeveloped countries and those of industrialized nations.

Some human attitudes are closely tied to economic interests, but wildlife values include both conventional economic categories which normally involve an expenditure in time, money, or participation and the less tangible economic categories in which the valuation of an experience involving wildlife is more difficult. Wildlife values include a number of categories that either do not enter a market (or national) economy, cannot readily be estimated in comparative market terms, or both. In recent years economists, sociologists, psychologists, and wildlife managers have all attempted to develop means of measuring wildlife values. Giles (1978) provides a list of twenty-five ways to measure them, and Steinhoff (1980) reviews several conceptual systems used for this purpose (Table 2.2). The classification or typology used by a wildlife manager will often depend upon the type of evaluation, survey, or report involved. We consider the King (1947) classification to be easily understood and applied. According to King, all wildlife values and services may be included under six general headings as follows:

1. Commercial values: income derived from sales of wild animals or their products or from direct and controlled use of wild animals and their progeny
2. Recreational values: monies expended in the pursuit of wildlife in connection with sports and hobbies
3. Biological values: the worth of the services rendered man by wild animals
4. Social values: values accruing to communities from the use of wildlife and values associated with organizations that exist because of a common interest in wildlife
5. Esthetic values: the values of objects and places possessing beauty, affording inspiration and opportunities for communion, contributing to the arts, etc.
6. Scientific values: values realized through the use of wildlife as a means of investigating certain fundamental and widespread natural phenomena

All of these value categories can be subdivided and some values can be considered to be both social and scientific (the use of animals in cancer research can create medical and social benefits); or biological and social (intensive feeding of predators on pest species is a biological value that translates into commercial and/or social benefits). Other groupings may be apparent to you as well; however, there are often problems of precise definition in dealing with

Table 2.2—Classification Systems Proposed for Wildlife Values or Related Values.[1]

Author of System	Basis of System	All[2] Values	All[3] Uses	Exclusive[4]	Categories of Value		
King (1947)	The Experience	Yes	Yes	Yes	Recreational Aesthetic Educational	Biological Social Commercial	
Hendee (1969)	The Experience	No	No	Yes	Appreciative Consumptive Passive Free-Play	Sociable Learning Active-Expressive	
Hendee (1974)	The Experience	No	No	Yes	Back-Country Hunt General Season Party Hunt Meat Hunt Special Skills Hunt Fly Only	Cast-drifting Boat-drifting Plunking etc. (example)	
Shaw (1974)	The Experience	No	No	Yes	Utility or Nuisance Consumptive Recreational Aesthetic or Existence		
Hendee (1974)	Elements of Experiment	No	No	Yes	Solitude Companionship Escapism Nature Appreciation	Outdoor Skill Trophy Exercise	
Nobe & Steinhoff (1973)	Economic Interest	Yes?	No	No	Direct Users Primary Beneficiaries Secondary Beneficiaries Alternative Resource User	Vicarious User Altruist Environmentalist Option Holder	

Swartzman & Van Dyne (1975)	Quality of Life	Yes	No	Yes	I. Economic 1. Income per capita 2. Employment stability 3. Net regional product change 4. Income distribution II. Ecological 5. Ecological degradation 6. Environmental quality index 7. Percentage use of renewable resources 8. Annual percent usage of non-renewable resources 9. Man-initiated energy consumption III. Sociocultural 10. Population size 11. Social differentiation 12. Cultural heterogenity 13. Sociopsychological 14. Information advantage IV. Political 15. Scope of governmental services 16. Uses of governmental services 17. Political participation 18. Property tax base 19. Political power advantage 20. Dollar investment
Kellert (1978)	Attitude	Yes	No	No	Utilitarian Naturalistic Dominionistic Humanistic Ecologistic Moralistic Negativistic Knowledge of Animals Scientistic Aesthetic
More (1973)	Attitude	No	No	Yes	Display Esthetic Affiliation Pioneering Kill Exploration Challenge

Man and Wildlife • 59

		²Includes all wildlife values?	³Applicable to all uses?	⁴Mutually exclusive categories?	

Langford & Cocheba (1978)	Sources of Activities	No	No	Yes	I. Current Period Values
					  A. Sensory Perception Values
					    1. Recreational hunting activity
					    2. Non-hunting recreational activity
					      a. Wildlife-based activities
					      b. Wildlife-related activities
					      c. Endemic-wildlife activities
					      d. Recording-based wildlife activities
					  B. Existence Values
					    1. Contemplative wildlife activities
					II. Future Period Values
					  A. Option Values
					    1. Option demand activities

Rolston (1979)	Philosophical Criteria	Yes	Yes	Yes	Economic Life-support Recreational Scientific Aesthetic Life-intelligibility Plurality-unity Stability-freedom Dialectical-environmental Sacramental
Raths et al. (1966)	Educational Criteria	No	No	No	Money Friendship Love and sex Religion and morals Leisure Work Family Maturity Character traits Politics and social origin

[1] Steinhoff, H. W. 1980. Analysis of major conceptual systems for understanding and measuring wildlife values. In W. W. Shaw and E. H. Zube, eds. Wildlife values. USDA Forest Service.

[2] Includes all wildlife values?
[3] Applicable to all uses?
[4] Mutually exclusive categories?

any value classification. Some people feel for instance that the term consumptive for a resource entity removed from its habitat and nonconsumptive for those entities used and enjoyed but not removed from their natural settings are not definitive. The argument here is that human impact on certain nonconsumptive natural areas so alters the setting that the original condition no longer exists. It is effectively "consumed."

Obviously, it is necessary to have some means of evaluating wildlife in order to measure elements of the resource and compare the measurements with those of other resources. It is also important in many instances to combine all values: the economic (or commercial, including recreational in King's classification), the sociocultural (including esthetic), ecological (including biological), and the scientific in order to provide the kinds of measurements and demand data needed to develop the required management activities, which when supported by adequate budgets provide for effective wildlife programs. In other instances, competitive agencies or interest groups may be involved in planning for the same land unit. Then not only the total wildlife resource must be measured, including the vegetation, fishes, and invertebrates, but also the land itself relative to its productive capacity for both wildlife and competitive resources.

Besides consideration of the wildlife itself, the evaluation process has other significant aspects. Appraisal of relevant wildlife activities and programs is important in maintaining budgets and personnel and effecting changes in research and management policy. In Canada, the federal government has a built-in evaluation methodology within the annual budget process for each wildlife branch and program. In addition, regular in-depth evaluations are made every few years. Similar annual assessments requiring rationalization and budget defense are also commonplace in most other North American jurisdictions.

The many questions of attitudes, ethics, cultures, and socioeconomic values relative to wildlife encompass far too broad an area for all wildlife managers to keep fully abreast of. Perhaps the best most of us can do is to be aware of the literature, the professions, and interest groups in the complex. We can then tap the resources and expertise required each time we are faced with human social problems we may not be adequately trained or equipped to deal with.

Many issues wildlife managers face today deal with areas of conflict among people or between animals and people. They may not be whale-fish-human conflicts or utilization versus preservation issues, but will often involve justification of management activity. This justification is usually required of senior wildlife officials, but it may also be necessary for biologists and managers to communicate with several publics that benefit from, or are interested in, the resource in question. Since some groups may disagree relative to policy decisions concerning use, the wildlife manager may have to defend his activities to various groups of preservationists and sportsmen at the same time. Your task will not be easy even though you may have an effective education and information unit to work through. The manager's role will be increasingly difficult as attitudes become more widely expressed and politicized.

Wildlife managers must prepare themselves on a broader front than they have done in the past. Today, in addition to the scientific, biological, or ecological data bases they normally obtain for their management project they must also:

- possess an understanding of their responsibility to the species, the population and the ecosystem in question.
- understand the views of all interest groups, including those benefiting directly and those not benefiting directly but who might influence policies through expression of a moralistic, humanistic or other attitude.
- have an understanding of the various values of the resource and how the resource and the policy toward it complement the total management program.

Culture, Conflicts, and Values—An Example

Although we do not usually have to search to find conflict in relation to management of wildlife, there are some areas of our globe where the cultural and economic bases of conflict are exemplified better than in others. One such area, until recently, has been Trinidad and Tobago, two islands close to the coast of Venezuela, that obtained independence from Great Britain in 1962. There are about 400 species of birds, and 100 mammals (including 58 bats), 600 butterflies and some 70 reptiles present on the islands. Trinidad was once a part of the South American mainland and reflects a continental, rather than a Caribbean flora and fauna.

In 1984, there was a six-month open season on legally hunted animals, no bag limit, and either sex could be harvested. Wild meat was sold legally during the open season and often illegally during the closed season. Trap guns, deadly devices ostensibly set for game, were "hunting" 365 days of the year and often killed dogs and maimed humans. Birds were frequently trapped and sold inside the country and sometimes smuggled into the United States. The Game Ordinance placed all bats on the vermin list to be killed anytime by any means. Many reptiles, birds, and other mammals were also declared vermin. The legal game species included cayman (alligator), all lizards, the agouti, the paca, the deer (*Mazama* sp.) the armadillo, and peccary. Among the legal birds were scarlet ibis (a national emblem), plovers, ducks, herons, cormorants, rails, sandpipers, and many more. Some species of cage birds are already extinct, others are endangered and some mammals including the ocelot and tayra are threatened.

There are four groups with specific interests in wildlife in Trinidad. One is a small group of naturalists and scientists sharing a common desire to protect and conserve. A second is a much larger group of legal hunters, a minority of whom are conservation oriented and a third group is the commercial hunters who obtain a part of their living from wildlife. A fourth and highly influential group is composed of the consumers of wild meat. In 1984, there was a ready market for wild meat which brought as much as $40 a kilo in local currency. In addition, farm land abandonment, illegal squatting in the forests, hillside quarrying, and deforestation had reduced diversity and altered habitats.

In relation to the enforcement of wildlife regulations, the following letter written by a local warden tells something about the problems existing in 1984:

> "Firearm owners lend their guns for other hunters to shoot wildlife. Today there is a rapid increase of unlicensed and trap guns in the forest. Yet as a Forest Officer or Game Warden you've no authority to request the person's Firearm User's License. This is a very dangerous situation.
>
> It is the Game Wardens and Forest Rangers who are exposed to all the dangers of traps, hunter's menacing threats and yet we are not armed. Many officers are often abused and insulted, they know of a lot of illegal activities practiced out there but fear and frustration keep them quiet. As an officer you have authority to charge people for committing an offense. Yet, I believe it is very tough for an unarmed officer to even try to talk to four men carrying unlicensed guns. You see, in the law of nature, self preservation is first. Morale in staff is very low. There is no spirit of brotherhood among the senior personnel of Wildlife Staff and that gap seems to be growing wider daily. Certain young officers also believe that they are superior to men with fifteen to twenty years service by virtue of their E.C.I.A.F. training. I have been threatened, a man tried to chop me and I was not armed. I had to run to save my life. Obviously this has me very depressed and sad.
>
> Too many people are granted permission to hunt. This should be controlled and all hunters should be members of a bona-fida Association. There should be complete ban on the sale of wild meat at the present time. However, if there are projects such as Wildlife farming, then the sale of wild meat from the farms should be approved. All of the protected species such as porcupines, red howler monkeys and mataperro (anteater) are shot ruthlessly and sold to unsuspecting consumers as the more popular species.
>
> Rapid exploitation of the forest such as squatting for short term crops, marijuana cultivation, quarrying and felling of Forest fruit trees have also affected Wildlife populations tremendously.
>
> Providing that hunting continues at the present rate for two decades, our wildlife will become very negligible. Despite all the appeals by the few interested Naturalists and a bunch of frustrated officers, very few hunters hear the voices "crying in the wilderness." Their sole intention is to kill and eat. They shoot preg-

nant deer, agoutis and lappes. I have heard reports of an extremely greedy hunter killing thirty pregnant agoutis during the close season.

Parents allow their children to trap robins, use laglee to trap birds. I know of mothers who are so greedy for wild meat that they allow their sons to hunt for hours during the night for manicous (opossum), deer, tatoo (armadillo) and lappe (paca). They usually serve visitors with the wild meat favors.

It is time administration make a positive decision or allow the Wildlife to become extinct by greedy hunters out there."

Progress in conservation has been made in Trinidad in recent years, however, conditions elsewhere in many poorly developed tropical jurisdictions such as Haiti, are still resulting in the rapid loss of populations, habitats, systems, and species. Challenges to change often include problems of poverty, demographics, greed and fear that reduce the effectiveness of even the most basic conservation approaches.

Bibliography

Burley, D. V. 1981. Rapid culture change and the fur trade: A case for the Micmacs of northeastern New Brunswick. Parks Canada, Winnipeg (unpubl.).
Clutton-Brock, J. 1981. *Domesticated Animals from Early Times*. Univ. of Texas Press, Austin, TX.
Dodds, D. G. 1976. Evolution of wildlife harvesting systems in Africa. Trans. Fed.-Prov. Wildl. Conf. 40:106–113.
Dodds, D. G. 1982. Micmac food resources at European contact. Can. Ethno. Soc., Vancouver, British Columbia (unpubl.).
Enos 1:3.
Federal-Provincial Wildlife Conference. (1981). *Report of the Federal Provincial Committee for Humane Trapping*.
Genesis 10:8–9.
Giles, R. H., Jr. 1978. *Wildlife Management*. W. H. Freeman, San Francisco,CA.
Gray, P. A., P. Boxall, R. Reid, F. Filion, E. DuWors, A. Jacquemont, P. Bouchard, and A. Bath. 1993. The importance of wildlife to Canadians: results from three national surveys. Proc. Intl. Union Game Biol. XXI Congr., Can. For. Serv., Chalk River, Ontario, Vol. 1:151–157.
Griffin, D. R. 1976. *The Question of Animal Awareness*. The Rockefeller University Press, New York, NY.
Howell, F. C. 1971. *Early Man*. Time-Life Books, New York, NY.
John 21.
Kellert, S. R. 1979. Public attitudes toward critical wildlife and natural habitat issues. U.S. Fish and Wildlife Service, Washington. Report by National Technical Information Service, U.S. Dept. Commerce, Springfield, VA.

Kellert, S. R. 1980a. Contemporary values of wildlife in American society. In: W. Shaw and E. Zube (eds.), *Wildlife Values,* pp. 31–60. Institutional Ser. Report No. 1, Center for Assessment of Non-commodity Material Resource Values. Rocky Mt. Forest and Range Exper. Stn., U.S. Forest Service.

Kellert, S. R. 1980b. Activities of the American public relating to animals. U.S. Fish and Wildlife Service, Washington. Report by National Technical Information Service, U.S. Dept. Commerce, Springfield, VA.

Kellert, S. R. and J. K. Berry. 1980. Knowledge, affection and basic attitudes toward animals in American society. U.S. Fish and Wildlife Service, Washington. Report by National Technical Information Service, U.S. Dept. Commerce, Springfield, VA.

Kertzer, M. N. 1955. What is a Jew? In: *A Guide to the Religions of America,* pp. 65–72. Simon and Schuster, New York, NY.

King, R. T. 1947. The future of wildlife in forest land use. Trans. N. Am. Wildl. Conf. 12:454–467.

Langford, W. and D. Cocheba. 1978. The wildlife evaluation problem: a critical review of economic approaches. Can. Wildl. Serv. Occ. Paper 37.

Leviticus 17:15.

Luke 5.

Luke 24:41–43.

Lutz, W. 1993. Attitudes toward nature conservation and hunting. Proc. Intl. Union Game Biol. XXI Congr., Can. For. Serv., Chalk River, Ontario, Vol. 1. 210–214.

Mark 6:38–44.

Martin, C. 1978. The war between the Indians and the animals. Nat. Hist. 87:92–96.

Saint-Exupery, A. de. 1943. *The Little Prince*. Harcourt, Brace and World, Inc., New York, NY.

Small, M. F. 1996. These animals think, therefore . . . Nat. Hist. 105:26–30.

Steinhoff, H. W. 1980. Analysis of major conceptual systems for understanding and measuring wildlife values. In: W. Shaw and E. Zube (eds.), *Wildlife Values*. pp. 11–21. Institutional Ser. report No. 1 Center for Assessment of Non-Commodity Natural Resource Values. Rocky Mt. Forest and Range Exper. Stn., U.S. Forest Service.

Whisker, J. B. 1981. *The Right to Hunt*. North River Press, Croton-on-Hudson, NY.

Recommended Readings

Clutton-Brock, J. 1981. *Domesticated Animals from Early Times*. Univ. of Texas Press, Austin, TX. A well documented account of how animals came to be used by man; when and where it first happened.

Griffin, D. R. 1976. *The Question of Animal Awareness*. The Rockefeller University Press, New York, NY. An eminent scientist considers animal resources from a balanced, unemotional viewpoint.

Herscovici, A. 1985. *Second Nature*. CBC Enterprises, Toronto, Ontario. An examination of the animal-rights controversy as exemplified by sealing and trapping issues in Canada.

Stanford, C.B. 1999. *The Hunting Apes, Meat Eating and the Origins of Human Behavior*. Princeton Univ. Press, Princeton, NJ. A discussion of meat sharing within primate and early human societies and its possible effects on today's social structures and human behaviors.

Whisker, J. B. 1981. *The Right to Hunt*. North River Press, Croton-on-Hudson, NY. An unabashed defense of hunting including an unusual review of the literature.

3 NATIVE AMERICAN ACCESS TO WILDLIFE

For the past 30 years the "rights" of native Americans to wildlife have been the subject of news articles throughout North America. This is a complex issue of considerable importance to wildlife managers as indigenous peoples have historical and cultural ties to wildlife and because a number of treaties between native bands, or tribes, and governments in North America, or elsewhere, representing colonists recognized that native peoples had certain rights when the treaties were written. For many natives, "ownership" is not a concept they can apply to wildlife. Many accept that they have a stewardship responsibility to wildlife and reject a jurisdiction's right to regulate. This has created both problems and opportunities for wildlife managers whether they work for state, provincial, or tribal agencies.

Before the Europeans

The first humans to live in North America were hunter-gatherers who moved across a tundra landscape from the Kamchatka Peninsula of northeastern Asia into Alaska. The isthmus was wide, perhaps 1000 km, and the people were in search of large animals for food. They were technologically advanced, using projectile points to kill, making fire for warmth and cooking, and tailoring skins for clothing. According to Snow (1996), available evidence would have these people we refer to as Paleo-Indians entering Alaska around 14,000 B.C., although there may well have been "failed" crossings previously. By 10,000 B.C. Paleo-Indians were widespread in the Americas and the retreating glaciers had eliminated the great isthmus that had allowed American continental settlement to occur. There also is the possibility that the populating of the Americas was assisted by Polynesian contacts in South America. For perhaps 4000 years the hunter-gatherers populated the continent, killing animals and foraging for edible plants as landscapes warmed, new forest communities developed, and plants and animals moved northward. When the initial expansion slowed,

68 • *The Philosophy and Practice of Wildlife Management*

restricted primarily by ocean waters, their movements lessened and groups "settled" in large well defined hunting territories. The evolution of North American cultures is understood to a great extent from the interpretation of artifacts as they related to various strategies for the inhabited environments, such as strategies for killing large creatures safely, for catching large numbers of fish, and for surviving critical winter periods, spawning run failures, or changes in caribou migration times. All such strategies developed over time and allowed increased longevity and infant survival.

Several Archaic cultures followed the Paleo-Indians, evolving from the hundreds of established Paleo-bands and adapting further within all the environments that North America offered, from maritime to desert, over the period between 8000 B.C. to sometime after 4000 B.C. The first cultivation, a tending of wild plants in situ, probably began in the Southwest as early as 3000 B.C. and by 1500 B.C. cultivation of sunflowers, squash, lamb's quarters, pigweed and other plants was becoming widespread south of the 120 days frost-free line. In the warm Southwest, maize (corn) and beans were farmed as early as 1500 B.C. and over the centuries maize gradually became the most important cultivated staple for native North American cultivators. It was important to natives in the central United States by 200 A.D. and was a primary food in the east by 800 A.D. The evolution of agrarian communities did not, however, eliminate hunting and gathering although regular harvests of foodstuffs did supplement wild food diets. Hunting, fishing, and gathering remained important to all Native Americans and, along with trade goods from tribes to the south, supported all northern peoples. Seasonal food strategies involving hunting, fishing, gathering and storage of domestic crops as they probably occurred prior to European contact are detailed throughout the literature for areas and tribes across the continent.

At the time of the Columbus voyages which were designed to seek trade routes across water from Europe to the Orient, native North Americans were a highly diverse people with complex social, political, and economic systems interacting with one another. Their lives revolved around seasonal subsistence cycles and a number of attendant ceremonies and feasts such as the potlatch ceremony in the Northwest, the sundance in the Southwest, and the many hunting, war, wedding, and funeral feasts. There were major population and trade centers but most people lived in one or more permanent or seasonal villages and were intimately familiar with the hunting and gathering areas of both the lands and waters they travelled. Conflicts among tribes sometimes occurred and territories often expanded or contracted as a result. Slavery occurred too, often based on economics and the importance of seasonal labor. Complex representative systems of government were in place and across the continent the diversity of social systems and practices was great. Native numbers at the time of Columbus are not known precisely but conservative estimates for the area north of Mexico suggest between 1,213,000 and 2,639,000 while liberal estimates range up to 7,000,000 (Trigger and Swagerty, 1996).

After European Contact

The first Europeans to set foot on the North American continent, whether Norsemen or Spaniards, were themselves descendants of hunter-gather societies. In 1500, they lived in a world of commerce, competition, and political intrigue. Their hunting and gathering "roots" had been forgotten and they exhibited a technical level of development totally different from what they encountered in North America. It was as if two foreign worlds had suddenly been joined. Evolving from the same origins, their different paths had brought these groups of humans together again after thousands of years of separate social development. According to Trigger and Swagerty (1996) overt reaction of the Indians ranged from "terror and flight to fascination and reverence" and the newcomers were sometimes viewed as spirits, as being supernatural, or as powerful shamans. But it took only a few years of contact for the Indians to understand that Europeans were, indeed, human beings. The early reactions were individual and specific to the tribal group yet as varied as their societies. Ironically, most whites did not recognize Indians as human beings and specific language was necessary in treaty councils to demonstrate that the white negotiators viewed them as equals.

Perhaps the most important effect of early European contact on native peoples was the introduction of diseases for which the Native Americans had little resistance. There were, of course, endemic diseases and parasites affecting Native Americans prior to European contact. Dysentery, influenza, pneumonia, arthritis, tuberculosis, cancer, syphilis, and numerous ecto- and endo-parasites brought illness and death. The average pre-contact life expectancy of Native Americans has been estimated at twenty-five to thirty years. The Europeans and their animals brought different diseases that prevented great longevity in Europe but proved far more devastating to the new hosts. Smallpox, measles, diptheria, scarlet fever, other diseases and disease organisms caused epidemics and a few pandemics. The Europeans, however, encountered venereal syphilis and carried it back to Europe where it resulted in thousands of deaths. In parts of northeastern North America, changes in diet following contact also caused illness and death. With so many dying, native social structures were weakened, their ability to provide food for themselves was reduced and many native populations declined drastically in the sixteenth and seventeenth centuries, then more gradually until the nineteenth century. Of course population declines were not always tied to biological causes like diseases and dietary change. For some, like the Cayuga and Seneca tribes of the Iroquois Nation who were unfortunate enough to be caught in the struggle between the colonists and the British, it was conflict. According to General Sullivan's report, "Forty Indian towns (in the Genessee region), the largest containing 128 houses were destroyed. Corn—160,000 bushels shared the same fate—fruit trees were cut down—Indians were hunted like wild beasts, till neither house nor fruit tree, nor field of corn, nor inhabitant remained . . . " (Kim,1900).

For some 400 years, Indians in North America dealt with the British, French and Spanish governments, and their colonists, and then with colonial powers by appeasement, accommodation, aggression and escaping across borders first of territories and then from the United States into Canada. Their trading goods changed following contact and their traditional trade routes were disrupted. Their intertribal socialization also was lost, often completely, but they clung to many aspects of their various cultures and they also continued to use wildlife. They hunted and fished, as did the colonists, even to the point of destroying total populations when wildlife entered a market economy to feed the growing numbers of colonists and provide the Europeans with furs. In a few instances, natives sought protection through treaties or other agreements when their people were being attacked or when they suffered for lack of food. Treaties and Royal proclamations have recently provided the bases for judicial decisions and have affected government policy relative to natives and wildlife.

Government Policies in the United States and Canada

Following the American Revolution, government policy, in Canada, was directed toward efforts to bring native peoples out of a subsistence economy into a general market economy. Governments' aims were to "civilize" the people and reduce native occupation of their original domains or, at least, to reduce the sizes of the lands occupied by Indians. In the United States, the policy was to separate Indians and whites and the policy continued until the allotment period. Until late in the nineteenth century, Native Americans still depended heavily on farming, trade goods, and wildlife. By 1900, a combination of a subsistence economy supplemented by earnings from employment outside reserves (Canada) and reservations (United States) was being threatened by reduced wildlife resources, increased numbers of non-native people and a reduction of the Indian land base. Most Indians in both countries were surrounded by non-natives in 1900. According to Hoxie (1996), "In Canada, 100,000 natives (1.3 percent of the population) lived on five million acres (2.02 million ha) of land divided into 1,500 reserves. In the United States, approximately 250,000 natives lived among 76,000,000 other people on 200 reservations, occupying nearly 72 million acres (29.14 million ha)."

Throughout the early part of the twentieth century, government efforts to suppress native rituals and force natives into the mainstream of North American society continued but in the 1930s the Indian Reorganization Act (1934) and several programs included in Franklin Roosevelt's New Deal (The National Recovery Act) began to assist the Indian people, albeit in a patronizing way. Native Americans had access to better education through contracts with public school districts. Public service employment was provided through the Indian Conservation Work Program. Several Indian enterprises were government supported and some 2.75 million acres (1.1 million ha) were added to the Indian domain in the United States (Hoxie, 1996). Indians in Canada received fewer benefits than their American counterparts during the 1930s but native political agitation did increase and their cultural concerns gradually began to get attention from the federal government.

Only in the past few years have the attitudes of the federal governments of both countries shifted from dominating paternalism and a welfare approach to natives on Indian land. Since 1960, and especially in the last two decades, the pace of change has quickened. Self government has been evolving and legislative and policy changes have affected health care, education, and entrepreneurship positively. Combined with this, tribal governments were taking actions to enhance wildlife populations and regulate recreational activities on their lands. The need for the Mescalero decision mentioned later in this chapter was a consequence of such action. Unfortunately the federal monies provided for programs did not always produce the intended results and the management of development programs by native officials often was poor but progress has been made. Indeed the processes of change have been so positive that Washburn (1996) entitled his survey of recent native history "The Native American Renaissance." His "Renaissance" includes political as well as religious, scholarly, literary, artistic, and economic renaissances. One small element that touches the religious and economic renaissance with political connotations is the native use of wildlife for food, religious (cultural) purposes, and sale.

The wildlife manager may become involved in native/wildlife issues in any of several ways: helping develop, manage, monitor, or administer comanagement programs; devising means to measure harvests where population data are required but where native harvests are conducted outside provincial or state jurisdiction; providing data to help guide policy relative to the species harvested, the harvest levels, and the sex and age classes to be harvested; working with nonnative user groups in explaining the legal and biological aspects relative to native harvesting of one or more species; teaching prospective native wildlife managers ; working for tribal organizations as a wildlife biologist or manager; or a variety of other opportunities many of which are still emerging. Even if you do not become directly involved with native/wildlife concerns, an understanding of the cultural and historical bases of native aspirations relative to wildlife and past and present government policy can help you deal with native and nonnative wildlife user publics you may encounter.

Some Examples of the Native American–Wildlife Complex

In September 1980, James Matthew Simon, a registered Indian under the "Indian Act" and an adult member of the Shubenacadie Band in Nova Scotia, was driving on a public road adjacent to the Shubenacadie Indian Brook Reserve. Mr. Simon was stopped by the Royal Canadian Mounted Police and searched. He was found to be in possession, during a closed season, of a shotgun and shells of a type not permitted by the Provincial Lands and Forests Act. Mr. Simon was charged with offenses under the Act. He pleaded in his defense the applicability of the Treaty of 1752. Mr. Simon was convicted by the Nova Scotia Provincial Court. His appeal of that conviction was dismissed by the Nova Scotia Supreme Court, Appeal Division. Ultimately, he was granted leave to appeal to the Supreme Court of Canada. The decision of the Supreme Court was rendered November 21, 1985. What James Matthew Simon v. The Queen (1985) 2 S.C.R. 387 decided is the subject of debate and interpretation, but includes the following:

1. The Treaty of 1752 was a binding and enforceable agreement between the British and the Mi'kmaq people. "Both the Governor and the Micmacs entered into the Treaty with the intention of creating mutually binding obligations which would be solemnly respected."
2. The Treaty contained a positive recognition of the existing and continuing Indian right to hunt and constituted a source of protection against infringement of these hunting rights.
3. The courts would interpret the Treaty provisions in a liberal and flexible manner which would accord with modern day practice. For example, hunting would not be restricted to the use of spears and handmade knives, as argued by the Attorney General of Nova Scotia. The interpretation of the hunting right also would not be restricted to hunting only for noncommercial purposes.
4. The Supreme Court of Canada held that there was no evidence to indicate that the Treaty of 1752 had been terminated by hostilities and that the burden of proving any such termination would be upon the Attorney General of Nova Scotia.
5. The Court held that, at a minimum, the Treaty was still available to protect hunting on treaty lands and, as a result, Mr. Simon would have the

free right to transport guns and munitions, in a safe manner, to these areas to carry on hunting activities. Consequently, his possession of the firearms was not contrary to the Lands and Forests Act.
6. The Court ruled that if it is legally possible to extinguish treaty rights to certain lands by nonIndian occupation, the onus is on the Crown to prove it
7. The Court would not require proof of descendancy of any present day Indian to any particular person in 1752, provided the person concerned could establish a significant connection with those included in the original treaty.

Because the Treaty of 1752 was a valid and enforceable treaty available to Mr. Simon, it overrode the provisions of the Provincial Lands and Forests Act. Mr. Simon was acquitted as a result.

Regardless of what questions may be posed, the most important aspect to the wildlife manager is the challenge to incorporate whatever manner of harvesting may result from such decisions into a system which will allow for population management. This is a concern not dealt with by the courts and it seems clear, both from recent court decisions in Canada and a review of trapping by aboriginal people by Peter Hutchins (1987), that the fate of wild populations and perhaps even species normally will not be considered in court judgements. This may well mean that the wildlife manager will have to provide wildlife population and habitat data to all involved and to appeal to native people that wildlife should not become a pawn in what are important political issues. People will come first, most certainly, both natives and other user groups, but sacrificing wildlife in power struggles for political gain makes little sense either for the people who stand to gain or the wildlife that stand to lose. Contrary to the Canadian situation, the U.S. Supreme Court, especially in the Puyallup cases, ruled that the states could not regulate hunting and fishing for natives except when conservation of a species was required. Nonetheless, compromise and mutual respect will still be necessary to ensure that wildlife interests are not subjugated to other agenda.

In another recent decision, The Queen v. Thomas Chevrier, an Ontario District Court granted an appeal of Chevrier's conviction for hunting moose out of season. Chevrier, a nonstatus Metis, had his conviction quashed and was acquitted. The bases of the appeal decisions were:

1. The accused inherited the right to hunt granted to his ancestors by the Crown in exercise of its jurisdiction over Indians. The province has no jurisdiction to take away that right even though the present holder may not be an Indian.

2. Even though the province can not negotiate treaty rights, the federal government may still retain the power to regulate their exercise.
3. The Provincial Court judge erred in holding that the accused's right was a community right that could only be upheld by representative action.

The District Court judge added the following comments concerning wildlife and rights to wildlife:

"To those who are concerned that this decision may lead to the destruction of our wildlife resources I can only say:

1. This decision will not lead to unrestricted hunting by everyone claiming to have an Indian ancestor because relatively few, other than status Indians, will be able to prove their descent from a signatory tribe.
2. The whites who rely upon these resources must do so recognizing that these resources have come to them subject to prior claims.
3. Although provincial law cannot negate these treaty rights the federal government may still retain the power to regulate the exercise of these rights for the good of everyone.
4. In the final analysis, everyone with a legitimate interest in the continuation of our wildlife resources must agree upon the proper management of these resources. If this is not done we may see animals such as the moose melt from our forests as has the woodland caribou."

Meanwhile, native people in Nova Scotia recently have been acknowledged to have a right to "any surplus" in relation to salmon and both the Crown and native people are considering the inclusion of all wildlife under provincial jurisdiction (Denny, Paul, and Sylliboy v. The Queen, March 5, 1990). The question of whether native Americans may market wildlife may be most important for wildlife professionals. In Canada there are existing and pending legal decisions relative to the marketing of wildlife that have resulted in great contention. In British Columbia a commercial native fishery exists and land claims settlements are at hand. Rights to wildlife may well be important in the establishment of precedents relative to the larger issues of self government, land claims and social development of native communities. For those involved in wildlife management in Canada, the prospect of wildlife taken outside of the normal regulatory framework being marketed has become a reality.[1] This reality is exceedingly complex as suggested by a November 1997 New Brunswick court decision. That decision upheld native rights of unlimited access to trees on Crown (public) lands in both New Brunswick and Nova Scotia because at the time of the treaty on which the decision was based, these provinces were a single jurisdictional entity. The government of New Brunswick has appealed the decision and the matter may be decided eventually by the Supreme Court of Canada. If the decision is upheld, the social, economic, political, and ecological ramifications of such a precedent will command the attention of the forestry and wildlife professions for the foreseeable future. More than 90 percent of Canada's productive forest lands are publically owned and the forest industry which provides more than a million jobs and is the backbone of the country's resource-based economy is highly dependent on those lands. What accommodations may be necessary or even possible to combine all the interests in such a circumstance will require understanding and compromise on a scale seldom envisioned anywhere. At a time when sustainable forest and wildlife management are in their infancy with no assurance of survival to adolescence, it may be possible that sustainable management initiatives and native rights will evolve side by side. A Supreme Court of Canada decision in December of 1997 ruled that the claims of native peoples to vast areas of Canadian territory and the natural resources they contain are far broader than the current law recognized. In effect, native peoples have the right to occupy and control activities on traditional (nontreaty) lands. This places them in a strong bargaining position for land claims negotiations. The ruling overturned a British Columbia court ruling that rejected oral histories and traditions as supporting evidence for traditional land use. The activities that can be controlled are not limited to traditional uses such as hunting and fishing.

In April 1999 a new territory was carved out of Canada's Arctic. Nunavut's population of slightly over 27,000 is 85 percent Inuit. The land mass is 1,994,000 square kilometers (769,887 square miles), most of it above treeline. Land claims settled or in progress all provide elements of resource control, either full or partial such as benefits from profits derived from mineral resources on crown lands as in Nunavut. In British Columbia the Nisga'a treaty (1999) provides self-government, resource control, and a cash settlement to the people on 2,000 square kilometers (772 square miles). This treaty provides a possible model for about 30 more initiated claims in British Columbia. A 1999 Saskatchewan Court of Queen's Bench ruling was based on land area, however, when the Court granted 120 acres for each tribal member. Thus, land claim settlements in Canada, numbering some 450, vary with circumstances of history including early treaties, proclamations, and agreements as well as geography, culture, and tradition.

In addition to the questions of subsistence and religious rights, there has been friction in the United States between state wildlife agencies and tribes over who has the authority to regulate non-Indians on reservations. In 1983, the U.S. Supreme Court ruled on a case between the Mescalero Apaches and the State of New Mexico. The tribe filed suit because the state was attempting to enforce the state's hunting and fishing regulations on nonmembers of the tribe when they were hunting or fishing on the reservation. Many of the tribal regulations were more liberal than the state's. The ruling stated that Mescaleros had, with federal (Bureau of Indian Affairs, Bureau of Land Management) assistance, established a comprehensive scheme for managing the fish and wildlife resources on the reservation. New Mexico's laws were preempted on the reservation by federal law.

At the time of the decision, a number of tribes were locked in battle with state agencies over regulation of nontribal members on reservation lands. The Mescalero decision paved the way for compromise and probably more effective wildlife management on the reservations. Cooperation became imperative if the states were to retain any ability to regulate or even monitor hunting activities on tribal lands.

Perhaps the most controversial decision to date in the United States related to resource allocation to native peoples was that of Judge Boldt (United States v. Washington) in 1974. Although the "Boldt Decision" concerned the rights of a number of northwestern Indian tribes to fish in the Puget Sound and Olympic Peninsula watersheds and adjacent off shore waters, it appeared to place a serious limitation on the state faunal ownership theory. But what the judgement did was recognize that during the territorial period the federal government held wildlife in trust for all the people. When the territories became states the property interests of Native Americans as stated in the treaties were still held by the natives and not transferred to the states. The judge interpreted the treaty language "in common with" to mean sharing equally the opportunity to take fish at traditional fishing areas. The tribes were given the chance to harvest 50 percent of the fish that would be available at the traditional sites above the population necessary for minimum conservation purposes. A few months later, Judge Belloni, in Oregon, made a decision to apply the same 50 percent apportionment to the treaty Indians who fished in the Columbia River system (including both Oregon and Washington). Ironically, the decisions came at a time when the value of the apportionment was muted by the diminishment of salmon stocks as a result of the dams on the Columbia and overfishing on the high seas.

Wildlife managers should not overlook the possibility that similar decisions could be handed down in the future for any wildlife resource. Consider the implications of a decision that would allocate 25 percent of the elk "surplus" on public lands in the western states to Native Americans. This would have a far broader impact than the current right to take game in traditional hunting areas.

One of the most difficult areas for adjudicators is the previously noted Indian religious rights and the use of wildlife in ceremonial activities. While U.S. courts have indicated certain limited government authority over native hunting rights derived from treaties, the killing of wildlife for native religious purposes presents a more difficult issue because the guarantee of religious freedom in the First Amendment of the Constitution would seem to take precedence over state and federal conservation laws. According to Bean (1989) courts will try to reach balanced decisions by judging two critical factors, "how central a specific practice is to Indian religious belief and how compelling the government's justification is for its actions that infringe...." Similarly, the claims of natives relative to "sacred ground" that often confront government and industry doubtlessly will bring forth other balanced court decisions. For the wildlife manager, such conflicts will be apparent only when a native religious practice or the proposed violation of a sacred land or burial area affects actions designed to protect or otherwise manage wildlife. The protection of the bald eagle by the Bald Eagle Protection Act along with provisions for limited ceremonial use of the feathers is one example of the first instance. As to the second, the desire to establish trails or campgrounds for wildlife users could conflict with sacred lands and accommodation might prove difficult.

Wildlife managers and protection officers may also be caught between what they view as their responsibilities and what government policy allows them to do in regard to Native Americans. Following the Supreme Court of Canada decision in Simon v. The Queen, and after unsuccessful attempts by the Nova Scotia government to negotiate "hunting agreements" with natives, provincial conservation officers were forced to cease attempting to enforce any wildlife regulations in regard to Indians except those involving safety. Similarly, biologists who depended on kill data as one information set used in determining seasons and bag limits were no longer able to obtain those data. It is possible that this will change in the future as cooperative efforts between governments and wildlife managers for Native Americans start to take hold.

Into the Future

In 1990, Indians, Aleuts, and Inuit in the United States numbered a little over two million people or about 0.5 percent of the total population of the country. Long range projections of demographic trends suggest that fullblooded Native Americans will decline because 48 percent of American Indian men and the same percentage of women are married to non-Indians (Washburn, 1996). However, in many parts of Canada Native Americans represent the fastest growing segment of the population. For nearly 500 years, Indians have held on to their identity, culture, and life style. Further intermarriage may not change these aspects of the American Indian very much over the next 500 years.

In 1990 approximately a million of Canada's 27 million people were natives or nearly 4.0 percent of the population. Proposals and treaty settlements to date suggest that eventually dozens of "First Nations" jurisdictions will be essentially self-governing and will exercise control over most of the resources within their boundaries. In both countries, there is a true "Native American Renaissance." Public sympathy and political liaisons such as those of the environmentalists and native peoples across North America have not always helped progressive wildlife management causes but in some cases they have brought awareness to some issues that required attention. In the future, one would hope that ecological and economic data will be considered along with social and cultural data to allow just and sound decisions regarding policy relating to natives and wildlife. An example of how that can be done comes from the Yakama Reservation in Washington (McCorquodale et al., 1997).

Throughout North America, federal, state, and provincial jurisdictions will be struggling to effect cooperative wildlife management in the years ahead. A number of such structures have developed in Canada's northern regions since 1975, while more recently some have been structured in response to massive land claims settlements. Comanagement programs in states, provinces, and territories will differ according to the specific circumstances but their continued development will be necessary. This struggle will result in amendments to legislation, new management structures, and much commitment to experimental management. In that regard, it could be a healthy renaissance of wildlife management but it will depend on the good will of all the negotiating parties.

It will be essential that wildlife educational programs adjust their curricula to ensure a knowledge of native interests and approaches to natural resources management and provide their graduates with the capacity to be skilled negotiators.

Note

1. The Supreme Court of Canada (R.V. Marshall case) ruled concerning commercial eel fishing without license, and then clarified in November 1999, that natives did not have "a right to trade generally for economic gain, but rather a right to trade for necessities. The (1760) treaty right is a regulated right and can be contained by regulation within its proper limits." Whatever one may think of the term "regulated right" it is apparent that, for fish, natives do not have absolute right. The legal accepted interpretation is still being debated and natives are challenging the government's right to regulate fish and wildlife resources.

Bibliography

Attorney General of Canada. Undated. Policy respecting the prosecution of aboriginal persons for hunting offences under the Migratory Birds Convention Act and the Canada Wildlife Act. (unpubl.).

Bean, M. J. 1989. Recent court decisions affecting wildlife. In: Audubon Wildlife Report, pp. 155–175. Academic Press, San Diego, CA.

Burley, D. V. 1981. Rapid culture change and the fur trade: a case for the Micmac of northeastern New Brunswick. (unpubl.).

Chevrier v. The Queen. 1988. Ontario District Court (1 C.N.L.R. 128).

Cordell, L. S. and B. D. Smith. 1996. Indigenous farmers. In: *The Cambridge History of the Native Peoples of the Americas.* Vol. 1., (pt 1):201–266. Cambridge Univ. Press, New York.

Czech, B. 1995. American Indians and wildlife conservation. Wildl. Soc. Bull. 23:568–573.

Denny, David, Lawrence John Paul and Thomas Frank Sylliboy v.The Queen. 1990. Supreme Court of Nova Scotia, appeal division, (94 N.S.R., (2nd). 253).

Dodds, D. G. 1993. *Challenge and Response. A History of Wildlife and Wildlife Management in Nova Scotia.* Dept. Nat. Resour., Halifax, Nova Scotia.

Dodds, D. G. 1982. Micmac food resources at European contact. Can. Ethno. Soc., Vancouver, British Columbia (unpubl.).

Flett v. The Queen. 1987. Manitoba Provincial Court (5 W.W.R., 115).

Hoxie, F. E. 1996. The reservation period, 1880–1960.. In: *The Cambridge History of the Native Peoples of the Americas.* Vol. 1, (pt 2):183–258 Cambridge Univ. Press, New York.

Hutchins, P. W. 1987. The law applying to the trapping of furbearers by aboriginal peoples in Canada: a case of double jeopardy. In: M. Novak, J. A. Baker, M. E. Obbard, and B. Malloch (eds.). *Wild Furbearer Management and Conservation in North America.* pp. 31–48. Ont. Min. Nat. Resour., Toronto, Ontario.

Hutchins, P. W. 1989. Aboriginal and environmental responses to resource development: a love-hate relationship with promise. Envir. Law, Native Justice, Nat. Resour. and Energy Panel, Can. Bar Assoc. Annu. Meeting, Vancouver, British Columbia (unpubl.).

Kim, S. C. 1900. *The Iroquois. A History of the Six Nations of New York.* Pierre Danforth, Middleburgh, New York.

Langdon, S. 1982. Production, exchange and destruction in the Tlingit economic system. Can. Ethno. Soc., Vancouver, British Columbia (unpubl.).

Martin, C. 1978. The war between the Indians and the animals. Nat. Hist.87:92–96.

McCorquodale, S. M., R. H. Leach, G. M. King, and K. R. Bevis. 1997. Integrating Native American values into commercial forestry. J. Forestry 95:15–18.

Morrison, R. B., and C. R. Wilson (eds.). 1986. *Native Peoples: the Canadian Experience*. McClelland and Stewart, Toronto, Ontario.

Paul, D.H. 1993. *We Were Not the Savages*. Nimbus, Halifax, Nova Scotia.

Simon, James Matthew v. The Queen. 1985. Supreme Court of Canada (2 S.C.R. 387).

Smith, B. D. 1996. Agricultural chiefdoms of the eastern woodlands. In: *The Cambridge History of the Native Peoples of the Americas*. Vol. 1 (pt 1):267–324. Cambridge Univ. Press, New York.

Snow, D. R. 1996. The first Americans and the differentiation of hunter-gatherer cultures. In: *The Cambridge History of the Native Peoples of the Americas*. Vol. 1 (pt 1):125–200. Cambridge Univ. Press, New York.

Sparrow, Ronald Edward v. The Queen. 1990. Supreme Court of Canada (1 S.C.R., 1075).

Trigger, B. C. and W. R. Swagerty. 1996. Entertaining strangers: North America in the sixteenth century. In: *The Cambridge History of the Native Peoples of the Americas*. Vol. 1 (pt 1):325–398. Cambridge Univ. Press., New York.

Washburn, W. E. 1996. The native American renaissance 1960–1995. In: *The Cambridge History of the Native Peoples of the Americas*. Vol. 1 (pt 1):401–474. Cambridge Univ. Press, New York.

Willett, L.T. 1982. Making tabagie: a Micmac parallel to the potlatch? Can. Ethno. Soc., Vancouver, British Columbia (unpubl.).

Recommended Readings

Colson, D. C. 1998. *The Nez Perce Treaties*. Confluence Press, Lewiston, ID. An example of United States treaty making and legislative interpretations as they affect one tribe in the western United States.

Crosby, A. W. 1993. *Ecological Imperialism*. Canto edition. Cambridge University Press. Cambridge, UK. The impact of European imperialism on native peoples within the world's temperate zones is discussed in ecological terms.

Krech, S. III. 1999. *The Ecological Indian, Myth and History*. W. W. Norton & Co., New York, 318 pp. An exhaustive discussion of the American Indian as both conservationist and wasteful exploiter through time. If you are a wildlife or conservation biologist and have time for one book on the subject of Indians and resources, let it be this one.

Martin, C. 1978. *Keepers of the Game. Indian-Animal Relations and the Fur Trade*. Univ. of California Press, Berkeley, CA. An anthropologist's detailed interpretation of Amerindian-animal relationships.

Martin, C. 1999. The Way of the Human Being. Yale University Press, New Haven, 235 pp. The American Indian's understanding of this world in myth that borders on reality. Calvin Martin will help you understand and, perhaps, appreciate.

Patton, D. R. 1997. Wildlife habitat relationships in forested ecosystems on Indian reservations. In: *Wildlife Habitat Relationships in Forested Ecosystems.* Timber Press, Portland, OR. Government-native relationships in the United States with legislation and events concerning native people. Gives examples of wildlife management programs on reservations.

Trigger, B. C. and W. E. Washburn. 1996. *The Cambridge History of the Native Peoples of the Americas.* Vol. 1 (pts 1–2.) Cambridge Univ. Press, New York. A valuable background reference covering all aspects of native history and culture from pre-European contact to the present.

Wildlife Advisory Council. 1993. Native people and wildlife. Ch. III. In: *Living with Wildlife. A Strategy for Nova Scotia,* pp. 38–60. Dept. Nat. Resour., Halifax, Nova Scotia. The Native American/wildlife issue in one jurisdiction, the present status and recommendations.

Orion Nature Quarterly. 1990. Vol. 9 (No. 3). This issue of Orion looks at indigenous peoples from North and South America in political, cultural, and socioeconomic contexts. In some cases, clear parallels with European contact on this continent can be made.

4 SOME BIOLOGICAL BASES FOR AND APPROACHES TO MANAGEMENT

What biological data need to be known for effective management? Considerations such as numerical changes in populations-an understanding of birth and death rates and sex and age composition are important but equally important to the manager is a knowledge of the factors, such as habitat quality, social interactions, disease, and genetics, effecting these population changes.

Computer Models

Few models being used to aid in management decision making are sophisticated enough to incorporate all of the above parameters. This is critical because there has been a tendency in recent years to use computer modelling to aid in deciding which regulations to impose. The models use the available biological data and harvest trends and provide comparative scenarios for different regulatory decisions. To be useful the data bases must be sound. Unfortunately, many wildlife agencies do not have accurate survey records. For those which do, the data bases are short covering only a few years. Effective simulation depends on a long and accurate data record but it is possible to work with incomplete data. Such models can help determine how important the missing data are and which management plan to follow (Ralls and Starfield, 1995).

There are two types of models that can be used: (1) stochastic or (2) deterministic. The latter is based on complete information about a natural population change. Using the giant Canada goose as an example, De Angelis (1976) pointed out that natural fluctuations in wild populations are best described by difference equations with randomly varying parameters. This approach recognizes the interdependence of birth and death rates and is superior to a logistic approach, with independence assumed. A stochastic difference equation model is therefore better equipped to reflect the rapid fluctuations likely to occur in

most natural populations. Burgman et al. (1993) describe a number of stochastic population models of interest to wildlife managers.

Caughley (1977) considers that there are but three problems of population management: conservation, harvesting, and control. Invariably though, the decision of which management practice to apply to the situation is dependent on basic life table information for the population. No effective management can be applied without an understanding of the biology of the species and a population's variability, current trends, and composition. Knowing this, a manager can decide at what level of harvest the population will increase, decline, or provide a sustained yield (SY) where SY=HN and where H is the instantaneous rate of harvest (=rp), the rate of population increase were the population not harvested, and N is the average size of the population throughout the year. SY is but a small proportion of a large population so maximum sustained yield (MSY) has been used as the management objective by many wildlife agencies. By operating at MSY, the maximum output of individuals is reached and hunter satisfaction sustained by higher harvests. The problem with MSY is that if exploitation exceeds this value, precipitous population decline is a likely consequence, particularly with a low productivity species.

When an adequate data base exists, proposed management strategies can be modelled to predetermine their impact on a species. Unfortunately, such models sometimes have been misused in the past. To be effective, a model must be limited to the species and circumstances for which it was designed. The use of models has often been abused by overextending them or projecting data to inappropriate situations. The U.S. Fish and Wildlife Service's model for the mallard has frequently been used to predict impact of hunting regulations on other dabbling ducks, even though the biology and indeed the effect of the regulations are known to be different. Walters et al. (1974) developed a model of the population dynamics of the mallard duck in North America so that long-term predictions of population response to management activities related to habitat and harvest could be made. They concluded that the key value of their model was not its predictive power but its ability to serve as a focus for research needs. Starfield (1997) proposes that a model be considered as an hypothesis. By that he means that it represents what the system may be rather than represents what it actually is. The two might be the same but there is nothing wrong if it is not providing the model has the flexibility to respond to different assumptions about how the system actually behaves.

Generalized models such as ONEPOP are used for many different species whose attributes satisfy the logic of the model. Managers can model deer, elk, moose, or other big game species as long as certain basic information is known about the populations and harvest values are available. Populations are delineated by geographical boundaries which represent the year round range. This minimizes the problem of immigration and emigration as confounding factors. The function of the model is to simulate the dynamics of the real population so that the effects of various management decisions can be predicted.

The major advantage of modelling is that it forces the manager to decide which population factors are most critical and to define clearly how those factors interact. The ability to determine management effects on simulated populations without a time delay or involvement of personnel is an important money saver.

The major disadvantage of modelling is that few data bases are complete enough for an acceptable level of predictability. Managers also may view models as a cure-all for their problems when they really are only valuable tools.

To receive maximum benefit from a model, the user must be familiar enough with it to understand the underlying assumptions, its strengths, weaknesses, and accuracy. Familiarity is best achieved by being involved in the development of the model. The manager can interact with the modeler by providing the data base and assisting in development of the model's conceptual framework. To use a model for management purposes without understanding the underlying assumptions is dangerous and can lead to costly errors. Starfield and Bleloch (1991) provide examples of model development in wildlife management and lead the reader through the vagaries of inadequate data and other confounding factors in model construction.

The basic assumption of most population models is that either the environment is homogeneous or the population exists in a vacuum. Most models are not sophisticated enough to cope with the natural environmental heterogeneity of the real world. More truly ecological models are in the process of development and will include dispersal, immigration, species interactions, social behavior, critical reproductive habitat (refuges), and patterns of hunter distribution and harvest rate. The new models will also reflect time and space factors. Starfield (1997) believes it is possible to simplify ecological complexity by using qualitative variables and frame-based modeling to construct a number of simple models rather than depending on a model that is built from the ecosystem processes and thus by necessity starts off as complex. In his approach, the complex model is created ultimately from the component simple models. Once these ecological models are in place, managers will have the predictive capacity necessary to make wiser decisions.

Nutrition and Energetics

Animal species have nutritional requirements that vary with physiological and physical condition, age, sex, and social status. A certain amount of energy needs to be ingested to sustain the metabolic requirements of the individual. A true measure of environmental carrying capacity thus becomes the energy and nutritional elements available to a species within a given area related to the species requirements. While social constraints may provide additional limitations to population size, it is the available energy which determines the maximum number of individuals of given age and sex categories which can be accommodated at any particular time of year.

As Moen (1973) stated, "There is an obvious need—for knowledge of the requirements of an animal for maintenance and productive purposes before a meaningful biological appraisal of carrying capacity can be made." Fortunately, there is a growing body of information on the biological energy requirements of wild animals (Hudson and White, 1985; Robbins, 1993).

Basic questions that need answering include what foods are eaten by the animal, what energy is contained therein, and how much of this energy is utilized by the animal? Knowing the average energy requirement per animal per month and the assimilation efficiencies for foliage eaten, one can calculate the support capacity of the habitat per unit of time and the percent of available food utilized by the existing population as follows (after Harris, 1970):

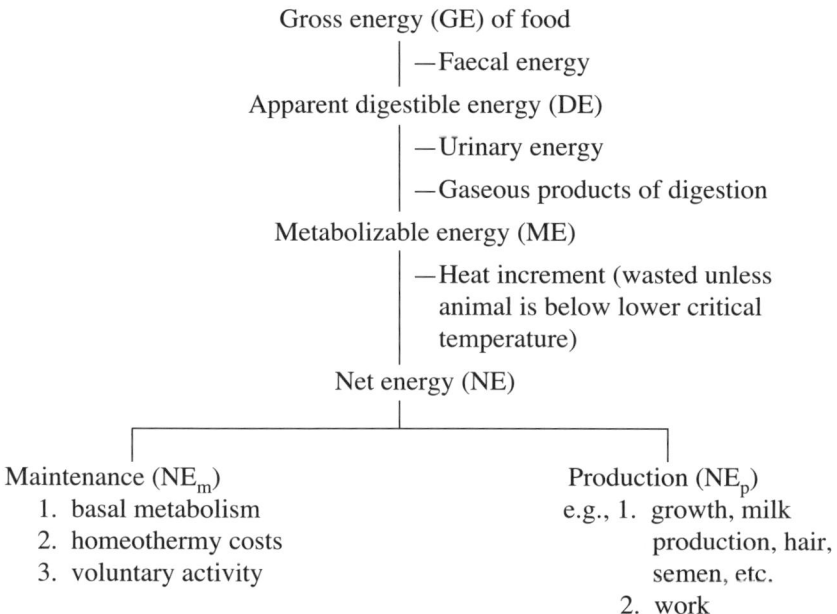

Such an energy analysis is actually a very simple expression of carrying capacity because it does not incorporate other important components such as water and protective cover. Complete energy budgets for free-ranging animals are difficult to construct as the estimates of energy requirements often are derived by respirometry in a laboratory setting.

Although we have considerable data for many species, these data are limited for the most part to basal metabolic measurements or to behaviors and activities restricted by a laboratory environment. The actual energy costs to the animal of food seeking, social interaction, and homeothermy maintenance in a natural setting have yet to be determined for virtually all wildlife species. What is required is an indirect measure of energy metabolism that does not constrain the animal and allows a complete and normal behavioral repertoire. Alternatively, complete time budgets for behavior in the wild need to be collected and the energy costs for each behavior then determined in the laboratory. While time budgets can be constructed for some vertebrates which are highly visible and have restricted home ranges, such studies generally require a high manpower input for a relatively limited data return. For example, Potvin and Huot (1983) devised a working model developed to estimate the carrying capacity of a white-tailed deer wintering area in Quebec based on measurements of surface area used by deer, forest cover types, browse accessibility, and nutritive requirements of deer expressed in energetic terms. Rather than measuring the time spent in various behaviors, the model predicted what should be the most energetically advantageous physiological response and how energy might be partitioned as a result of different snow covers. The model's primary weakness is the impreciseness of the data used to construct it. The biological data base is still too weak to allow refinement of such costs as locomotion in snow and measurement of the relative energy available to an animal selecting food within a given environment. Hanley and Rogers (1989) estimated carrying capacity of black-tailed deer in four hypothetical forest habitats in Alaska during summer and winter using the quantity and quality of available food and the species' nutritional requirements. Three sets of snow conditions and two levels of metabolic requirements for digestible energy and digestible protein were used. One of the limitations of the model was the persistence of the snow condition for the entire winter season and another that all available forage, within the constraints established, was eaten. Their algorithm nicely illustrated the fundamental importance of habitat, season, and nutritional status if carrying capacity is to be determined.

Doubly labeled water (deuterium and oxygen-18 isotopes) offers a method for measuring energy expenditure in free ranging animals. A slightly higher daily energy expenditure estimate is given than that based on activity budget analysis because the energy expended in food processing is included. The isotopes are injected into an animal. After they have equilibrated, the animal is released. Several days later, the animal is recaptured and a blood sample taken. The difference in the decay rate in body water and CO_2 gives a measure of energy expenditure. But it is a measure of total expenditure over the period of time between samples without regard to habitat use or the animal's behavioral patterns. Heart rate is one indirect measure of energy metabolism showing considerable promise. It seems to work for white-tailed deer, mule deer, caribou, marten, prairie dogs, green winged teal, black duck, and red squirrel among other species. Until such a method has been confirmed totally and is available for use with free-ranging wildlife, minimum habitat requirements and land-use allocations for wildlife species will continue to be determined on the basis of less definitive data. With a shrinking wildlife habitat base, more precision is required to ensure that critical mistakes are not made when stating habitat needs for any particular species.

Nutrient deficiencies can also cause problems in populations and managers should be aware of potential difficulties. Selenium is a particular problem in the Pacific Northwest where grazing ungulates may not be meeting their minimum requirements. Selenium deficiency has been shown to significantly lower reproductive rates in black-tailed deer in an area of northern California (Flueck, 1989). This is one suggested reason why ungulate species were never abundant in the intermountain Columbia Basin area of the Pacific Northwest. Acid precipitation can further reduce selenium availability by forming sulphur-selenium complexes, thus exacerbating naturally low selenium levels in eastern North America.

It is important to know the mineral availabilities in local soils and vegetation and to compensate, if necessary, for problems that may develop at the population level as a result of deficiencies or over-abundances of macro or micro elements. For instance, pheasants (*Phasianus colchicus*) have never become established south of the thirty-ninth parallel in eastern North America. Jones et al. (1968) and Anderson and Stewart (1969) presented evidence that pheasant density was related to the inorganic chemistry of the environment. It is known that pheasants, like other indeterminate layers, have high Ca-P requirements for egg shells (NRC, 1984) and their distribution in eastern North America is primarily limited to Ca rich till soils associated with recent glaciation. Pheasants can compensate by ingesting the calciferous shells of snails. Anderson and Stewart (1973) found a potential excess of Ba in unsorted grit from an area of low pheasant numbers. As Ca concentration in the environment and trace element absorption are inversely related, Ca poor soils would exacerbate any such trace mineral excesses. We might expect an impact on survival, particularly during fall and winter months when pheasants feed heavily on corn and other grains low in Ca and concurrently consume relatively small amounts of calcitic grit. Such a seasonal response of high fall-to-spring mortality on Ca poor soils may illustrate a toxic response to increased Ba concentration in body tissues.

Fraser et al. (1980) have shown that Na appears to be the major mineral sought by wildlife using mineral licks. Many herbivorous terrestrial mammals cannot meet their physiological needs for certain minerals from the vegetation normally available to them. In addition to mineral licks, some mammals, like moose, can utilize Na rich plants (aquatic vegetation) and perhaps store the Na for later mobilization. Another possible mechanism used to conserve Na is K substitution in salivary-rumen fluids when Na is deficient or when there is decreased sodium excretion in urine (Blair-West et al., 1968). Thus we see that wildlife may adapt to deficiencies (or excesses) of certain minerals in the natural environment.

Human activities can rapidly alter the availability of certain elements, disrupting the physiology of exposed animals, e.g., increased heavy metal (mercury) levels in granitic lakes, fish, and fish-eating wildlife due to acid precipitation and leaching from soils. This also occurs in soils and wildlife due to pesticides, seed dressings, and automobile emissions, or aerial fallout from smelters. Even the application of chloride salts to highways can affect wildlife, if only indirectly, by altering vegetation or by attracting wildlife to roadsides where they may be hit by vehicles.

It is important to remember that many nutrient deficiencies occur naturally. For example, vegetation quality measured by crude protein content, Ca, and P (and Ca-P ratio) was below that required for maintenance of white-tailed deer during winter in some Maine yarding areas (Abell and Gilbert, 1974). With N and P deficiencies, these animals are susceptible to winter mortality. The differential mortality observed in Maine deer wintering areas with similar vegetative composition and structure and snow and temperature conditions was likely a function of vegetational quality. Hobbs and Swift (1985), in developing a model to predict carrying capacity of burned and unburned mountain shrub habitat for mule deer and mountain sheep, emphasized that forage quality must be considered along with forage quantity as an integrated feature of habitat. Using their algorithm, a species stocking rate can be predicted based on the selected nutritional plane.

Nutrient deficiencies may become increasingly important as factors for wildlife managers to consider as more whole-tree harvesting and cutting of forests on erosion susceptible soils takes place. Such practices result in nutrient depletion, and without fertilization by forest managers the soils can be expected to produce poor forest crops and unthrifty wildlife populations, at least of herbivores. Switzer and Nelson (1972) showed considerable retention of nutrients in the woody, aerial portions of trees. White (1974) demonstrated that relatively large amounts of P and K were removed from a site when cottonwood (*Populus deltoides*) was whole-tree harvested. In such harvesting all portions of the tree are utilized and no "waste,"—branches, leaves, or stumps—is left behind. The implications to wildlife management of whole-tree harvesting of pulpwood species are that either artificial nutrient supplements will have to be provided, which may attract wildlife, and potentially increase regenerating tree damage as sometimes has occurred in Christmas tree plantations, or if no fertilization occurs, wildlife populations will drop with the decline in quality of the vegetation.

Wildlife species have functioned as sentinels of environmental degradation, forewarning the consequences of contamination by such chemicals as DDT, mercury, PCBs, and dieldrin. This prompted development of a wildlife model using the raccoon in Florida to provide information on environmental quality. The raccoon lives up to eight years and has omnivorous food habits. Its diet includes aquatic organisms which biomagnify waterborne diseases, toxins, and pollutants. Because of the habitat it uses, it is exposed to blood-sucking vectors of arboviral, rickettsial, and parasitic diseases. Raccoon serum can be used to detect St. Louis and Venezuelan equine encephalitis. Leptospirosis, tularemia, and some enteric bacteria and viruses can also be found in the raccoon. Pesticide residues in the omentum fat can be used to monitor industrial pollution. Radionuclide contamination will show up in whole body counts.

Raccoon hair can be used as biopsy material to detect Hg contamination. This species is thus well suited for use as a wildlife surveillance system. Any such system must include at least: (1) a rather common species that is sufficiently sensitive to indicate the diseases and environmental contaminants to be monitored, (2) sufficient data returns without adversely affecting the population status of the species, (3) adequate base line information to show the current situation with respect to physiological measures, reproductive processes, susceptibility to disease, parasite burdens, pesticide, and other pollutant concentrations in body tissues so that any changes can be detected, and (4) capability to integrate with data derived from other surveillance systems. Nutrient deficiencies prompted by environmental alteration should also be detectable with such a system.

Behavior

Social interactions within and between populations place additional constraints on the potential carrying capacity of an area. Many species exhibit density-dependent effects of population size. Intraspecific competition for increasingly limited resources results in physiological and behavioral responses influenced by the increased social contact between individuals in the population. The ensuing endocrine gland and hormone production changes can result in lowered reproduction, changes in sex ratio of offspring, increased mortality, or a state of physiological stress—all of which can affect population density.

Territoriality serves to space individuals or family groups within any given environment. The energy costs of territory delineation become greater as population density increases and/or territory size shrinks, although territory size may also shrink as a result of improved environmental conditions such as increased food supply. This means the territorial animal in a high density situation must spend more of its time and energy interacting with other individuals in a threat or defense posture. Thus less energy in a time-energy budget is available for reproduction, foraging, or other activities.

Behavior of the sexes may vary within species at different times of the year. This is particularly important to a wildlife manager attempting to obtain population data for such things as a life-table analysis. It means that erroneous values can be derived from surveys if spatial and temporal patterns of movement are different due to differing behavioral responses of the sexes. Trapping or visual observation of individuals in a population may well produce data that are useless unless the behavior of the sexes and age groups is known.

The behavior of ungulates and how it relates to management was the subject of an international symposium published as part of the International Union for the Conservation of Nature series. In all, 56 papers were presented which introduce the reader to how important wild ungulate behavior is to the process of making management decisions. The summary paper by Cowan (1974) should be required reading for all wildlife managers.

If we search for North American examples of how behavior affects populations of wild animals, perhaps one of the most overlooked is that of the plains bison, which was probably doomed by the advent of the railroad. The wanton destruction so often alluded to as causing the demise of the species was in part a result of their behavior. Thousands were shot from trains because they simply refused to cross the tracks, milled alongside them, and were easy targets.

Knowing the behavioral responses of wildlife to human disturbance is important in making correct management decisions. Bald eagles tend to be sensitive to disturbance during the nesting season and a buffer zone should be provided during that time of year. Elk and grizzly bears are disturbed by vehicular traffic and will abandon preferred habitat in roaded areas unless provision is made to close roads or control traffic. Road development for resource extraction which provides long-term access to areas and oil well installation can disrupt elk (Van Dyke and Klein, 1996) and grizzly bear. Caribou have been shown to avoid crossing sections of the trans-Alaska oil pipeline, and what was originally one population is becoming two separate populations with reduced management flexibility. Musk-oxen, caribou, and other arctic species react to low flying aircraft, and minimum flying levels have had to be introduced for oil and mining exploration work in the arctic and subarctic regions of Canada and Alaska to protect the resident wildlife from energy depleting harassment or disturbance. There is continuing controversy over NATO low level exercises in the Canadian arctic and their impact on caribou. There is evidence that wildlife will accommodate to such disturbance when it occurs with predictable frequency (e.g., Weisenberger et al., 1996) or adjust behavior to minimize the disturbance but not necessarily return to predisturbance or predevelopment levels (e.g., Morrison et al., 1995). The development of a town site (Churchill, Manitoba) in the seasonal migration path of the polar bear has caused considerable human-bear interaction and conflict. Female mule deer have been shown to have reduced reproductive performance following experimental harassment with an all-terrain vehicle. Wood ducks apparently do not habituate to aircraft disturbance (Conomy et al., 1998).

Other behavioral responses in wildlife may be more subtle. For example, cow moose apparently respond to antler size and configuration in the bull, and female elephants respond to tusk size of the bulls. Such mate selection by the female has ensured that the species maintains a genetic structure reflective of the most vigorous and strongest members of the male sex. Yet wildlife managers with dubious wisdom have instituted bulls-only seasons in many areas of moose range and poachers, of course, harvest male elephants with the largest tusks. While the productive segment of the population, the female may be protected, such disproportionate harvesting of males may lead to social and genetic disruption. Bubenik (1972, 1977) strongly advocated hunting seasons for moose which would protect the "primes" of the population and put pressure on the senile and juvenile segments. This means a selective hunt with each license being issued for a particular age and sex category. Although common to many European countries, such selective and restrictive hunting requirements until recently were used rarely in North America, but in the long run may well be in the best interests of the game populations and hunting public if adopted here. Ontario is an example of a jurisdiction which has such selectivity in its moose regulations.

Other behavioral characteristics important to managers include seasonal migration in many species from summer to winter range and back. The pattern of migration and the types of seasonal ranges required as well as the necessary connecting links between the seasonal ranges to allow migration to occur will dictate how land acquisition patterns should take place; will delineate critical seasonal habitats such as deer wintering areas; should help indicate where potential barriers (e.g., pipelines or roads) might be located to avoid conflict with species like caribou and elk.

These examples again illustrate how critical it is for the manager to have a full and thorough understanding of the basic biological responses of the species to be managed. Without such understanding and knowledge of the limitations and requirements of individual animals within wild populations, a manager is operating with a severe handicap. We can ill afford the luxury of inappropriate action in these times of generally dwindling habitats and stressed populations. Attempting to manage wild populations without first acquiring an adequate knowledge of the environmental constraints, ecological interactions, and zoological realities of the situation often can cause more harm than good. Although the immediacy of response required at times may necessitate action based on limited data, it is contingent on the manager to point out the deficiencies in the data base and strive to fill them as soon as possible so that potentially more appropriate actions can be taken in the future. We should know why an animal does what it does, and yet too often we do not even know what it does behaviorally in given situations.

What are the biological limitations of the environment that often dictate behavioral responses? Most are energetically based and reflected in basal metabolic rates, upper and lower critical temperatures, and the ability to extract energy from the environment. Animals move to shaded or generally cooler microclimates when continued solar exposure would cause them to exceed their upper critical temperature or they would show other adaptations to avoid using energy to maintain homeothermy. Conversely, if the lower critical temperature is exceeded, an animal must improve its insulation, as grouse do by using snow burrows; or moderate the microclimate as deer may do by huddling and having their respiration (which condenses overhead) reduce radiative heat loss, or as gray partridge do by maintaining coveys and roosting together. Otherwise they must use energy to maintain homeothermy.

One convenient way of expressing the relationship between organism and environment is by the use of climate-space diagrams (Porter and Gates, 1969). In this way the biologist is able to determine if the organism can remain homeothermic in a particular microhabitat at a particular time of day or season. Not only is the energetic relationship between the animal and its environment revealed, but by superimposing the climate-space diagrams of other species (e.g., predators), the probability and timing of interaction can be predicted as can the degree of niche separation. In essence, a climate-space diagram predicts when an animal must thermoregulate by behavioral means. Let us imagine a ruffed grouse in the winter having a choice of three forest stands (hemlock, white pine, and hard maple) or an open field. It should choose the environment which would minimize its exposure to heat sinks and where the net radiant energy exchange is positive. Looking at these four particular environments at T ranging between -10 and $+10°$ C under clear skies and using an abbreviated Gates' equation

$$M + Q_a = \epsilon A_i \sigma T^4$$

where

M = basal metabolic rate
Q_a = average radiation absorbed by all surfaces of the animal
A_i = surface areas presented to streams of radiation (assume 1/2 of grouse's body exposed to sky hemisphere and 1/2 to ground)
ϵ = emissivity of animal's surface to long-wave radiation
σ = Stefan Boltzmann constant (8.2×10^{-11}/y °K^{-1} min^{-1})
T = surface temperature (°K) of animal (feather surface temperature $\approx$ ambient temperature [T_A])

and by plugging in the appropriate values, the energy balance for a medium-sized grouse (500–700 grams) is likely to be positive in the two conifer stands and negative in the hardwood stand and open field. It would therefore be disadvantageous for the grouse to roost in the latter situations when night skies are clear and the T_A less than 10 °C. We can predict the behavioral response of a species by knowing the energy flow in the available habitat and its climate-space diagram. Needless to say, grouse behavior coincides with the prediction (Rasmussen and Branden, 1973).

Weather plays an important role in mediating energy exchange, and at no time is this more apparent than in winter in the temperate and arctic zones. The average specific gravity of snow ranges from 0.056 for large fluffy flakes to 0.135 for small flakes. After lying on the ground for a time, or due to slow melting, the snow becomes compacted into ice which has a specific gravity of about 0.920. A key physical relationship for an animal is the direct proportionality of the coefficient of the thermal conductivity of snow to the square of specific gravity: if specific gravity $=0.10$, the thermal conductivity $= 0.004$; if specific gravity $= 0.20$, the thermal conductivity $= 0.0162$. Snow is important in providing thermal insulation for biological organisms, in determining the availability of food, and in affecting the locomotion of animals both at and below its surface. These characteristics are all dependent on the hardness, granularity, and thickness of the snow cover. Wildlife most highly adapted for movement on snow are lynx, snowshoe hare, ptarmigan, and caribou. Even these species can be handicapped by deep, fluffy snow (lynx) or heavily crusted snow (ptarmigan and caribou). The hindering of movement is not only dependent on the characteristics of the snow cover but also the morphology and track-load values for the appendages of the wildlife.

Subnivean species can suffer extensive mortality if there is little or no snow cover and the ambient temperature is below 10 °C. Even with just a moderate snow cover (10 cm+) there can be great thermal advantages. Marchand (1982) demonstrated that temperature stability in subnivean environments could be achieved with a single snowfall of 15–20 cm of low specific gravity.

The advantage of hibernation in deep burrows becomes apparent with the increasing soil temperature at increasing depth. However, the effect of snow cover is usually lost 100 cm below the surface except in extreme winters with no snow cover.

There are energy costs associated with feeding during periods of snow cover. Caribou dig craters in the snow to access ground vegetation. As snow hardness increases, the average daily cost of digging craters increases. An Alaskan study showed that energy costs to an adult caribou for digging craters increased from 46.6 kcal/day in early winter to 148.5 kcal/day in late winter. During this same time period snow hardness increased from 2300 mg/cm^2 to 9000 g/cm^2. This 50 percent increase in energy expense must be met by increased forage intake or use of stored energy reserves (Thing 1977).

Ruffed grouse have very small energy reserves. Increased glycogen levels in the liver and pectoral muscles may enhance shivering thermogenesis so the response to cold temperature is not lipolytic as it is in many species but rather is glycolytic. Because the fat reserves only provide about two days' energy for a fasting grouse, the birds must reduce energy expenditures to survive extended periods of inclement winter weather. Snow depth and compaction are important determinants of winter survival. The animals have three winter roosting alternatives: (1) snow burrows, (2) snow bowls, and (3) tree roosts. If sufficient soft snow is available (minimum 20–25 cm uncrusted snow), they prefer to burrow. With temperatures rarely falling below −2 °C under 15 cm of snow, even with an air temperature of −20 °C, the birds are capitalizing on the insulation provided by the snow. The birds use thermogenesis to warm the air spaces in their plumage after burrowing into the snow to further restrict heat loss (Thomas et al., 1975).

Wind is another important climatic factor. For mammals, fur conductance (hc) determines the convective and radiative heat flow (H) through the fur and surrounding air for a given set of skin (T_s) and ambient air (T_a) temperatures:

$$H = hc\,(T_s - T_a)$$

Fur conductance is also a function of wind speed. For example, caribou calves one to three days old have

$$h = 3.49 + 0.03 V^{1-6}$$

where

h = conductance (W/m² · °C)
V = wind speed (m/sec)

When wind chill (combination of V and T_a) values exceed 1100 kg · cal/m² per hour or metabolic rate approaches 25 cal/hour/kg for calves, exposure of long duration usually means death.

As T_a passes out of the thermoneutral zone of a homeothermic animal, the individual will increase its insulation, i.e., decrease conductance by reducing evaporative heat loss, increasing tissue insulation by peripheral vasoconstriction, increasing fur or feather insulation by piloerection, and reducing its heat dissipating surface by curling up or otherwise attempting to attain spherical form. The effect of wind increases the rate of heat flow from an animal and raises the effective lower critical ambient temperature by reducing the depth of the thermal boundary layer. One of the offsetting environmental factors to wind chill is solar radiation, and in addition to seeking solar exposed locations, animals select bedding sites or habitats which effectively reduce or eliminate the effects of wind.

White-tailed deer in northern parts of their range use cedar, hemlock, or other conifer stands during the winter. By so doing they reduce the effective wind chill and energy loss due to radiation and convection. Moose will use isolated trees or brushy cover in cut-over areas to minimize the effects of wind when bedding (McNicol and Gilbert, 1978). In western North America, altitudinal migration occurs for the same reason. Deer and elk will move from higher altitudes to more forested lower elevations for the thermal advantages afforded there during the winter months (see Loveless, 1964).

An area not as thoroughly explored for wildlife species is that of heat stress. Behavioral adjustments are used here also: seeking shade, nocturnal versus diurnal activity, limiting movements to areas with an adequate water supply, and water baths. These behavioral adjustments often complement physiological and morphological adaptations that are evolutionary factors developed to cope with a predictably xeric and/or hot climate. These include (1) radiative appendages—ears of desert jackrabbit, long legs of coyote, (2) an ability to maintain a higher body temperature—antelope with protective heat exchangers to maintain a cooler brain temperature, and (3) conservation of water. Management recommendations for ungulate species such as elk and deer include summer thermal cover. However, there is increasing evidence that these species can behaviorally accommodate to high temperatures even in the absence of such cover when there is minimal disturbance (Petron, 1987). The actual need is, thus, for hiding not thermal cover. Cook et al. (1998) confirmed in an Oregon study that thermal cover has erroneously been advanced as critical to elk in both summer and winter situations.

Animals have evolved adaptive mechanisms as a result of millennia of exposure to relatively predictable environmental conditions. Unfortunately, many current environmental changes are outside the adaptive capabilities of most wildlife species. The effects of toxic chemicals on organisms were graphically displayed by the rapid decline of predatory birds and the encouraging population response upon cessation of use of chemicals such as DDT. Yet we continue to contaminate the environment with chemicals which will continue to have biological effects generations hence. Acid precipitation now poses a considerable threat to entire ecosystems in northeastern North America, and species such as otter and loons which are dependent on aquatic organisms for food, are consequently threatened. Other chemicals mimic sex steroids and cause declines in male testosterone and sperm counts and can even result in sex changes in some species of fish. The effects on reproductive capacity can be enormous. What we ultimately will reap from our insults to the environment unfortunately remains to be determined. Not the least of the currently visible effects are modifications in behavior which can indirectly increase mortality. For example, dieldrin has been shown to reduce vigilance behavior in sheep

(Sandler et al., 1969), operant behavior in bobwhite quail (Gesell et al., 1979; Kreitzer, 1980), and a number of behaviors in second and third generation pheasants (Dahlgren and Linden, 1974). Parathion reduced incubation time in laughing gulls (White et al., 1983), and a number of other chemicals including methylmercury, toxaphene, and endrin have caused behavioral alterations in wildlife species (see Heinz, 1979, King et al., 1991). Dioxins and furans have been linked to reproductive impairment in wood ducks (White and Seginak, 1994). What else we will discover in the future is likely to be both acute and subtle and unfortunately there is little evidence that we are prepared to deal with the consequences of the direct and synergistic effects of new chemical combinations in the environment.

Populations

The unit most often considered by managers is the population, which is a group of organisms of the same species inhabiting a particular geographic area at a particular time. It is defined in terms of both time and space, a fact sometimes ignored by wildlife managers. To manage populations requires an understanding of their dynamics. We examine some examples of population interaction: predator-prey relationships and resource partitioning by similar sympatric species.

A classic wildlife example of predator-prey interaction is the Isle Royale moose-wolf situation. Originally the short-term data base suggested a dynamic equilibrium situation in which the predator and prey populations were in balance. With continued study it became apparent that the wolf population expanded as the moose population increased beyond the limitation of predator pressure. A moose population decline occurred in the late 1970s and early 1980s and the expanded wolf population then exerted heavy predator pressure on the declining prey. Both species since have seen precipitous population declines. In essence, a predator population may be able to suppress a prey population's growth or hasten a decline when it is at a low point in a cyclical pattern that may have been initiated by other factors such as hunting. Stephenson (1973) suggested such a relationship existed between white-tailed deer and wolves north of Montreal, Quebec, and predicted deer populations would not increase unless either hunting or predators were controlled. The type of response that might be expected with different prey species interacting with wolves is summarized by Seip (1995).

A northern Alberta study showed that one pack of timber wolves annually consumed about 15 percent of the yearling and older moose within their territory (Fuller and Keith, 1980). With an estimated annual recruitment of new yearlings of about 19 percent, it meant that predation was closely approximating annual recruitment. The study occurred in an area of proposed industrial development which would have resulted in an increase in human population and thus increased hunting pressure. The obvious management conclusion under those circumstances was that without control of the wolves the moose population would decline.

Densities of white-tailed deer have been found to be higher in the areas of overlap between wolf pack territories compared to territory centers. Wolves apparently avoid hunting in the areas of territory overlap until they are unable to catch prey elsewhere. While such refuges are only important in the summer range of the deer, it does ensure reservoirs for deer repopulation of core wolf areas as wolf populations decline due to reduction in deer populations.

Predator removal studies in Minnesota, Texas, and South Dakota have shown that populations of prey species will increase dramatically under such circumstances. In the case of waterfowl, this is probably because land-use practices such as mowing up to the edges of potholes have made the prey vulnerable to predators like raccoons that concentrate their efforts in these remnant habitats.

Modern research has repudiated a general philosophy which pervaded wildlife management for many years. While compensatory mortality has some validity for extremely prolific species such as bobwhite quail and muskrat, overhunting and overtrapping of even these species can occur. We must recognize that humans, acting either as predators (hunters) or as eliminators of habitat, constitute an additive factor. In reality most mortality in natural populations is indeed additive. Certainly, populations generally produce a surplus annually, but this may be small or even locally nonexistent under natural conditions. Even small additions to mortality may precipitate a decline in numbers of individuals in future generations of species with low productivity.

Management recommendations for areas where predators exist and are to be sustained or where new predator populations are developing, for example, the western United States now include limitations on hunter harvests to support species such as wolves (Boyce, 1995; Vales and Peek, 1995).

The presence of buffer species must be accounted for in management decisions. Beaver often serve this function by maintaining high wolf populations even while the primary prey species (moose or deer) may be declining in number. An interesting example of management models for a wolf-ungulate system based on biological simulation is presented by Walters et al. (1981) and is recommended for review.

Resource Partitioning

Food habit studies of herbivores and carnivores have shown considerable overlap in diet between certain species. The Canadian National Park Service conducted an elk reduction program in some of the western parks when there was evidence that elk were causing range deterioration which was affecting deer and moose populations. There has been speculation that closely related wildlife species such as mink and otter or coyotes and timber wolves may be competitors. Though diet overlap occurs between these species, each is adapted to exploit more efficiently (in an energetic sense) particular prey. This means that resource partitioning among species is the norm, and the presence or absence of particular species is what defines the amount of niche separation or overlap among species. Nonetheless, larger species will dominate smaller and sometimes force them out of their territories or foraging areas. As an example, red fox in Maine establish their home ranges outside coyote territories or in the boundary areas between adjacent coyote groups in order to minimize interaction (Harrison et al., 1989).

This type of discussion rapidly leads into theoretical ecology, and the best operative philosophy for a manager often is to ignore most competition between wildlife species as a factor and only react to species' interactions when domestic livestock or introduced species occupy the same area as native wildlife. There are relatively few situations where competition among wildlife species is a problem. Instead of initially suspecting competition, it is usually more productive to seek other causative factors except when domestic or exotic species are involved.

This chapter intentionally has not been comprehensive in its treatment of wildlife biology because there are many texts available that concentrate solely on the subject. Instead we have chosen to exemplify situations that might aid the manager in determining when to consider particular biological factors in management decisions.

Bibliography

Abell, D. H. and F. F. Gilbert. 1974. Nutrient content of fertilized deer browse in Maine. J. Wildl. Manage. 38:517–524.

Adamczewski, J. Z., C. C. Gates, B. M. Soutar, and R. J. Hudson. 1988. Limiting effects of snow on seasonal habitat use and diets of caribou (Rangifer tarandus groenlandicus) on Coats Island, Northwest Territories, Canada. Can. J. Zool. 66:1986–1996.

Anderson, W. L. and P. L. Stewart. 1969. Relationships between inorganic ions and the distribution of pheasants in Illinois. J. Wildl. Manage. 33:254–270.

Anderson, W. L. and P. L. Stewart. 1973. Chemical elements and the distribution of pheasants in Illinois. J. Wildl. Manage. 37:142–153.

Anderson, B. L., R. D. Pieper, and V. W. Howard, Jr. 1974. Growth response and deer utilization of fertilized browse. J. Wildl. Manage. 38:525–530.

Andrews, R. V., R. W. Belkap, J. Southard, M. Lorinez, and S. Hess. 1972. Physiological, demographic and pathological changes in wild Norway rat populations over an annual cycle. Comp. Biochem. Physiol. 41A:149–165.

Baker, D .L., D. E. Johnson, L. H. Carpenter, 0. C. Wallmo, and R. B. Gill. 1979. Energy requirements of mule deer fawns in winter. J. Wildl. Manage. 43:162–169.

Balser, D. S., H. H. Dill, and H. K. Nelson. 1968. Effect of predator reduction on waterfowl nesting success. J. Wildl. Manage. 32:669–682.

Beasom, S. L. 1974. Relationships between predator removal and white-tailed deer net productivity. J. Wildl. Manage. 38:854–859.

Belovsky, G. E. R. and P. A. Jordan. 1981. Sodium dynamics and adaptations of a moose population. J. Mammal. 62:613–621.

Bigler, W. J., J. H. Jenkins, P. M. Cumbie, G. L. Hoff, and E. C. Prather. 1975. Wildlife and environmental health: raccoons as indicators of zoonoses and pollutants in Southeastern United States. J. Am Vet. Med. Assoc. 167:592–597.

Boyce, M. S. 1995. Anticipating consequences of wolves in Yellowstone: model validation. In: L. N. Carbyn, S. H. Fritts, and D.R. Seip. (eds.), *Ecology and Conservation of Wolves in a Changing World.* pp. 199–209 Can. Circumpolar Inst., Edmonton, Alberta.

Bubenik, A. B . 1972. North American moose management in light of European experiences. N. Am. Moose Conf. 8:276–295.

Bubenik, A. B. 1977. Moose and man. Recent studies and their relationship to hunting. Annu. Conv. Ont. Fed. Anglers and Hunters. Toronto, Ontario, Canada. (unpubl.).

Bubenik, A. B., 0. Williams, and H. R. Timmerman. 1978. Some characteristics of anthrogenesis in moose—a preliminary report. N. Am. Moose Conf. 14:157–177.

Burgman, M. A., S. Ferson, and H. R. Akcakaya. 1993. Risk Assessment in Conservation Biology. Chapman and Hall, London, UK.

Buskirk, S. W. and S. L. Lindstedt. 1989. Sex biases in trapped samples of Mustelidae. J. Mammal. 70:88–97.

Cameron, R. D., K. R. Whillen, W. T. Smith, and D. D. Roby. 1979. Caribou distribution and group composition associated with construction of the trans-Alaska pipeline. Can. Field-Nat. 93:155–162.

Carbyn, L. N. 1982. Coyote population fluctuations and spatial distribution in relation to wolf territories in Riding Mountain National Park, Manitoba. Can. Field-Nat. 96:176–183.

Caughley, G. 1977. *Analysis of Vertebrate Populations.* John Wiley Sons, New York, NY. 234 pp.

Christian, J. J. 1963. Endocrine adaptive mechanisms and the physiologic regulation of population growth. In: M. V. Mayer and R. G. Van Gelder (eds.), *Physiological Mammalogy.* pp.189–353. Academic Press, New York.

Christian, J. J. 1971. Fighting, maturity and population density in *Microtus pennsylvanicus.* J. Mammal. 52:556–567.

Christian, J. J. 1978. Neurobehavioral endocrine regulation of small mammal populations. In: D. P. Snyder (ed.). *Populations of Small Mammals Under Natural Conditions.* pp.143–158. Pymatuning Lab of Ecology, University of Pittsburgh, PA. Spec. Publ. Ser. Vol. 5.

Clark, D. R., Jr. 1979. Lead concentrations: bats vs. terrestrial small mammals collected near a major highway. Environ. Sci. Tech. 13:338–341.

Coggins, V. 1976. Controlled vehicle access during elk season in the Chesuimnus area, Oregon. In: S.R. Hieb (ed.) *Proceedings of the Elk-logging-roads Symposium.* pp. 58–61, For. Wildl. Range Expt. Stn., University of Idaho, Moscow, ID.

Conomy, J. T., J. A. Duborsky, J. A. Collazo, and W. J. Fleming. 1998. Do black ducks and wood ducks habituate to aircraft disturbance? J. Wildl. Manage. 62:1135–1142.

Cook, J. G., L. I. Irwin, L. D. Bryant, R. A. Riggs, and J. W. Thomas. 1998. Relations of forest cover and condition of elk: a test of the thermal cover hypothesis in summer and winter. Wildl. Monogr. 141.

Cowan, I. McT. 1974. Management implications of behavior in the large herbivorous mammals. In: V. Geist and F. Walther (eds.). *The Behavior of Ungulates and its Relation to Management.* Vol. 2. IUCN Publ. New Ser. No. 24:921–934. Morges, Switzerland.

Dahlgren, R. B. and R. L. Linder. 1974. Effects of dieldrin in penned pheasants through the third generation. J. Wildl. Manage. 38:320–330.

Dale, F. H. 1954. Influence of calcium on the distribution of the pheasant in North America. Trans. N. Am. Wildl. Conf. 19:316–322.

De Angelis, D. L. 1976. Application of stochastic models to a wildlife population. Math. Biosci. 31:227–236.

Diamond, J. M. 1975. The island dilemma: lessons of modern biogeographic studies for the design of natural reserves. Biol. Conserv. 7:129–146.

Downing, R. L., E. D. Michael, and R. J. Poux, Jr. 1973. Monthly differences in the accuracy of sex and age ratio counts of white-tailed deer. Trans. Northeastern Deer Study Group. 9:37–53.

Duebbert, H.F. and J.T. Lokemoen. 1980. High duck nesting success in a predator-reduced environment. J. Wildl. Manage. 44:428–437.

Dunks, J. H., R. E. Tomlinson, H. M. Reeves, D. D. Dolton, E. Braun, and T. P. Zapatka. 1982. Migration, harvest, and population dynamics of mourning doves banded in the central management unit, I967–77. U.S. Fish Wildl. Serv., Washington, D.C., Spec. Sci. Rep. Wildl. No. 249.

Ellenberg, H., J. Dietrich, M. Streppler, and H. W. Nurnberg. 1985. Environmental monitoring of heavy metals with birds as pollution integrating biomonitors. I. Introduction, definitions and practical examples for goshawk (*Accipiter gentilis*), In: D. T. Lekkas (ed.). *Heavy Metals in the Environment.* pp. 724–726. CER Consultants Ltd., Edinburgh, Scotland.

Environment Canada. 1983. Canadian Wildlife Service, Annual Review 1982–1983. Can. Wildl. Serv., Ottawa, Ontario.

Erlinge, S. 1972. Interspecific relations between otter, *Lutra lutra,* and mink, *Mustela vison,* in Sweden. Oikos 23:327–335.

Fancy, S. G. and R. G. White. 1986. Predicting energy expenditures for activities of caribou from heart rates. Rangifer Special Issue 1:123–130.

Fancy, S. G. and R. G. White. 1987. Energy expenditures for locomotion by barren-ground caribou. Can. J. Zool. 65:122–128.

Fisher, R., F. F. Gilbert, and J. D. Robinette. 1987. Heart rate as an indicator of oxygen consumption in the pine marten (*Martes americana*). Can. J. Zool. 65:2085–2089.

Flook, D. R. 1964. Range relationships of some ungulates native to Banff and Jasper National Parks, Alberta. In: D. J. Crisp (ed.). *Grazing in Terrestrial and Marine Environments.* pp. 119–128 Symp. Br. Ecol. Soc. No. 4. Blackwell, Oxford.

Flueck, W. T. 1989. The effect of selenium on reproduction of black-tailed deer (*Odocoileus hemionus columbianus*) in Shasta County, California. PhD. thesis, (unpubl.), Univ. Calif., Davis, CA.

Frank. R., K. Ishida, and P. Suda. 1976. Metals in agricultural soils of Ontario. Can. J. Soil Sci. 56:181–196.

Fraser, D., E. Reardon, F. Dieken, and B. Loescher. 1980. Sampling problems and interpretation of chemical analysis of mineral springs used by wildlife. J. Wildl. Manage. 44:623–631.

Fraser, D. and E. R. Thomas. 1982. Moose-vehicle accidents in Ontario: relation to highway salt. Wildl. Soc. Bull. 10:261–265.

Freddy, D. J. 1984. Heart rates for activities of mule deer at pasture. J.Wildl. Manage. 48:962–968.

Fritzell, E. K. 1978. Habitat use by prairie raccoons during the waterfowl breeding season. J. Wildl. Manage. 42:118–127.

Fuller, T. K. and L. B. Keith. 1980. Wolf population dynamics and prey relationships in northeastern Alberta. J. Wildl. Manage. 44:583–602.

Gaines, K. F., C. G. Lord, C. S. Boring, I. L. Brisbin Jr., M. Gochfeld, and J. Burger. 2000. Raccoons as potential vectors of radionuclide contamination to human food chains from a nuclear industrial site. J. Wildl. Manage. 64:199–208.

Geist, V. and F. Walther (eds.). 1974. Symposium on the behavior of ungulates and its relation to management. (2 vols.). Int. Union Conserv. Nat. Natur.Resour. Morges, Switzerland. IUCN Publ. News. Ser. No. 24.

Gerrard, J. M., P. Gerrard, W. J. Maher, and D. W. A. Whitfield. 1975. Factors influencing nest site selection of bald eagles in northern Saskatchewan and Manitoba. Blue Jay 33:169–176.

Gesell, G. G., R. J. Robel, A. D. Dayton, and J. Frieman. 1979. Effects of dieldrin on operant behavior of bobwhites. J. Environ. Sci. Health B14:153–170.

Gjessing, E. T., A. Henriksen, M. Johannessen, and R. F. Wright. 1976. Effects of acid precipitation on fresh water chemistry. In: F. H. Braekke (ed.). *Impact of acid precipitation on forest and Fresh water ecosystems in Norway.* pp. 64–85, Sum. Rep. on Res. Results from Phase I (1972–75) of SNSF Proj. Oslo, Norway.

Gordon, A. G. and E. Gorham. 1963. Ecological aspects of air pollution from an iron-sintering plant at Wawa, Ontario. Can. J. Bot. 41:1063–1078.

Gray, B. T. and H. H. Prince. 1988. Basal metabolism and energetic cost of thermoregulation in wild turkeys. J. Wildl. Manage. 52:133–137.

Green, B., J. Anderson, and T. Whateley. 1984. Water and sodium turnover and estimated food consumption in free-living lions (*Panthera leo*) and spotted hyenas (*Crocuta crocuta*). J. Mammal. 65:593–599.

Grenier, P. A. 1974. Orignaux tues sur la route dans le Parc des Laurentides, Quebec, de 1962 a 1972. Nat. Can. 101:737–754.

Halls, L. K. 1978. White-tailed deer. In: J. L. Schmidt and D. L. Gilbert (eds.). *Big Game of North America: Ecology and Management.* pp. 43–65. Stackpole Books, Harrisburg, PA.

Hanley, T. A. and J. J. Rogers. 1989. Estimating carrying capacity with simultaneous nutritional constraints. USDA For. Serv. Res. Note PNW-RN-485.

Harper, J. A. and R. F. Labisky. 1964. The influence of calcium on the distribution of pheasants in Illinois. J. Wildl. Manage. 28:722–731.

Harris, L. E. 1970. *Nutrition Research Techniques for Domestic and Wild Animals.1*. Intl. Agric. Serv., Utah State Univ., Logan, UT.

Harrison, D. J., J. A. Bissonette, and J. A. Sherburne. 1989. Spatial relationships between coyotes and red foxes in eastern Maine. J. Wildl. Manage. 53:181–185.

Heinz, G. H. 1979. Methylmercury: reproductive and behavioral effects on three generations of mallard ducks. J. Wildl. Manage. 43:394–401.

Hobbs, N. T. and D. M. Swift. 1985. Estimates of carrying capacity incorporating explicit nutritional constraints. J. Wildl. Manage. 49:814–822.

Holter, J. B., W. E. Urban, Jr., H. H. Hayes, and H. Silver. 1976. Predicting metabolic rate from telemetred heart rate in white-tailed deer. J. Wildl. Manage. 40:626–629.

Hudson, R. J. and R. G. White (eds). 1985. *Bioenergetics of Wild Herbivores*. CRC Press, Boca Raton, FL.

Hurst, R. J., M. L. Leonard, P. D. Watts, P. Beckerton, and N. A. Oritsland. 1982. Polar bear locomotion: body temperature and energetic cost. Can. J. Zool. 60:40–44.

Hutchinson, F. E. and B. E. Olson. 1967. The relationship of road salt applications to sodium and chloride ion levels in the soil bordering major highways. Highway Res. Rec. 193:1–7.

Hutchinson, T. C. and L. M. Whitby. 1976. The effects of acid rainfall and heavy metal particulates on a boreal forest ecosystem near the Sudbury smelting region of Canada. In: *Proceedings of the First International Symposium on Acid Precipitation and the Forest Ecosystem*. pp. 745–765. USDA For. Serv. Gen. Tech. Rep. NE-23.

Jenkins, D. 1956. Chick survival in a partridge population. Anim. Health. 7:6–10.

Jones, R. L., R. F. Labisky, and L. W. Anderson. 1968. Selected minerals in soils, plants and pheasants: an ecosystem approach to understanding pheasant distribution in Illinois. Ill. Nat. Hist. Surv. Biol. Notes 63.

Jordan, P .A., D. B. Botkin, A. S. Dominski, H. S. Lowendorf, and G. E. Belovsky. 1973. Sodium as a critical nutrient for the moose of Isle Royale. Proc. N. Am. Moose Conf. 9:13–42.

Kautz, M. A., G. M. VanDyne, L. H. Carpenter, and W. W. Mautz. 1982. Energy cost for activities of mule deer fawns. J. Wildl. Manage. 46:704–710.

Kelsall, J. P. 1968. The migratory barren-ground caribou of Canada. Can. Wildl. Serv. Monogr. 3.

King, K. A., T. W. Custer, and J. S. Quinn. 1991. Effects of mercury, selenium, and organochlorine contaminants on reproduction of Forster's terns and black skimmers in a contaminated Texas bay. Arch. Envir. Contam. Toxicol. 20:32–40.

Krefting, L. W., A. B. Erickson, and V. E. Gunvalson. 1955. Results of controlled deer hunts on the Tamarac National Wildlife Refuge. J. Wildl. Manage. 19:346–352.

Kreitzer, J. F. 1980. Effects of toxaphene and endrin at very low dietary concentrations on discriminatory acquisition and reversal in bobwhite quail, *Colinus virginianus*. Environ. Pollut. A23:217–230.

Labisky, R. F., J. A. Harper, and F. Greeley. 1964. Influence of land use, calcium, and weather on the distribution and abundance of pheasants in Illinois. Ill. Nat. Hist. Surv. Biol. Notes 51.

Leopold, A. 1931. *Report on a Game Survey of the North Central States*. Sporting Arms and Ammunition Manufacturers' Institute. Madison, WI.

Litvaitis, J. A. and W. M. Mautz. 1980. Food and energy use by captive coyotes. J. Wildl. Manage. 44:56–61.

Loveless, C. M. 1964. Some relationships between wintering mule deer and the physical environment. Trans. N. Am. Wildl. Nat. Resour. Conf. 29:415–431.

Loveless, C. M. 1967. Ecological characteristics of a mule deer winter range. Colorado Game, Fish and Parks Dept. Tech. Publ. No. 20.

Luck, B. R. and R. G. White. 1986. Oxygen consumption for locomotion by caribou calves. J. Wildl. Manage. 50:148–152.

Lund, G. F. and G. E. Folk, Jr. 1976. Simultaneous measurements of heart rate and oxygen consumption in black-tailed prairie dogs (*Cynomys ludovicianus*). Comp. Biochem. Physiol. 55A:201–206.

Lyon, J. L. 1979. Influences of logging and weather on elk distribution in western Montana. USDA For. Serv. Res. Paper INT-236.

Lyon, J. L. 1980. Coordinating forestry and elk management. Trans. N. Am. Wildl. Nat. Resour. Conf. 45:278–281.

Mack, R. N. and J. N. Thompson. 1982. Evolution in steppe with few large hooved animals. Am. Nat. 119:757–773.

Marchand, P. J. 1982. An index for evaluating the temperature stability of a subnivean environment. J. Wildl. Manage. 46:518–520.

Mautz, W. W. and J. Fair. 1980. Energy expenditure and heart rate for activities of white-tailed deer. J. Wildl. Manage. 44:333–342.

McLean, B., K. Jingfors, and R. Case. 1986. Abundance and distribution of muskoxen and caribou on Banks Island, July 85. Renewable Resources, Government of the Northwest Territories, File Report 64:45.

McLellan, B. N. and D. M. Shackleton. 1988. Grizzly bears and resource extraction industries: effects of roads on behavior, habitat use and demography. J. Appl. Ecol. 25:451–460.

McNicol, J. G. and F. F. Gilbert. 1978. Late winter bedding practices of moose in mixed upland cutovers. Can. Field-Nat. 92:189–192.

Mierau, G. W. and B. E. Favara. 1975. Lead poisoning in roadside populations of deer mice. Environ. Pollut. 8:55–64.

Miller, F. L. and A. Gunn. 1979. Responses of Peary caribou and muskoxen to helicopter harassment. Can. Wildl. Serv. Occ. Paper No. 40.

Miller, F. L., F. W. Anderka, C. Vithayasai, and R. L. McClure. 1974. Distribution, movements and socialization of barren-ground caribou radiotracked on their calving and post-calving areas. Proc. Intl. Reindeer/caribou Symp. 1:423–435.

Moen, A. N. 1973. *Wildlife Ecology*. W.H. Freeman Co., San Francisco. CA.

National Research Council (NRC). 1984. *Nutrient Requirements of Poultry*. Nat. Acad. Sci., Washington, DC.

Newsome, A. E. 1971. The ecology of house-mice in cereal haystacks. J. Anim. Ecol. 40.1–15.

Owen, R. B., Jr. 1969. Heart rate, a measure of metabolism in blue-winged teal. Comp. Biochem. Physiol. 31:431–436.

Owen, R. B., Jr. 1970. The bioenergetics of captive blue-winged teal under controlled and outdoor conditions. Condor. 72:153–163.

Ozoga, J. J. and L. W. Gysel. 1972. Response of white-tailed deer to winter weather. J. Wildl. Manage. 36:892–896.

Parker, K. A., C. T. Robbins, and T. A. Hanley. 1984. Energy expenditures for locomotion by mule deer and elk. J. Wildl. Manage. 48:474–488.

Pauls, R. W. 1980. Heart rate as an index of energy expenditure in red squirrels (*Tamiasciurus hudsonicus*). Comp. Biochem. Physiol. 67A:409–418.

Pimlott, D. H., J. A. Shannon, and G. B. Kolenosky. 1969. The ecology of the timber wolf in Algonquin Provincial Park. Ontario Dept. Lands For., Toronto, Ontario.

Porter, W. P. and D. M. Gates. 1969. Thermodynamic equilibria of animals with environment. Ecol. Manage. 39:227–244.

Potvin, F. and J. Huot. 1983. Estimating carrying capacity of a white-tailed deer wintering area in Quebec. J. Wildl. Manage. 47:463–475.

Ralls, K., and A. M. Starfield. 1995. Choosing a management strategy: two structured decision-making methods for evaluating the predictions of stochastic simulation models. Conserv. Biol. 9:175–181.

Rasmussen, G. and R. Branden. 1973. Standard metabolic rate and lower critical temperature for the ruffed grouse. Wilson Bull. 85:223–229.

Rautenstrauch, K. R. and P. R. Krausman. 1989. Influence of water availability and rainfall on movements of desert mule deer. J. Mammal. 70:197–201.

Regelin, W. L., C. C. Schwartz, and A. W. Franzmann. 1985. Seasonal energy metabolism of adult moose. J. Wildl. Manage. 49:388–393.

Renecker, L. A. and R. J. Hudson. 1986. Seasonal energy expenditures and thermoregulatory responses of moose. Can. J. Zool. 64:322–327.

Robbins, C. T., Y. Cohen, and B. B. Davitt. 1979. Energy expenditure by elk calves. J. Wildl. Manage. 43:445–453.

Robbins, C. T. 1993. *Wildlife Feeding and Nutrition.* (2nd ed.). Academic Press. Inc., San Diego, CA.

Rogers. L. L., L. D. Mech, D. K. Dawson, J. M. Peek, and M. Korb. 1980. Deer distribution in relation to wolf pack territory. J. Wildl. Manage. 44:253–258.

Rose, G. B. 1973. Energy metabolism of adult cottontail rabbits, *Sylvilagus floridanus,* in simulated field conditions. Am. Midl. Nat. 89:473–478.

Rose, G. B. 1974. Energy dynamics of immature cottontail rabbits. Am. Midl. Nat. 91:473–477.

Rose, G. A. and C. H. Parker. 1982. Metal content of body tissues, diet items and dung of ruffed grouse near the copper nickel smelters at Sudbury, Ont. Can. J. Zool. 61:50–511.

Sandler, B. E., G. A. VanGelder, D. D. Elsberry, G. G. Karas, and W. B. Buck. 1969. Dieldrin exposure and vigilance behavior in sheep. Psychon. Sci. 15:261–262.

Schitoskey, F., Jr. and S. R. Woodmansee. 1978. Energy requirements and diet of the California ground squirrel. J. Wildl. Manage. 42:373–382.

Schroeder, H. A. 1965. The biological trace elements or peripatetics through the periodic table. J. Chronic Dis. 18:217–228.

Seip, D. R. 1995. Introduction to wolf-prey interactions. In: L. N. Carbyn, S. H. Fritts and D. R. Seip (eds.). *Ecology and Conservation of Wolves in a Changing World.* pp. 179–186. Can. Circumpolar Inst., Edmonton, Alberta.

Sheffy, T. B. and J. R. St. Amant. 1982. Mercury burdens in furbearers in Wisconsin, J. Wildl. Manage. 46:1117–1120.

Shield, J. 1972. Acclimation and energy metabolism of the dingo, *Canis dingo* and the coyote, *Canis latrans.* J. Zool.(Lond.)168:483–501.

Singer, F. J. 1979. Habitat partitioning and wildfire relationships of cervids in Glacier National Park, Montana. J. Wildl. Manage. 43:437–444.

Stalmaster, M. V. and J. A. Gessaman. 1982. Food consumption and energy requirements of captive bald eagles. J. Wildl. Manage. 46:646–654.

Stephenson, A. B. 1973. Deer management in the North Montreal region, Quebec Min. Tourism Fish and Game, Min. Agric. and Colonization. (unpubl.).

Stokes, A. W. 1954. Population studies of the ring-necked pheasants on Pelee Island, Ontario. Ontario Dept. Lands For., Tech. Bull. Wildl. Ser. No. 4.
Stokes, A. W. and D. F. Balph. 1965. The relation of animal behavior to wildlife management. Trans. N. Am. Wildl. Nat. Resour. Conf. 13:401–410.
Switzer, G. L. and L. E. Nelson. 1972. Nutrient accumulation and cycling in loblolly pine (*Pinus taeda* L.) plantation ecosystems: The first twenty years. Soil Sci. Soc. Am. Proc. 36:143–147.
Teubner, V.A. and G.W. Barrett. 1982. Bioenergetics of captive raccoons. J. Wildl. Manage. 47:272–274.
Thiessen, D. D. 1966. Role of physical injury in the physiological effects of population density in mice. J. Comp. Physiol. Psychol. 62:322–324.
Thing, H. 1977. Behavior, mechanics and energetics associated with winter cratering by caribou in northwestern Alaska. Biol. Pop. Univ. Alaska No.18.
Thomas, V. G., H. G. Lumsden, and D. H. Price. 1975. Aspects of the winter metabolism of ruffed grouse (*Bonasa umbellus*) with special reference to energy reserves. Can. J. Zool. 53:434–440.
Udevitz, M. S., C. A. Howard, R. J. Robel and B. Curnutte. 1980. Lead contamination in insects and birds near an interstate highway. Kansas Environ. Entom. 9:35–36.
U. S. Environmental Protection Agency. 1979. EPA Research Summary-Acid Rain. Off. Res. Develop. EPA-600/8-79-028.
Vales, D. J. and J. M. Peek. 1995. Projecting the potential effects of wolf predation on elk and mule deer in the East Front Portion of the northwest Montana wolf recovery area. In: L. N. Carbyn, S. H. Frits and D. R. Seip (eds.), *Ecology and Conservation of Wolves in a Changing World.* pp. 211–221 Can. Circumpolar Inst., Edmonton, Alberta.
Verme, L. J. 1973. Movements of white-tailed deer in upper Michigan. J. Wildl. Manage. 37:545–552.
Vikoren, T. and G. Stuve. 1996. Fluoride exposure in cervids inhabiting areas adjacent to aluminum smelters in Norway. II. Fluorosis. J. Wildl. Dis. 32:181–189.
Wagner, F. H. and LC. Stoddart. 1972. Influence of coyote predation on black-tailed jackrabbit populations in Utah. J. Wildl. Manage. 36:329–342.
Walters, C. J., M. Stocker and G. C. Haber. 1981. Simulation and optimization models for a wolf-ungulate system. In: C. W. Fowler and T. D. Smith (eds.), *Dynamics of Large Mammal Populations.* pp. 317–337. John Wiley Sons, New York, NY.

Westermark, T., T. Odsjo and A. G. Johnels. 1975. Mercury content of bird feathers before and after Swedish ban on alkyll mercury in agriculture. Ambio. 4:87–92.

White, D. H., C. A. Mitchell, and E .F. Hill. 1983. Parathion alters incubation behavior of laughing gulls. Bull. Environ. Contam. Toxicol. 31:93–97.

White, D. H. and J. T. Seginak. 1994. Dioxins and furans linked to reproductive impairment in wood ducks. J. Wildl. Manage. 58:100–106.

White, E. H. 1974. Whole-tree harvesting depletes soil nutrients. Can. J. For. Res. 4:530–535.

Wooley, J. B., Jr. and R. B. Owen, Jr. 1978. Energy costs of activity and daily energy expenditure in the black duck. J. Wildl. Manage. 42:739–745.

Wren, C., H. MacCrimmon, R. Frank, and P. Suda. 1980. Total and methyl mercury levels in wild mammals from the Pre-Cambrian Shield area of south central Ontario, Canada. Bull. Environ. Contam. Toxicol. 25:100–105.

Yarmolov, C., M. Bayer, and V. Geist. 1988. Behavior responses and reproduction of mule deer (*Odocoileus hemionus*) does following experimental harassment with an all-terrain vehicle. Can. Field-Nat. 102: 425–429.

Zwickel, F. C. and J. F. Bendell. 1972. Blue grouse, habitat and populations. Proc. Intl. Ornith. Congr. 15:150–169.

Additional Readings

Carbyn, L. N. 1983. Wolves in Canada and Alaska. Can. Wildl. Serv. Rep. Ser. No. 45. A collection of papers by wolf researchers and managers given at an lUCN/Species Survival Commission meeting on wolves in Edmonton, Canada.

Geist, V. and F. Walther (eds).—see Bibliography. An excellent collection of papers on ungulate behavior as it relates to management.

Fowler, C. W., and T. D. Smith. 1981. Dynamics of large mammal populations. John Wiley Sons, New York, NY. Population biology of large mammals examined from a dynamic aspect. Many fine simulation models are explained.

Hudson, R. J. and R. G. White (eds.). 1985. Bioenergetics of wild herbivores. CRC Press Boca Raton, FL. The contributions in this book cover such topics as energy and supplies, energy expenditures, foraging behavior and thermoregulation of wild herbivores.

5 PARASITES, DISEASE, AND WILDLIFE

For many years most wildlife biologists have discounted the influence of disease and parasites on large wild mammal populations. The accepted theory has been compensatory mortality and few considered that populations could be controlled by disease organisms despite the ample evidence from human history. Wildlife managers have considered disease and parasitism as mortality factors of little consequence in the dynamics of the populations they studied and managed.

Evidence has mounted that given the right circumstances disease organisms can have major impacts on populations of mammals. This chapter will be restricted to mammals considered medium to large in body size. This encompasses the largest rodents, e.g. beaver, and most carnivores but largely focuses on the ungulates. Although the reader will be introduced to the disease organism in sufficient detail to understand its biological attributes an attempt has been made to focus on the impact(s) the disease or parasite has had on mammalian populations.

According to Yuill (1986), there are three situations where direct mortality is the expected consequence of disease. The first is when the death of the host enhances disease transmission, the second is when the disease organism will infect many different species and thus have many hosts for aiding disease transmission and the third is when the host populations have a wide distribution and transmission is spread over a long period of time. Yuill provides examples of each of these cases, namely rabies, the screw-worm fly (*Cochliomyia hominivorax*) and yellow fever. The population effects vary in these three situations with the first two often resulting in significant population losses while in the third case the mortalities seldom cause any noticeable population effect.

Major population impacts usually occur when a disease organism infects an aberrant or maladapted host. *Parelaphostrongylus tenuis* exemplifies this situation but disease can be indirect in its effect on populations by affecting the

112 • *The Philosophy and Practice of Wildlife Management*

condition and behavior of individuals, the natality rate, or the cost of existence. More difficult to document, such subtle influences become increasingly important as wildlife habitats decrease in size and their resident wildlife populations become more isolated and thus self-sustaining. It also suggests that there may be disproportionate harvesting of diseased or parasitized animals during hunting seasons.

Rabies

Chalmers and Scott (1969) detail situations where the rabies virus has reached a dynamic balance with host species and thus mortality has become minimal or nonexistent in the host populations. However, such mutually nondetrimental associations derive from long standing ecological systems and humans have disrupted most such climax associations. Mustelidae, desmodontid vampire bats, viverridae and domestic dogs of northern equatorial Africa and possibly myomorphic rodents are examples of climax associations which serve as reservoirs for the disease and transmission sources to more susceptible species. Transmission to susceptible hosts can occur from predation and bites or otherwise coming into contact with the body fluids of the infected individual. Transmission within reservoir hosts may be facilitated by agonistic behavior during times of high population density. Latent rabies may be reactivated by such a stress factor or by physiological stressors such as parturition, lactation, or starvation.

In North America, the red fox is the predominant wildlife species sustaining rabies epidemics but other, perhaps independent cycles of infection may reinforce this source, e.g., arctic fox (*Alopex lagopus*), wolves, raccoon, skunks, and insectivorous bat species (e.g., *Eptesicus fuscus*). By contrast in Europe, the fox seems to be the susceptible species, perhaps the only species able to sustain an epidemic. Control measures are therefore focused on the fox and the epidemiology can be predicted somewhat from fox population data. Moegle et al. (1974) concluded that, for the portion of the German Federal Republic they studied from 1963–1971, in those areas where foxes were shot at >0.5 animals/km^2/yr rabies spread at a rate independent of the population density. The incidence of disease had no influence on the advance of the epidemic, rather it was a function of the behavior of young animals prior to, and during, dispersal. Frequencies of rabies infections and population densities of foxes were related but not proportional.

Because fox populations in Europe turn over at a very rapid rate (~ 2/3 per year), the reduction in population due to rabies is relatively short-lived. In contrast, badger populations in many areas of central Europe were reduced to 10–50 percent of previous levels by rabies (Moegle and Knorpp, 1978; Wachendorfer and Schwierz, 1980).

In other parts of the world, other species become predominant in sustaining rabies epidemics by being the primary reservoir, for example, the raccoon-dog (*Nyctereutes procyonoides*) in far eastern Asia and dogs and jackals (*Canis* spp.) in southwest Africa. In the latter case, kudus (*Tragelaphus strepsiceros*) were seriously affected in one area with about 10,000 killed in a three year period. It was suggested that high population densities of kudus and jackals facilitated the outbreak with possible spread of the disease occurring within kudu populations due to social and feeding factors. The jackal appears to be the main wildlife reservoir in all of southern Africa.

Ontario took a different approach to a potential outbreak of rabies in raccoons by creating a buffer zone of vaccinated animals in areas where infected raccoons were expected to enter the province from the United States (Rosatte et al., 1997). Should rabies be detected in an area, raccoons would be killed in a 4 km radius of the detected case and vaccinated in a buffer zone 4–8 km from the site. It will be interesting to see how successful this strategy will be.

A stochastic spatial simulation model developed for skunks in the midwestern United States showed that rabies persisted for at least twenty years in the majority of the simulations. The biological importance of this is the support for the ability of some populations and species to maintain enzootic and even epizootic rabies without external influence of other species, or disease reintroduction, or an internal latency period (Bremillion-Smith, 1986). The interepizootic interval averaged eight years and the author concluded that rabies can control populations under given conditions that will not cause extinction of either the disease or the host species. Similar conclusions could likely be reached for all the prime vectors of rabies such as the red fox, jackal, or arctic fox. Johnson (1970) disputes this suggestion and considers the canids to be aberrant hosts of the virus with the mustelids being the natural and thus sustaining hosts for the disease.

Rabies thus is a disease which can cause precipitous population declines in susceptible species which for various reasons facilitate transmission of the disease to most members of the population (e.g., badgers and kudus). For those species which sustain the disease (e.g., certain canids and skunks) the effect is similar to "waves" of the virus occurring periodically with relatively short-lived population declines followed by rebounds to pre-epizootic numbers. In effect, a cyclical pattern is established which is loosely tied to population density and species behavioral patterns.

Canine Distemper

Domestic dogs are routinely vaccinated against canine distemper virus. The virus commonly shows up in antibody tests of wild canids though there is little evidence of population effects. Indeed Guo et al. (1986) suggest distemper in Texas coyotes is enzootic and the coyote population may serve as a reservoir for the disease in domestic dogs.

Pathologic findings associated with canine distemper include disruption of the immune system. This primary symptom of canine distemper disease led to the discovery that a similar virus (*Morbillivirus*) was responsible for a substantive die-off of harbor seals (*Phoca vitulina*) in the North Sea area during 1988. There is documented evidence of >8000 seal mortalities in the North Sea region in a few months of 1988. For example, over two-thirds of a colony of 6000 animals on the west coast of Sweden died (Kirwin, 1988). Infected seals died within a week of infection, usually from pneumonia. The virus also has had an effect on dolphins in both the Atlantic and Pacific Oceans in recent years (De Swart et al., 1995; Reidarson et al., 1998).

The disease is a significant mortality agent in Mustelidae with the last known population of the black-footed ferret (*Mustela nigripes*) in the wild, which numbered about 125 individuals, almost succumbing totally (reduced to about seventeen animals) to distemper (May and Anderson, 1987). Monson and Stone (1970) suggest that the disease may be a significant mortality factor of many species of wild carnivores.

These recent findings suggest that common diseases in domestic animals which are currently controlled by immunization may be having profound effects on wild populations. It is only in circumstances where the effects are obvious, such as seen in seals, or where a population is being closely monitored, such as with ferrets, that we are ever witness to the event.

Hemorrhagic Disease

Epizootic hemorrhagic disease (EHD) is a virus believed responsible for epizootics in white-tailed deer (*Odocoileus virginianus*) populations in the southeastern United States. At the same time as a 1971 southeastern U.S. outbreak, another epizootic was observed in southwestern North Dakota. Epizootics have been reported in Alberta and South Dakota as well. In fact until the southeastern outbreaks, it was considered a disease of the northern United States and Canada.

Midges such as *Culicoides* are believed to be vectors of EHD and the severity of epizootics appears to be density related. In a Kentucky study, 62 percent of a captive white-tailed deer colony of 104 animals died (Roughton, 1975). At the same time, losses were occurring in the free-ranging deer population but the exact extent was not documented. The 1962 Alberta epizootic caused the deaths of at least 440 white-tailed deer, 18 mule deer and 13 pronghorn antelope (*Antilocapra americana*) (Chalmers et al., 1964) while documented losses in North Dakota the same year were >200 white-tailed deer, 18 mule deer (*Odocoileus hemionus*), and 11 pronghorn antelope (Richards, 1963). It was estimated that at least 2000 white-tailed deer died during the 1970 outbreak in southwestern North Dakota (Hoff et al., 1973).

Experimental studies have demonstrated white-tailed deer mortality rates >90 percent (Fay et al., 1956; Karstad et al., 1961; Shope et al., 1968). Thus, in situations of high densities of susceptible host and vector, significant population losses might be expected. The epizootics appear to occur in midsummer to early fall and were preceded by long periods of very hot, dry weather in the Dakotas which may have facilitated development of vector populations.

Bluetongue

Desert Bighorn sheep (*Ovis canadensis*), and deer have been affected by bluetongue virus. *Culicoides* gnats are the vectors and in Texas, white-tailed deer and domestic sheep are apparent reservoirs. Bluetongue is thought to be a limiting factor to bighorn sheep populations in that state. During a 1976 dieoff in the Missouri River basin at least 4100 pronghorn antelope and deer were lost in eastern Wyoming. A population of about 400 deer on the Turnbull Wildlife Refuge in Washington State was reduced to about 100 animals by the disease in 1987 (Boone, pers. comm.).

The virus is common in many parts of Africa and the Middle East where the *Culicoides* vectors exist but has primarily impacted domestic cattle, sheep, and goats. The pathology of bluetongue is similar to that for EHD and the hemorrhagic syndrome in white-tailed deer is considered to be a result of either EHD or bluetongue viruses. Virus isolation and identification need to be done to differentiate the infections.

Bluetongue can cause early fetal absorption or spontaneous abortion in white-tailed deer. Thus, as well as direct mortality losses, there is the potential for productivity losses.

Arboviruses

About 200 or so arboviruses are known and wildlife serve as reservoirs for most of these insect borne infections. The term arbovirus is a truncation of arthropod-borne virus and by definition an arbovirus infects vertebrate animals and multiplies within the arthropod vector. Most arboviruses fall within four antigenic groups. Group A contains American equine encephalitis viruses and Group B St. Louis encephalitis, yellow fever, dengue, and the Russian spring-summer complex viruses.

Most of these viruses are endemic in wildlife populations and only receive notoriety when humans or their domestic animals are affected. Although birds and small mammals are the most common wildlife hosts for the viruses, the one arbovirus known to commonly occur in large mammals is vesicular stomatitis virus. It is characterized by vesicles on the mucous membranes of the mouth, the skin of the muzzle, the teats, and coronary bands of the feet. Karstad et al. (1956) found natural infections in deer, raccoon and bobcats (*Felis rufus*). The disease is limited to the Americas and is rarely fatal. Thus it has minimal influence on wild populations.

Nonetheless there is strong serological evidence that many arboviruses do infect large wild mammals. Bighorn sheep, white-tailed deer, bison (*Bison bison*), red fox, caribou, grizzly bear (*Ursus horribilis*), moose, and Dall sheep (*Ovis dalle*) have all tested sera positive for antibodies. The high prevalences found in some of these species (40–49 percent) suggests that they may be important reservoirs for the viruses and function in the epizootiology of the diseases. No evidence in the literature exists related to population effects on large mammals for any of the arboviruses.

Rinderpest

One of the most famous (or infamous) panzootics on record was the rinderpest virus pandemic in Africa at the end of the nineteenth century. Millions of cattle and innumerable wild animals died. In East Africa, cattle succumbed first followed by cape buffalo (*Syncerus caffer*), eland (*Taurotragus* ssp.), warthogs (*Phacochoerus* ssp.), giraffe (*Giraffa camelopardalis*), kudu (*Strepsiceros strepsiceros*), roan antelope (*Hippotragus equinus*), bushbuck (*Tragelaphus scriptus*), and finally wildebeest (*Connochaetes* ssp.). In South Africa, buffalo were hardest hit, herds of eland were lost, and warthogs and bushpigs (*Potamo choerus*) died in substantive numbers. Other species affected included blesbok (*Damaliscus dorcus*), bushbuck, duiker (*Cephalophus*), gemsbok (*Oryx*), kudu, reedbok (*Redunea*), springbok, steinbok (*Raphicesrus*), and waterbuck (*Kobus*) (Theiler, 1897). It was thought that some species such as wildebeest and cape buffalo in East Africa were almost exterminated but numbers recovered even though distribution patterns have been changed. The panzootic was so virulent that it is believed to have left no foci of infection in southern Africa. In east and west Africa though, it became enzootic and periodic epizootics have occurred. In the 1960 epizootic in Uganda and northern Kenya, it was estimated that 60 percent of the eland, kudu and warthogs, 50 percent of the buffalo, bushbuck and giraffe, 40 percent of the impala, and many oryx died (Grimwood, 1961).

Rinderpest also decimated wildlife outside Africa. The population decline of the European bison (*Bison bonasus*), a deer (*Cervus*) dieoff in the United Kingdom in 1865, deer (*Odocoileus*) losses in Brazil, deer, wild buffalo (*Bubalis bubalis*), and wild boar losses in India, wild pig die-offs in Sri Lanka, and losses of wild pigs, bonteng (*Bibos*), buffalo, gaur (*Bos gaurus*), antelope, and deer in Southeast Asia have all been ascribed to rinderpest.

Transmission is through the air and requires close contact between sick and healthy animals. Carrion feeding can also transmit the disease to species like wild pigs. Presently rinderpest is almost gone from Africa with only Mauritania, Mali and Ethiopia still experiencing epizootics (1974 was the last year that any wild mammal was found with rinderpest in any of the other African nations). Coincident with this decline was the elimination of rinderpest from domestic ungulates. This suggests that the virulent panzootic and recurring epizootics may have been initiated in domestic species and wildlife are not a reservoir. Another possibility put forward by Provost (1981) is that a hypovirulent strain of the virus occurs in wildlife, especially buffalo, and the virulence is induced when cattle are infected. Rinderpest is the best example of a disease that decimated large mammal populations to the extent that distributional patterns of a number of species were permanently affected because local populations never recovered.

Canine Parvovirus

A recent pandemic was caused by canine parvovirus. This virus is thought to be a mutation from the feline panleukopenia virus (Parrish et al., 1985) and it was first noticed in the late 1970s in domestic dogs. It caused mortality and morbidity in dogs and a vaccine was rapidly developed which provided adequate protection.

Serologic surveys of wild animals suggested that wild canids contracted the virus about the same time as did the domestic canids. In Minnesota, the first positive samples from wolves were found in 1976 (Goyal et al., 1986) and in Alaska from 1979 (Zarnke and Ballard, 1987).

In Sweden, the first case of canine parvovirus was detected in 1979. Wierup (1983) estimated that 17 percent of the domestic dog population developed clinical infections. An epidemic was calculated to be initiated when the density of susceptible dogs was $>12/km^2$ and to cease when the density was $<6/km^2$. The infection has been widespread in Korea since 1982 (Choi et al., 1986) and now has spread globally (Parrish et al., 1985). Mortality is concentrated in pups that have been recently weaned.

While most serological surveys have shown widespread antibody titres to the disease in wild canids, foxes tested in France showed only 4 of 142 sera with canine parvovirus antibodies (Schwers et al., 1983). It was concluded that the fox was not a vector of the disease and certainly not the origin of the infection in dogs in Europe.

Canine parvovirus demonstrates the possibility that new disease organisms may develop with virulent effects. Humans have lived with this knowledge of viruses and bacteria for most of the twentieth century and have seen some serious influenza epidemics and the development of bacteria resistant to antibiotics. It appears that canine parvovirus had few detrimental effects on wild wolf populations although individuals undoubtedly died or were taken ill. Such a statement appears in Zarnke and Ballard (1987) but once again this is a subjective observation for as in so many other disease situations there simply are no data available from affected wild populations. Foreyt (1991, pers. comm.) noted that survival of coyote pups in 1979–80 was minimal based on pelts taken by trappers and hunters in Washington and Idaho and attributed this to canine parvovirus.

Pasteurellosis

Epizootics of pasteurellosis have occurred in many different parts of the world and the causative bacteria *Pasteurella* spp. have worldwide distribution. *Pasteurella multocida* is usually implicated in pasteurellosis of wild mammals. Bighorn sheep (*Ovis canadensis*), pronghorn antelope, bison, white-tailed deer, black-tailed deer, mule deer, caribou, elephant, elk, kangaroo (*Macropus* spp.), sea lion (*Zalophus californicus*), wild swine and water buffalo (*Bubalus* spp.) have all demonstrated the disease in the wild. The bacteria contaminate the environment in the saliva and feces of infected individuals. Mortalities in mule deer and bighorn sheep have been reported but often the observed pneumonia is attributed to lungworm even when *Pasteurella* is present; although Post (1962, 1971), Thorne et al. (1982), and Foreyt (1989) believe that pasteurellosis was responsible for the historical die-offs of bighorn sheep which resulted in substantive population reductions. Corroborative evidence includes the loss of wild and captive bighorn sheep to pasteurellosis.

It is not unusual to have respiratory viruses such as parainfluenza-3 virus and respiratory syncytial virus as well as lungworms present and involved in the pathogenesis of a pasteurellosis infection in domestic animals. However, Nordkvist and Karlsson (1962) found no evidence of myxovirus in a reindeer pasteurellosis outbreak in Norway.

The most important impact of pasteurellosis does appear to be on bighorn sheep populations. Onderka et al. (1988), Onderka and Wishart (1988), and Foreyt (1989) all recommend strongly that domestic sheep and possibly cattle not be allowed access to bighorn sheep range because of the susceptibility of bighorns to *P. haemolytica* infections and the high mortality rates which might be anticipated.

Tularemia

Although tularemia is usually found in smaller mammals it can be transmitted to large mammals including foxes, coyotes, beaver, bobcat, badger (*Taxidea taxus*), and deer in North America. *Francisella tularensis,* a bacterium, causes an acute, moderately severe infectious septicaemia and is transmitted by a number of ectoparasites as well as by contact with a contaminated environment. Ticks, mites, deer flies (Tabanidae), fleas, lice, midges, mosquitoes, and bed bugs (*Cimex lectularius*) are all known transmitters of the disease.

There are reports of massive die-offs of beaver resulting from tularemia. In the early 1950s, epizootics were reported in the Upper Peninsula of Michigan, Minnesota, Wisconsin, Manitoba, and northwestern Ontario. In the latter situation, beaver all but disappeared from the area. The populations rebuilt rapidly aided by transplanting programs so that by 1970 they were at or above former levels and authorities in Ontario were anticipating a possible recurrence of the disease. The epizootic spread from Manitoba and Ontario to Minnesota, Wisconsin, and finally the Upper Peninsula of Michigan taking about four years to spread this distance. Virulence seemed to diminish as the disease spread south and east but appeared to be positively related to population densities.

The bacterium was first isolated from beavers in Wyoming during a 1938 die-off. It has been implicated in other die-offs in the western United States and central Canada since that time. It remains a potentially serious threat to beaver populations in high density areas where an epizootic in a small rodent, such as, *Microtus* may cause aquatic contamination and initiate a beaver epizootic.

Brucellosis

Brucellosis is well known in domestic cattle where it causes abortions during the latter half of gestation, female sterility, and pathological changes in the male genital tract. It is highly contagious and the causative agents are bacteria of the genus *Brucella* (in cattle, *Brucella abortus*). The disease occurs in such species as bison, elk, moose, Dall's sheep (*Ovis dalli*), caribou, chamois (*Rubicapra rupicapra*), foxes, and European deer.

Brucella is only transmitted orally but contamination of the eyes, wounds, or genital tract can also result in infection. The only important population effect would be through decreased productivity as a result of fetal losses and reduced fertility. Although most wildlife populations of the species mentioned above test positive to some degree to brucellosis, there is little information on the actual impacts on the populations. Thorne et al. (1978) examined the effects of the disease on sixty adult elk and over seventy-two of their offspring for sixty-five months. They artificially infected twenty-seven adults and 96 percent of the twenty-four naturally exposed adults also became infected. The most important effects on the population were abortion and birth of nonviable calves. Fourteen of twenty-nine cows lost their first calf following infection. One of nine cows followed through a second pregnancy lost her calf and one of five lost a third calf. The authors thought that the rate of abortion was actually higher than they reported as several fetuses suspected to have aborted were not found. Although high, the abortion rate in elk was lower than that normally found in cattle. Peterson et al. (1991) looked at different bison-brucellosis management schemes involving vaccination combined in one case with removal of infected animals and deduced that for Grand Teton National Park the best that could be achieved would be a reduction in seropositive animals to 20 percent from the then level of 69 percent. This would have little effect on disease transmission. Compared to elk, brucellosis may be more virulent in moose and result in death of the infected individual. If so, there is the potential for greater direct losses but transmission may be more difficult in this species as it tends to be solitary compared to the elk.

Anthrax

The bacterium, *Bacillus anthracis,* has caused considerable mortality within populations of large mammals. Choquette et al. (1972) described the Canadian situation where anthrax was first diagnosed in 1962 in bison in the Northwest

Territories. In 1972 there were 12,000–14,000 bison in Wood Buffalo Park and adjacent areas. During the various outbreaks from 1962–1972 good data were available on carcasses found relative to local population size so we can assess the virulence of the disease and its potential population effects. The outbreak area in 1962 encompassed a population of about 1300 bison and 281 carcasses were found during August. In 1963, some 269 carcasses were found in an area where about 2500 bison were resident. Although smaller outbreaks occurred through 1971, 1964 marked the last major outbreak. However, in 1971, there were thirty-one carcasses found in an area where a population of 130 bison existed. This would represent a population loss of 24 percent. Young animals were rarely affected but death was concentrated in the adult male segment of the population. This is a pattern similar to that found in anthrax outbreaks in kudus in Kruger National Park, South Africa. It also means that effects on the population are not as severe as the reproductive females are less vulnerable and thus reasonable productivity can be maintained.

The disease is disseminated by wallowing in infected mud, by carrion feeders, excreta (including birds, mammals, and insects), and by wind. As might be expected, anthrax has almost universal distribution. It is reported in red deer, fallow deer (*Dama dama*), elk (*Alces alces*), roe deer (*Capreolus capreolus*), and wild boar (*Sus scrofa*) in Europe and was believed to be responsible for an 1874 epizootic in Prussia where 2000 red and fallow deer died. Anthrax has been found in wild pigs, wild dogs, and dingoes in Australia and in elephants in Asia. It is prevalent in Africa where sporadic outbreaks have involved zebra (*Equus burchelli*), hartebeest (*Alcelaphus caama*), wildebeest (*Connochaetes gnu*), springbok (*Antidorcas marsupialis*), red hartebeest (*Alcelaphus buselaphus*), black wildebeest (*Connochaetes taurinus*) and kudu. Choquette (1970) lists another twenty-five or so wildlife species that have been diagnosed as dying from anthrax in Africa.

The South Africa situation clearly demonstrates the impact this disease can have on populations. There anthrax has been the single most important cause of large-scale mortality in certain antelope species and zebras over the period 1960–1980. Pienaar (1961) reported that at least 1195 animals died of anthrax in the northern part of Kruger National Park from 1959–1961. In three separate outbreaks (1959, 1960, and 1970) at least eighty-three of an estimated population of 250 roan antelope (*Hippotragus equinus*) died of anthrax. The park's roan antelope population represented 80% of the total South African population of this species. A vaccination program seems to have helped the roan antelope rebuild somewhat between 1971 and 1980. Ebedes (1976) reported that between 1967 and 1974 anthrax caused 54 percent of the recorded wildlife mortality in Etosha National Park including 54 percent of the zebra and 35 percent of the blue wildebeest (*Connochaetes taurinus*) losses.

Berry (1980 as reported in Ebedes, 1981) documented 62 percent of the blue wildebeest and 43 percent of the zebra deaths in 1976–78 resulted from anthrax. The availability of carcasses caused an increase in lion (*Panthera leo*) and spotted hyaena (*Crocuta crocuta*) populations and resulted in increased predation on wildebeest and zebra. The loss of pregnant females or females recently undergoing parturition further depressed the herbivore populations as losses were concentrated on adult animals during the rainy season.

To illustrate the extent of population declines, the zebras in Etosha National Park dropped from an estimated 18,000 individuals in 1968 to 9166 in 1978. Wildebeest population estimates were 25,000–30,000 in 1913, 25,000 in 1954, 30,000 in 1965, 4000–5000 in 1968–1970 and by 1978, 2493. Population reductions of 50 percent within a decade occurred and the long-term prospects were not promising according to Ebedes (1981). However, Whyte and Joubert (1988) dismissed the disease hypothesis for the decline of blue wildebeest populations in Kruger National Park between 1969–1979 and instead suggested that adverse habitat conditions caused by above average rainfall and the severing of migration routes by erection of a fence were responsible.

A white-tailed deer anthrax epizootic on a 3600 acre island in Arkansas resulted in a 60–90 percent herd reduction (Kellogg et al., 1970). Common factors in many outbreaks of the disease are high density populations, previous history of anthrax in the area and low food supply. Herbivores may be forced to graze closer to the soil surface thus facilitating ingestion of anthrax spores. Once the disease is initiated, biting flies may sustain and spread the infection.

It is interesting to note that Whyte and Joubert (1988) so easily dismissed the anthrax evidence presented by Ebedes (1981). As zebra and blue wildebeest populations increased (1979–1986), they even hinted that earlier estimates of population declines may have been in error because of sampling techniques. It is this type of contention that can lead to easy dismissal of disease as a population regulating mechanism.

Yersinosis or Pseudotuberculosis

Yersinia pseudotuberculosis is a bacterium that causes different disease symptoms in a number of species. It is a known mortality agent in domestic or captive wild ungulates. Transmission is oral-fecal but exposure is not considered sufficient to cause disease. Avian, rodent, and lagomorph species are suggested reservoir hosts and susceptible species are made vulnerable by stressors such as overcrowding, poor physical condition, or environmental factors and pathogenesis results.

Blake and McLean (1988) relate a particularly well-documented example of population effects of this disease organism on the muskoxen (*Ovibos muschatus*) of northern Banks Island in the Canadian arctic. The Thomsen River area once supported a "substantial muskox population of undetermined size" based on archaeological findings. By 1900, these muskoxen had disappeared from the region and it has taken most of the twentieth century to build a Banks Island population estimated at 25,000 animals with the highest density found in the Thomsen River area. In the summers of 1986 and 1987, acute yersinosis was identified in the muskoxen. Animals were dying despite being in excellent physical condition with abundant fat stores. The muskox population density in a 200 km^2 study area around Thomsen River was 2.3/km^2. Fifty-eight animals were pathologically affected by yersinosis over a thirty day period. This represented 12.5 percent of the population. It appears that the high population density must have been facilitating disease transmission in an otherwise healthy population.

It will be interesting to have observations of the long-term impacts of this disease on muskox populations on Banks Island and to see if there is a cyclical pattern that may extend over 100 or more years.

Chronic Wasting Disease

Chronic wasting disease (CWD) has been found in mule deer and elk in Colorado. It was found first in a herd of captive mule deer in Fort Collins and diagnosed as CWD and also found in captive elk in Fort Collins and a Wyoming research facility. It has been recognized in free ranging wild deer and elk since 1981. A 1996 fall survey of brain tissue from hunter harvested animals showed 6 percent of the bucks in northeastern Colorado (Larimer County) were affected compared to 1 percent of the elk. Comprehensive monitoring continues and 1997 results showed 5 percent of the deer and a lower percent of the elk sampled were infected. The disease is similar to bovine spongiform encephalopathy found in British cattle and is a member of a group of diseases known as Transmissible Spongiform Encephalopathies. The disease seems to be confined to north central Colorado and south central Wyoming but one wonders whether the extensive sheep ranching which used to occur in the area was responsible for CWD (which may be derived from scrapies in sheep) emerging as a problem in wildlife populations.

Ectoparasites

While anemia is not an unexpected result of heavy infestations of ectoparasites, death seems an unlikely consequence of such events. Yet Brunetti and Cribbs (1971) reported that the African blue louse (*Linognathus africanus*) was the likely causative agent of periodic die-off of mule deer on the Tejon

Ranch in California. They found thirty-four carcasses of deer in the areas of greatest population density in December 1969. Although no evidence of infectious disease was found, the deer had massive infestations of the louse and cause of death was exsanguination anemia. The emaciation associated with such continuous loss of blood predisposed the animals to death from hypothermia during normal winter snowstorms. The animals had no fat reserves with which to withstand the seasonal cold and snow.

A similar scenario on a province-wide scale has been determined for moose in Alberta, Canada. The winter tick, *Dermacenter albipictus,* reaches numbers of 25,000 to 50,000 per individual moose in the central part of Alberta. Many animals showed tick-induced loss of winter coat. More importantly, moose with hair loss have been found dead and it appears that in severe winters, die-offs can be precipitated by the loss of thermal insulation. The hair loss seems to occur as the moose attempts to rub the ectoparasites off its body. Glines and Samuel (1984) summarize the relevant life history of the winter tick in its association with the moose. The exact population effects are yet to be determined.

Tick paralysis was described in springbok from South Africa (Fourie and Horak, 1987). The Karoo paralysis tick (*Ixodes rubicundus*) was responsible and paralyzed animals had average tick burdens of thirty compared to healthy animals with an average of ten. A neurotoxin is believed to cause the paralysis and the affected animal usually succumbs to respiratory failure. Apparently numerous springbok carcasses have been found at localities where *I. rubicundis* is present suggesting it could be an important mortality factor.

Scabies has been implicated in bighorn sheep losses. In 1988, the Cottonwood Creek (Washington State) sheep herd declined from seventy-two to twenty-eight animals (a 60 percent reduction). Poor range conditions contributed to the affliction. The disease is caused by mites of the genus *Psoroptes* and is widespread in bighorn sheep, elk and white-tailed deer. The mites cause abrasion of the skin and elicit an immune response at the site of the abrasions. Sheets of epidermis may be shed, which causes loss of physical condition and makes the animal vulnerable to mortality from normal environmental conditions. The disease historically has been linked with large scale mortalities throughout bighorn sheep range. For example, in 1881 thousands of sheep in the Upper Greybull River of Wyoming were lost and the population was reduced to a few individuals. In New Mexico, a 1978–79 epizootic resulted in the loss of 60–70 percent of a Mexican desert bighorn sheep herd but because contagious ecythema was also involved, scabies could not be confirmed as cause of death.

Ectoparasites can be key vectors for disease organisms, particularly the arboviruses, which have been discussed earlier in this chapter.

Lyme Disease

A spirochete (*Borrelia burgdorferi*) is responsible for a disease which was first reported in the United States in Wisconsin in 1969. White-tailed deer and the white-footed mouse (*Peromyscus leucopus*) are the primary reservoirs of the spirochete. The infection was given the name Lyme disease because children in Lyme, Connecticut, suffered an outbreak of arthritis. Ticks serve as vectors for the disease. The disease can be asymptomatic in some species but will cause ocular and arthritic reactions in others. The disease has spread or been detected from much of the United States including California. It first appeared in coyotes in Texas in 1984 and wolves have been shown to be susceptible.

Most of the current concern centers on the effects on *Homo sapiens* and little is known about the disease's potential to affect wildlife populations. If it causes similar symptoms in affected wildlife it is easy to conjecture that it could indirectly influence populations by making individuals more susceptible to predation. As lyme disease is apparently widespread in North America and as yet little more than sera antibody titres have been examined in wildlife, this is another disease which bears watching.

Elaeophorosis

Hibler and Adcock (1971) describe the etiology of elaeophorosis. The disease is a result of infections of adult filarial *Elaeophora schneideri* in the common carotid and internal maxillary arteries of their ungulate hosts. *Odocoileus* is the usual host for the parasite and seldom shows clinical signs of the infection although experimental infection of a white-tailed deer fawn resulted in mortality (Titche et al., 1979) and another case of a debilitated female was reported (Prestwood and Ridgeway, 1972). Tabanidae serve as the intermediate hosts and third-stage larvae enter the final host's bloodstream when the flies are feeding.

Pathology is common in abnormal definitive hosts and in elk bilateral blindness is a common clinical feature. The disease affects calves and yearlings predominantly and mortality is a common result. Examination of the Gila Forest elk herd in New Mexico showed that at least 90 percent of the cows calved, yet by late summer and early fall only 15–20 percent of the calves were alive. Elaeophorosis may also be an important mortality factor for this species in other areas of New Mexico as well as Colorado, Arizona, and Wyoming.

This disease is potentially important in moose as well although only four cases from Montana have been reported. Worley (1975) reported on an epizoological survey of Montana and northwestern Wyoming ruminants where 15 percent of twenty mule deer and 4 percent of seventy-four moose were infected. There was no evidence of neurological disorder, including blindness, in these animals.

Sika deer (*Cervus nippon*) in Texas were found infected with *Elaeophora schneideri*. White-tailed deer were thought to be the reservoir host and the effects of elaeophorosis on Sika deer was considered serious enough by the authors for them to recommend it be considered in the management of that species in Texas.

The impact of elaeophorosis is likely a function of density of susceptible hosts and of the intermediate hosts. As most mortality occurs in young animals, clinical evidence of the disease may not be found in older animals even though they are infected. Thus, Worley's (1975) survey reveals little of the pathogenesis which might be occurring in remote high density host populations.

Parelaphostrongylosis

Parelaphostrongylus tenuis is a metastrongyloid, a group of nematodes usually found in the lungs, which has the white-tailed deer as its normal definitive host. In this cervid species the adult worms are found associated with the meninges of the brain in the cranial subdural space and venous sinuses. Although infection rates in white-tailed deer populations can be high they seldom cause any pathology. Terrestrial gastropods which contact the first stage larvae are penetrated in the foot region or the snail or slug may ingest the larvae in which case the gut is penetrated. The third stage larvae are infectious and the final host is infected by ingesting the gastropod.

Hosts other than white-tailed deer also ingest infected gastropods and these cervids, including moose, elk, and caribou, suffer different fates than the white-tailed deer. The parasite is geographically restricted to eastern North America and it is not coincidental that there is no overlap between caribou and infected white-tailed deer ranges, that moose populations are reduced and distributed in patterns related to intensity of the parasite in white-tailed deer, and that elk populations within the infection zone are small, isolated, and unthrifty. White-tailed deer typically have high infection rates even in low population density areas.

The rate of infection is in fact more dependent on the intermediate host populations and deer behavior than definitive host densities. Although there are only a few reports of clinical signs of neurologic disease or pathology in white-tailed deer, other deer including fallow deer and black-tailed deer (*Odocoileus hemionus columbianus*) have succumbed when introduced to an infected range. Llamas (*Lama guanicoe*) have also been affected.

The most important population effects have been demonstrated, or implied, for the eastern North American cervids which share or formerly shared, their distributional ranges with the white-tailed deer. In these species, the parasite causes excessive trauma to the central nervous system and depending on the migratory pathway, or end point of migration, can result in blindness, paraplegia, functional disorders, or behavioral abnormalities which in turn often result in death. Perhaps the clearest indicator of this relationship is the virtual disappearance of caribou from areas of infected white-tailed deer and the reintroduction failures in places like Maine and Nova Scotia. Trainer (1973) reported that fourteen woodland caribou released on a Wisconsin game preserve occupied by white-tailed deer all died as a result of *P. tenuis* within six months. Anderson (1971) reported a similar result with reindeer introduced to Ontario. In essence, these findings support the contention that extinct caribou populations in the southern portion of their former distributional range succumbed to *P. tenuis* mortality as the white-tailed deer expanded its range northward over the last century.

A similar case can be made for elk. Severinghaus and Darrow (1976) suggested that the failure of elk to establish populations in the Adirondacks of New York state after substantial reintroductions with successful reproduction was a function of mortality caused by *P. tenuis*. Carpenter et al.(1973) found neurologic disease in elk transplanted from the Wichita Mountains Wildlife Refuge in southwestern Oklahoma to areas of eastern Oklahoma. Annual losses to the disease currently are estimated at 10 percent for elk herds in Oklahoma. A captive elk herd in western Pennsylvania had endemic cerebral nematodiasis caused by *P. tenuis* (Woolf et al., 1977). Although not a major mortality factor in this case, the greater susceptibility of the calf and yearling age classes would have limited recruitment and thus population growth. Such "unthrifty" elk populations in white-tailed deer range have also been reported in Ontario, Michigan, and elsewhere.

Moose present a somewhat more complicated picture. Again, mortality has been associated with *P. tenuis* and the so-called "moose-sickness" is caused by the parasite. Fenstermacher (1934) was actually the first to describe the infection in moose in Minnesota. Karns (1967), Lankester (1974), Gilbert (1974), and Whitlaw and Lankester (1994) noted that the prevalence of *P. tenuis* in moose was directly related to deer density in Minnesota, Manitoba, Maine, and Ontario respectively. Parker (1964) made a similar observation for Nova Scotia and all authors indicated that increases in moose populations would have to be linked to decreases in deer populations because of the effect of *P. tenuis*. Severinghaus and Jackson (1970) recommended against any attempt to reintroduce moose to the Adirondacks of New York state because *P. tenuis* would doom it to failure. Several attempts to introduce moose to a Wisconsin reserve did fail under similar circumstances. In Nova Scotia both Parker (1964) and Telfer (1965) noted that moose populations in the Cobequid Hills and Pictou-Antigonish highlands seemed to be unaffected by declines which occurred elsewhere in the province. These populations escaped the disease because they were physically separated from the deer and the infected range at critical transmission periods. This is, at least partially, related to snow conditions which are tolerable to moose but not deer, a situation resulting in documented ecological separation of the two species in Fundy National Park in New Brunswick (Kelsall and Prescott, 1971). In essence, the moose exist in refugia and sustain healthy populations within the refugia but the expansion of range and numbers is limited by the surrounding deer populations. Kearney and Gilbert (1976) were able to show that moose and deer could coexist in the same geographical area in Ontario by not using the same habitats during the transmission period. This was facilitated by a complex environmental situation with considerable habitat diversity. A similar situation may have been responsible for the results noted by Saunders (1973) where similar deer densities in two parts of northern Ontario had different relationships to moose densities. In the area of higher moose population logging had occurred more recently and the habitat was more viariable. Whitlaw and Lankester (1994) caution that the effect of the parasite on moose populations is more subtle than most believe and that further study is required to ascertain its importance as a mortality agent for this species.

The relationship between moose, deer, and *P. tenuis* is more complex and variable than noted for other susceptible species. Vagaries of climate, elevation, habitat diversity, and gastropod populations will dictate the population effects. In some areas where long periods of suitable transmission conditions occur, moose probably can not survive. In other areas where populations are only minimally affected, some resistance may build up in the populations or other factors such as habitat condition and weather may be operative.

This resistance development may be facilitated by predation and hunting because infected animals are more vulnerable to both types of mortality. This is so because they tend to seek out open areas and do not readily flee a predator or hunter. In fact, examinations of moose heads collected after hunting seasons were opened in New Brunswick and Maine after long periods of closure confirmed that prevalence of *P. tenuis* in the hunted sample was higher than that in the moose population generally. This finding is similar to that made by Rau and Caron (1979), that moose heavily infected with *Echinococcus granulosis* were more susceptible to harvest by hunters.

Elaphostrongylosis is a comparable disease found in reindeer in northern Europe and Russia caused by *Elaphostrongylus cervi rangifer*. Neurologic disorders are among the clinical signs but often the lungs are affected and considerable mortality of young animals may occur. It appears the parasite may be present in North America and resident in moose and caribou populations although no pathology has been reported related to the parasite on this continent. Halvorsen et al. (1989) believe that several species of nematode may be involved and suggested that *Elaphostrongylus rangifer* occurs in reindeer and *E. cervi* in red deer. Stein et al. (1989) named *E. alces* as unique to moose.

Lungworm-Pneumonia Complex

The metastrongyloidea are adapted to the respiratory system of mammals. Some, such as *Parelaphostrongylus tenuis,* invade other tissues but the respiratory tract remains the means of elimination from the host. The larvae move up the bronchial escalator, are swallowed and passed out with the feces. *Protostrongylus stilesi* and *P. rushi* which are found in the lung parenchyma and the bronchioles, respectively, of bighorn sheep are often associated with fatal pneumonia especially in lambs. Land molluscs serve as the intermediate hosts and transplacental infections in sheep have been documented. Numerous die-offs of bighorn sheep have been associated with the lungworm-pneumonia complex. Buechner (1960) considers the disease to be the mechanism for preventing overpopulation in bighorn sheep and hence preventing habitat destruction. However, sheep are gregarious and will develop patterns of preferential use of the habitat which predisposes them to infection by lungworms. They do this in the absence of major predators such as wolves and mountain lions (*Felis concolor*) which through their predatory activities probably induced greater use of the range as well selecting out the most infected sheep. In such circumstances, the fitness of the population can be maintained but in the absence of predation there is a risk of not only die-offs but loss of the entire population. There is no definitive evidence that the lungworm-pneumonia complex has adaptive value in such circumstances despite Buechner's (1960) conjecture to the contrary.

Die-offs have been recorded throughout bighorn sheep range in North America attributable to this disease. Because many of these populations were relatively small and somewhat isolated there is good documentation on the extent of the losses (Table 5.1). Fortunately, there are drug treatments which have proven practical and effective in reducing or eliminating this disease problem (Miller et al., 1987; Schmidt et al., 1979).

The possibility that other species have lungworm-pneumonia complexes exists as lungworms seem to be quite ubiquitous in distribution. As an example, McColl and Spratt (1981) reported a parasite pneumonia caused by *Marsupostrongylus* in a koala (*Phascolaretas cinereus*) from Victoria, Australia. It is interesting to speculate what conditions of environment, and koala behavior

Table 5.1 — Recorded Die-offs of Bighorn Sheep in the United States and Canada (Much of table from Forrester 1971)

Estimated Area	Date	Reported Losses	Cause of Deaths	Reference
Alberta				
Waterton Park Area	1982–1983	65% loss	Pneumonia-pasteurellosis	Onderka and Wishart, 1984
British Columbia				
East Kootenay	1965	95% loss in some herds	Lungworm-pneumonia complex	Straight, 1966
East Kootenay	1981–1982	65% loss in some herds	Lungworm-pneumonia complex	Schwantje, 1986
Kootnenay	1966–1968	75% loss	Pneumonia-lungworm disease	Stelfox, 1973
California				
Santa Rosa Mts.	1920s	"large scale die-off"	Scabies	Jones et al., 1957
Colorado				
Tarryall	1885	"severe decimation"	Scabies	Pillmore, 1958b
Sapinero Creek	1902	75 dead animals counted	Scabies	Potts, 1938
Pikes Peak	1911	"losses"	Hemorrhagic septicemia	Pillmore, 1958b
Tarryall	1921–1924	"near extermination"	Hemorrhagic septicemia	Pillmore, 1958b

Pikes Peak	1930	"a number of sheep died"	Lungworm infection	Dikmans, 1931
Tarryall, Pikes Peak and Kenosha	1952–1953	1,000 of 1,500 died	Verminous pneumonia	Hunter and Pillmore, 1954
Idaho				
Salmon River (Middle Fork)	1905–1906	"mass mortality"	Unknown	Buechner, 1960
Salmon River (East Fork)	1956	21 of 150 died	Pneumonia	Buechner, 1960
Montana				
Sun River	1924–1925	15 of 100 died	Verminous pneumonia	Marsh, 1938
Sun River	1924–1925	186 of 250 died	Pneumonia	Rush, 1927
Glacier Park	1927	26 of 150 died	Verminous Pneumonia	Buechner, 1960
Glacier Park	1936–1937	33 of 84 died	Verminous pneumonia and hemorrhagic septicemia	Bond, 1936, 1937
Glacier Park	1983		Pneumonia	Onderka and Wishart, 1984
Wyoming		269 of 346 died	Verminous pneumonia	
Yellowstone	1927–1928	602 of 1,205 died	Pneumonia	Buechner, 1960
Crystal Creek	1934–1936			Honess and Winter, 1955

might combine (or change) to make this species susceptible to a disease induced dieoff. As most mammals harbor some species of lungworm, similar conjecture can be made as to what might induce virulence or an epizootic in each situation.

Strongyloidosis

Forrester et al. (1974) reported a situation where strongyloidosis caused mortality in 39 percent of 251 white-tailed deer fawns born into a captive herd on the campus of the University of Florida. Astronomical *Strongyloides* burdens were noted with one typically infected fawn having 50,000 female parasites in its small intestine. Egg counts from feces of infected fawns varied from 200 to 286,000 eggs/g feces. Intrauterine transmission of the parasite also occurred. Because the mortality was concentrated in young fawns, the authors speculated that livestock grazing on areas supporting white-tailed deer herds might be responsible for strongyloidosis in the cervids. They suggested that further studies of the parasite in wild populations are needed.

This example is included because it is pertinent to some of the statements made in the summary related to our knowledge of disease in wild mammalian populations.

Summary

Perhaps the most obvious finding made while reviewing the literature and writing this chapter was the astounding lack of documentation on population effects. Wild mammal populations are difficult to assess at the best of times and often surveys were not taken prior to the outbreaks of disease. Most work is done in response to an unpredicted die-off, therefore data are not always complete. Furthermore, counts of mortalities are incomplete in most cases and thus often estimates of losses were all that the literature offered. Despite the lack of quantitative data it should be apparent that disease can and does play a key role in determining population size and dynamics. There may be factors which predispose a population to an epizootic, be it population density (of the host and/or vector), condition of individuals within the population (range quality), behavior or geographical isolation. All relate to the ease of transmission and hence morbidity.

It is a rare event when disease results in extermination of a population, at least within our historical frame of reference. The reported loss of 95 percent of the wildebeest and cape buffalo in East Africa in the 1890s due to rinderpest is the most modern example we have. There are tantalizing hints that such events have taken place earlier as certain species may no longer occur in parts of their former range because of disease (caribou-*Parelaphostrongylus tenuis*, bighorn sheep-lungworm-pneumonia complex). The potential exists for some new disease factor, more virulent than the canine parvovirus pandemic, to cause havoc among our increasingly reduced, in area and size, large mammal populations (see Daszak et al., 2000). Anderson and May (1979), May (1983), and May and Anderson (1987) present mathematical models that examine the critical components of disease outbreaks in epizootics and human epidemics. These models not only permit tracking the progression of disease but also delimit the critical factors influencing population dynamics of infectious disease. Not surprisingly, virulence of the pathogen and ease of transmission within host populations are the two key factors in the equation. While we have stressed direct mortality and secondarily productivity through natality impacts, a species that survives an epidemic and rebounds numerically may have acquired some characteristics which have evolutionary significance. The survivors of an epizootic should express greater resistance to the responsible pathogen, and their allele frequencies at other loci genetically linked to loci affecting resistance would also be affected. These latter gene frequency changes would be otherwise unrelated to pathogen resistance. In severe pandemics, a large population may be reduced sufficiently to result in significant inbreeding and thus the resultant population bottleneck would cause reduced genetic diversity. The African wildebeest would be a good example of such a situation resulting from disease but others have been noted in both captive cheetahs (*Acinonyx jubatus*) (O'Brien et al., 1985) and wild lions (Wildt et al., 1987). The black-footed ferret's susceptibility to canine distemper virus is yet another example. O'Brien and Evermann (1988) also discuss the host populations' genetic defense mechanisms and the consequences of disease outbreaks on pathogen evolution. A disease epidemic may result in a major genetic change in the disease organism that can be expressed as a change in host range, a change in virulence, or any of the myriad biological components involved in a disease episode. Here the canine parvovirus pandemic illustrates two such major genetic changes. The original outbreak in canids was thought to be caused by a mutation of the feline panleukopenia virus (or mink enteritis virus) called canine parvovirus-2 (CPV-2). A new strain CPV-2a appeared in 1981 and replaced CPV-2 within two years and persists in domestic dogs worldwide (Parrish et al., 1985). Because of these realities, it was discouraging to find wildlife biologists and managers who still often discount the importance of disease in controlling wildlife populations. The old arguments of reduced habitat quality, external factors, and compensatory mortality are presented in the face of countervailing evidence of disease. We need better communication and understanding between wildlife managers, biologists, and researchers on the issue of disease and its effects on large mammal populations.

Our knowledge base is still weak and many more detailed studies are needed before we can predict with any degree of certainty the impact a disease may have on a particular species under any particular set of circumstances. The only exception to this seems to be for captive populations where we can predict some of the risks and therefore manage to reduce them.

Many organisms such as anthrax, brucellosis, epizootic haemorrhagic disease in large mammals, fowl cholera, botulism, enteritis, and leucocytozoonosis in waterfowl appear to have a density-dependent relationship. Transmission and the opportunity for epidemics are enhanced in many cases by management practices which concentrate wildlife populations. The waterfowl refuge system now fights a continual and apparently losing battle against diseases such as avian cholera in overwintering duck and goose populations. With 90 percent of the waterfowl populations of the Pacific flyway concentrated in California refuges, it is not difficult to envision the potential for massive die-offs if a disease becomes established. Efforts to disperse the birds premigration are ineffective as the animals simply have nowhere else to go. The advantage lies with many small refuges instead of a few large ones, contrary to current biogeographic theory, but this is because of the specialized habitat requirements of overwintering waterfowl.

One relatively new management technique is the use of abomasal worm counts in white-tailed deer to determine the physical condition of the population. This was developed by the Southeastern Cooperative Wildlife Disease Study. Three classifications of populations: overpopulated, optimum, and suboptimum were correlated to average worm counts giving high (1000), moderate (500–1000), and low (500) categories of parasite incidence. The technique appears to have general applicability for the southeastern United States where it provides an evaluation of deer herd health as related to range conditions. In areas of more severe cold, there is a winter diapause in the parasites' life cycles which makes the technique useless. Demarais et al. (1983) also cautioned that abomasal counts for Mississippi deer collected in December and January did not consistently indicate either current or future trends in herd health compared to measurements such as body weight and antler beam diameter. They suggested the technique, based on summer parasite counts, may not have validity outside that season even in the Southeast. However, Schultz et al. (1993) found that abomasal parasite counts and eggs per gram in feces were highly associated and that further work on the relationship between herd health and fecal egg counts was warranted.

One final observation on the consequences of disease and parasites in wild mammals relates to how we deal with such when the afflicted wildlife pose a potential threat to human interests. If domestic cattle are threatened it is not uncommon to have population control imposed on the wildlife species that sometimes has resulted in massive mortality numbers. This was the case in Africa with control for tsetse and rinderpest and to a lesser extent in North America with control for anthrax and brucellosis. Even today bison that cross the boundary of Yellowstone Park are "fair game" because of the threat to livestock interests.

Bibliography

Alibasoglu, M., D. C. Kradel, and H. W. Dunne. 1961. Cerebral nematodiasis in Pennsylvania deer (*Odocoileus virginianus*). Cornell Vet. 51:431–441.

Amundson, T. E. and T. M. Yuill. 1981. Natural LaCrosse virus infection in red fox (*Vulpes fulva*), gray fox (*Urocyon cinereoargenteus*), racoon (*Procyon lotor*), and opossum (*Didelphis virginiana*). Am. J. Trop. Med. Hyg. 3:706–714.

Anderson, R. C. 1963. The incidence, development, and experimental transmission of *Pneumostrongylus tenuis* Dougherty (Metastrongyloidea: Protostrongylidae) of the meninges of the white-tailed deer (*Odocoileus virginianus borealis*) in Ontario. Can. J. Zool. 41:775–792.

Anderson, R. C. 1964. Motor ataxia and paralysis in moose calves infected experimentally with *Pneumostrongylus tenuis* (Nematoda: Metastrongyloidea). Northeast Wildlife Conf., Hartford, Connecticut (Jan; unpubl.).

Anderson, R. C. 1965. Cerebro-spinal nematodiasis (*Pneumostrongylus tenuis*) in North American ruminants. Proc. Intl. Congr. Parasit. 1:461–462.

Anderson, R. C. 1965. Cerebro-spinal nematodiasis (*Pneumostrongylus tenuis*) in North American Cervids. Trans. N. Am. Wildl. Nat. Resour. Conf. 30:156–157.

Anderson, R. C., M. W. Lankester, and U. R. Strelive. 1966. Further experimental studies of *Pneumostrongylus tenuis* in cervids. Can. J. Zool. 44:851–861.

Anderson, R. C. 1968. The pathogenesis and transmission of neurotropic and accidental nematode parasites of the central nervous system of mammals and birds. Helminthology Abs. 37(3):191–210.

Anderson, R. C. and U. R. Strelive. 1968. The experimental transmission of *Pneumostrongylus tenuis* to caribou (*Rangifer tarandus terraenovae*). Can. J. Zool. 46:503–510.

Anderson, R. C. 1971. Neurologic disease in reindeer (*Rangifer tarandus tarandus*) introduced into Ontario. Can. J. Zool. 49:159–166.

Anderson, R. C. 1972. The ecological relationship of meningeal worm and native cervids of North America. J. Wildl. Dis. 8:304–310.

Anderson, R. M. and R. M. May. 1979. Population biology of infectious diseases: Part 1. Nature 280:361–367.

Anderson, R. M., H. C. Jackson, R. M. May, and A. M. Smith. 1981. Population dynamics of fox rabies in Europe. Nature 289:765–771.

Baker, M. R. and R. C. Anderson. 1975. Seasonal changes in abomasal worms (*Ostertagia* sp.) in white tailed deer (*Odocoileus virginianus*) at Long Point, Ontario. Can. J. Zool. 53:87–96.

Barnard, B. J. H. and R. H. Hassel. 1981. Rabies in kudus (*Tragelaphus strepsiceros*) in southwest Africa/Namibia. J.S. Afr. Vet. Assoc. 52(4):309–314.

Becklund, W. W., 1964. Revised checklist of internal and external parasites of domestic animals in the United States and Possessions and in Canada. Am. J. Vet. Res. 25:1380–1416.

Behrend, D. F. and J. F. Witter. 1968. *Pneumostronglyus tenuis* in white-tailed deer in Maine. J. Wildl. Manage. 32:963–966.

Behrend, D. F. 1970, The nematode, *Pneumostrongylus tenuis,* in white-tailed deer in the Adirondacks. New York Fish Game J. 17(1): 45–49.

Benson, D. A. 1958a. Moose sickness in Nova Scotia, I. Can. J. Comp. Med. 22:244–248.

Benson, D. A. 1958b. Moose sickness in Nova Scotia, II. Can. J. Comp. Med. 22:282–286.

Bergerud, A. T. 1974. Decline of caribou in North America following settlement. J. Wildl. Manage. 38(4):757–770.

Berry, H. H. 1980. Behavioral and eco-physiological studies on blue wildebeest(*Connochaetes taurinus*) at the Etosha National Park. Unpubl. PhD Thesis, University of Capetown, South Africa.

Blake, J. E. and B. D. McLean. 1988. Epidemiology of Yersiniosis in muskoxen on Banks Island, Northwest Territories, Canada. (unpubl.).

Boas, D. A. 1983. The response of terrestrial snails to the presence of ungulate feces, a source of nematode larvae (Metastrongyloidea: Protostrongylidae). Can. J. Zool. 61(8): 1852–1856.

Bogaczyk, B. A. 1990. A survey of metastrongyloid parasites in Maine cervids. M.S. Thesis, University of Maine, Orono, Maine, USA.

Bosler, E. N., B. G. Ormiston, J. L. Coleman, J. P. Hanrahan, and J. L. Benach. 1984. Prevalence of the Lyme disease spirochete in populations of white-tailed deer and white-footed mice. Yale J. Biol. Med. 57:651–659.

Botvinkin, A. D., V. P. Savitskii, V. N. Sidorov, and V. G. Yudin. 1981. Role of the raccoon-dog, *Nyctereutes procyonoides,* in the epidemiology of

rabies in the far east of the USSR. Z. Mikrobiol. Epidemiol. I Immuniobiol. 1981(12):79–82. (Engl. Abs.).
Bowne, J. G., A. J. Luedke, N. M. Foster, and M. M. Jochim. 1966. Current aspects of bluetongue in cattle. J. Am. Vet. Med. Assoc. 148(10):1177–1180.
Brand, C. J. 1984. Avian cholera in the Central and Mississippi flyways during 1979–80. J. Wildl. Manage. 48:399–406.
Braverman, Y. and R. Galun, 1973a. The occurrence of Culiocoides in Israel with reference to the incidence of bluetongue. Refuah Vet. 30(3):121–127.
Braverman, Y. and R. Galun. 1973b. The occurrence of Culiocoides in Israel with reference to the incidence of bluetongue. Refuah Vet. 30(4):68–70.
Bristowe, J. S. 1866. Supplementary report on the morbid anatomy of the cattle plague, as it occurs in sheep, goats, and deer. Third report of the commissioners appointed to inquire into the origin and nature etc. of the cattle plague. H.M. Stationary Off., London.
Brown, J. E. 1983. *Parelaphostrongylus tenuis* (Pryadko and Bove) in the moose and white-tailed deer of Nova Scotia. Unpubl. M.Sc. Thesis, Acadia University, Wolfville, Nova Scotia, Canada.
Brunetti, O. and H. Cribbs. 1971. California deer deaths due to massive infestation by the louse (*Linognathus africanus*). Calif. Fish Game 57(3):162–166.
Buechner, H. K. 1960. The bighorn sheep in the United States, its past, present, and future. Wildl. Monogr. 4.
Burgdorfer, W., A. G. Barbour, S. F. Hayes, J. L. Benach, E. Grunwaldt, and J. P. Davis. 1982. Lyme disease-a tick-borne spirochetosis? Science 216:1317–1319.
Burgdorfer, W., R. S. Lane, A. G. Barbour, R. A. Gresbrink, and J.R. Anderson. 1985. The western black-legged tick *Ixodes pacificus:* a vector of *Borrelia burgdorferi*. Am. J. Trop. Med. Hyg. 34:925–930.
Burgess, E. C., T. E. Amundson, J. P. Daris, R. A. Kaslow, and R. Edelman. 1986a. Experimental inoculation of *Peromyscus* spp. with *Borrelia burgdorferi:* evidence of contact transmission. Am. J. Trop. Med. Hyg. 35:355–359.
Burgess, E. C., D. Gillette, and J. P. Pickett. 1986. Arthritis and panuveitis as manifestations of *Borrelia burgdorferi* infection in a Wisconsin pony. J. Am. Vet. Med. Assoc. 189:1340–1342.
Burgess, E.C. and L.A. Windberg. 1989. *Borrelia* sp. Infection in coyotes, black-tailed jack rabbits and desert cottontails in southern Texas. J. Wildl. Dis. 25:47–51.

Carpenter, J. W., H. E. Jordan, and B. C. Ward. 1973. Neurologic disease in wapiti naturally infected with meningeal worms. J. Wildl. Dis. 9:148–153.

Casals, J. and W. C. Reeves. 1965. The arboviruses. In: F. L. Horsfall and I. Tamm, (eds.). *Viral Rickettsial Infections of Man.* pp. 580–582. Lippincott, Philadelphia, PA.

Chalmers, A. W., and G. R. Scott. 1969. Ecology of rabies. Tropical Animal Health and Production 1:33–55.

Cheatum, E. L. 1951. Disease in relation to winter mortality of deer in New York. J. Wildl. Manage. 15:216–220.

Choi, D. Y., Y. S. Lyoo, C. H. Kwon, Y. H. Kim, and D. H. Kim. 1986. Incidence of canine parvovirus infection in Korea. Res. Rep. Rural Dev. Adm. 28:108–114.

Choquette, L. P. E., G. G. Gibson, and B. Simard. 1971. *Fascioloides magna* (Bassi, 1875) Ward, 1917 (Trematoda) in woodland caribou, *Rangifer tarandus caribou* (Gmelin) of northeastern Quebec, and its distribution in wild ungulates in Canada. Can. J. Zool. 49:280–281.

Choquette, L. P. E., E. Broughton, A. A. Currier, J. G. Cousineau, and N. S. Novakowski. 1972. Parasites and diseases of Bison in Canada. III. Anthrax outbreaks in the last decade in northern Canada and control measures. Can. Field-Nat. 86:127–132.

Christian, J. J., V. Flyger, and D. E. Davis. 1960. Factors in the mass mortality of sitka deer. Chesapeake Sci. 1:79–95.

Corner, A. H. and R. Connell. 1958. Brucellosis in bison, elk and moose in Elk Island National Park, Alberta, Canada. Can. J. Comp. Med. 22:9–20.

Cousineau, J. G. and R. J. McClenaghan. 1965. Anthrax in bison in the Northwest Territories. Can. Vet. J. 6:22–24.

Daszak, P., A. A. Cunningham, and A. D. Hyatt. 2000. Emerging infectious diseases of wildlife—threats to biodiversity and human health. Science 287:443–449.

Dauphine, T. C., Jr. 1975. The disappearance of caribou reintroduced to Cape Breton Highlands National Park. Can. Field-Nat. 89:299–310.

David, J. M. and L. Andral. 1982. Modelling of spatial evolution and population dynamics of healthy foxes subsequently infected with rabies. Commun. Immun., Micro. Infect. Dis. 5(1/3): 351–358.

Davies, F. G., T. Shaw, and P. Ochieng. 1975. Observations on the epidemiology of ephemeral fever in Kenya. J. Hyg. 75(2):231–235.

Davies, F. G. 1981, The possible role of wildlife as maintenance hosts for some African insect-borne virus diseases. In: L. Karstad, B. Nestel, and M. Graham (eds.), *Wildlife Disease and Research and Economic Development.* pp. 24–27. IDRC-179e, IDRC Publ., Ottawa, Canada.

Davis, J. W., L. H. Karstad, and D. O. Trainer. 1970. *Infectious Diseases of Wild Mammals.* Iowa State University Press, Ames, IO.
Davis, J. W. and R. C. Anderson, 1971. *Parasitic Diseases of Wild Mammals.* Iowa State University Press, Ames, IO.
Davis, J. W. and K. G. Libke. 1971. *In:* J. W. Davis, and R. C. Anderson (eds.), *Parasitic Diseases of Wild Mammals.* pp. 235–257. Iowa State University Press, Ames, IO.
Davoust, B., G. Muller, and G. Chappuis. 1985. Serological survey of canine parvovirus infections, Recl. Med. Vet. Ec. Alfort. 161(4): 323–328.
Demarais, S., H. A. Jacobson, and D. C. Buynn. 1983. Abomasal parasites as a health index for white-tailed deer in Mississippi. J. Wildl. Manage. 47: 247–252.
DeSouza Lopes, O. and De Abreau Sacchetta. 1974. Epidemiology of boraceia virus in a forested area in Sao Paulo, Brazil. Am. J. Epidem. 100(5):410–413.
DeSwart, R. L., T. C. Harder, P. S. Ross, H. W. Vos, and A. D. M. E. Osterhaus. 1995. Morbilliviruses and morbillivirus diseases of marine mammals. Infect. Agents Dis. 4:125–130.
DeVilliers, S. W. 1943. An outbreak of anthrax amongst kudus. J. S. Afr. Vet. Med. Assoc. 14:17–18.
Dieckerhoff, W. 1866. Geschichte der Rinderpest und Ihre Literatur. Ensslin, Berlin, Germany.
Drew, M. L. and W. M. Samuel. 1985. Factors affecting transmission of larval winter ticks, *Dermacenter albipictus* (Packard), to moose, *Alces alces* L., in Alberta. Proc. N. Am. Moose Conf. Workshop 15:303–348.
Dunn, F. D. 1965. Reintroduction of woodland caribou to Mt. Katahdin, Maine. Northeast Fish Wildl. Conf.
Dunn, F. D. and K. I. Morris. 1981. Preliminary results of the Maine moose season (1980). Alces 17:95–100.
Ebedes, H. 1976. Anthrax epizootics in wildlife in the Etosha National Park, Southwest Africa, In: *Wildlife Diseases,* Plenum Publ., New York, NY.
Ebedes, H. 1981. A new look at anthrax. In: *Wildlife Diseases of the Pacific Basin and Other Countries.* pp. 85–88 Depart. of Medicine, School of Vet. Med., Univ. Calif., Davis, CA.
Eckroade, R. J., G. M. Zurhein, and W. Foreyt. 1970. Meningeal worm invasion of the brain of a naturally infected white-tailed deer. J. Wildl. Dis. 6(4):430–436.
Edwards, M. A. and U. McDonnell. 1982. *Animal Disease in Relation to Animal Conservation.* Symp. Zool. Soc. London, No. 50, Academic Press, New York, NY.

Emerson, H. R. and W. T. Wright. 1968. The isolation of a *Babesia* in White-tailed deer. Bull. Wildl. Dis. Assoc. 4:142–143.

Emmons, R. W., J. Ruskin, M. L. Bissett, D. A. Lyeda, R. M. Wood, and C. L. Lear. 1976. Tularemia in a mule deer. J. Wildl. Dis. 12:459–463.

Enigk, K. and K. Friedhoff. 1962. *Babesia caprieoli* n. Sp. Beim Reh (*Capreolus capreolus* L.), Z. Tropenmed. Parasitol. 13:8.

Eve, J. H. and F. E. Kellogg. 1977. Management implications of abomasal parasites in Southeastern white-tailed deer. J. Wildl. Manage. 41:169–177.

Fay, L. D., Boyce, A. P., and Youatt, W. G. 1956. An epizootic in deer in Michigan. Trans. N. Am. Wildl. Nat. Resour. Conf. 21:173–184.

Fay, L. D. and Stuht, J. N. 1973. Meningeal worms in association with neurologic disease in Michigan wapiti. Wildl. Dis. Conf. Papers and Abstr. 24.

Fenstermacher, R. 1934. Disease affecting moose. The Alumni Quarterly (Minnesota). 22:81–94.

Fenstermacher, R. and W. L. Jellison. 1933. Diseases affecting moose. Minnesota Agric. Exp. Stat. Bull. 294.

Fenstermacher, R. and O. W. Olsen. 1942. Further studies of diseases affecting moose. III. Cornell Vet. 32:241–254.

Foreyt, W. J. 1989. Fatal *Pasteurella haemolytica* in bighorn sheep after direct contact with clinically normal domestic sheep. Am. J. Vet. Res. 50:341–344.

Foreyt, W. J. and D. A. Jessup. 1982. Fatal pneumonia of bighorn sheep following association with domestic sheep. J. Wildl. Dis. 18(2):163–168.

Forrester, D. J., W. J. Taylor, and K. P. C. Nair. 1974. Strongyloidosis in captive white-tailed deer. J. Wildl. Dis. 10:11–12.

Fourie, L. J. and I. G. Horak. 1987. Tick-induced paralysis of springbok. S. Afr. J. Wildl. Res. 179(4):131–133.

Gates, C. C. and W. M. Samuel. 1977. Prenatal infection of the Rocky Mountain bighorn sheep (*Ovis canadensis*) of Alberta with the lungworm *Protostrongylus* spp. J. Wildl. Dis. 13:248–250.

Geist, V. 1971. *Mountain Sheep: A Study in Behavioral Evolution*. Univ. Chicago Press, Chicago, IL.

Gilbert, F. F. 1973. *Parelaphostrongylus tenuis* (Dougherty) in Maine: I—The parasite in white-tailed deer (*Odocoileus virginianus*, Zimmerman). J. Wildl. Dis. 9:136–143.

Gilbert, F. F. 1974. *Parelaphostrongylus tenuis* (Dougherty) in Maine: II-Prevalence in moose. J. Wildl. Manage. 38:42–46.

Gleich, J. G., F. F. Gilbert, and N. P. Kutscha. 1977. Nematodes in terrestrial gastropods from central Maine. J. Wildl. Dis. 13:43–46.

Glines, M. V. and W. M. Samuel. 1984. The development of the winter tick, *Dermacentor albipictus,* and its effects on the hair coat of moose, *Alces alces,* of central Alberta, Canada. In: D. A. Griffiths and C. E. Bowman (eds.), Acarology VI., Vol. 2, pp. 1208–1214. Ellis Horwood Ltd, Chichester, PA.

Gothe, R. 1984. Tick paralysis, reasons for appearing during ixodid and argasid feeding. In: K. F. Harris (ed.), *Current Topics in Vector Research.* Vol. 2, Ch. 9, Praeger Publ., New York, NY.

Goyal, S. M., L. D. Mech, R. A. Rademacher, M. A. Kahn, and U. S. Seal. 1986. Antibodies against canine parvovirus in wolves of Minnesota: a serologic study from 1975 through 1985. J. Am. Vet. Med. Assoc. 189(9):1092–1094.

Gremillion-Smith, C. A. 1986. Skunk rabies: aspects of epizootiology and simulation modeling. PhD thesis (unpubl.), Southern Illinois Univ., Carbondale, IL.

Grimwood, I .R. 1961. Report of the Game Department, Kenya 1960, Govt. Print., Nairobi, Kenya.

Guo, W., J. F. Evermann, W. J. Foreyt, F. F. Knowlton, and L. A. Windberg. 1986. Canine distemper virus in coyotes: a serologic survey. J. Am. Vet. Met. Assoc. 189(9):1099–1100.

Hallen, J. H. B., K. McLeod, J. G. Charles, H. C. Kerr, and M. M. A. Jan. 1871. Report of the Indian Cattle Plague Commission with Appendices. Govt. Print., Calcutta, India.

Halvorson, O., A. Skorping, and K. Bye. 1989. Experimental infection of reindeer with *Elaphostrongylus* (Nematoda: Protostrongylidae) originating from reindeer, red deer, and moose. Can. J. Zool. 67:1200–1202.

Hammersland, H. L. and E. M. Joneschild. 1940. Tularemia in beaver. J. Am. Vet. Med. Assoc. 96:96–97.

Haynes, F. A. and A. K. Prestwood. 1969. Some considerations for disease and parasites of white-tailed deer in the southeastern United States. In: *Proceedings of the Symposium: White-tailed deer in the Southern Forest Habitat,* pp. 32–36. Nacogdoches, TX.

Hibler, C. P., J. L. Adcock, R. W. Davis, and Y. Z. Abdelboki. 1969. Elaeophorosis in deer and elk in the Gila Forest, New Mexico. Bull. Wildl. Dis. Assoc. 5:27–30.

Hibler, C. P. and J. L. Adcock. 1971. Elaeophorosis. In: (J.W. Davis and R.C. Anderson, eds.) *Parasitic Diseases of Wild Mammals.* Iowa State Univ. Press, Ames, IO.

Hibler, C. P., Lange, R. E., and C. J. Metzgar. 1972. Transplacental transmission of *Protostrongylus* spp. in bighorn sheep. J. Wildl. Dis. 9:389.

Hibler, C. P. and C. J. Metzger. 1974. Morphology of the larval stages of *Elaeophora schneideri* in the intermediate and definitive hosts with some observations on their pathogenesis in abnormal definitive hosts. J. Wildl. Dis. 10:361–369.

Hoff, G. L., S. H. Richards, and D. O. Trainer. 1973. Epizootic of hemorrhagic disease in North Dakota deer. J. Wildl. Manage. 37(3):331–335.

Howe, D. C., G. T. Woods, and G. Marquis. 1966. Infection of bighorn sheep (*Ovis canadensis*) with myxovirus para-influenza-3 and other respiratory viruses, results of serologic tests and culture of nasal swabs and lung tissue. Bull. Wildl. Dis. Assoc. 2:34.

Hudson, J. R. 1959. Pasteurellosis. In: A. W. Stableforth and I. A. Galloway (eds.). *Infectious Diseases of Animals*. pp. 413–436. Butterworths Sci. Publ., London.

Hussein, N. A., R. N. Sharma, and H. G. B. Chizyuka. 1984. Further review of the epidemiology of rabies in Zambia (1975–1982). Rev. Sci. Tech. Off. Int'l. Epizootiology 3(1):125–135.

Irmer, S., Kuttler, D., and H. L. Schlegel. 1981. The rabies in Lower Saxony and Hessen in the years 1970–1980; a critical evaluation about the combatting methods, especially about the activities of gassing to the populations of foxes and badgers and also just to the development of rabies. 88(6):248–252.

Issel, C. J., Trainer, D. O., and W. H. Thompson. 1972. Serological evidence of infections of white-tailed deer in Wisconsin with three California group arboviruses (LaCross, Trivittatus, and Jamestown Canyon). Am. J. Trop. Med. Hyg. 21:985–988.

Jellison, W. L., C. W. Fishel, and E. L. Cheatum. 1953. Brucellosis in a moose, *Alces americanus*, J. Wildl. Manage. 17:217–218.

Johnson, H. N. 1970. Keynote address: The ecological approach to the study of zoonotic diseases. J. Wildl. Dis. 6:194–204.

Karns, P. D. 1967. *Pneumostrongylus tenuis* in deer in Minnesota and implications for moose. J. Wildl. Manage. 31:229–303.

Karns, P. D. 1972. Minnesota's 1971 moose hunt: a preliminary report on the biological connections. N. Am. Moose Conf. Workshop. 8:115–123.

Karstad, L., E. V. Adams Jr., R. P. Hanson, and D. H. Ferris. 1956. Evidence for the role of wildlife in epizootics of vesicular stomatitis. J. Am. Vet. Med. Assoc. 129:95–96.

Karstad, L. 1957. Enzootic vesicular stomatitis. MS thesis (unpubl.), University of Wisconsin, Madison, WI.

Karstad, L. and R. P. Hanson. 1957. Vesicular stomatitis in deer. Am. J. Vet. Res. 18:162–166.

Karstad, L. 1970. Arboviruses. In: J. W. Davis, L. H. Karstad, and D. O. Trainer (eds.). *Infectious Diseases of Wild Mammals.* pp. 60–67. Iowa State Univ. Press, Ames, IO.

Kay, B. H., P. L. Young, I. D. Fanning, and R. A. Hall. 1981. Which vertebrates amplify Murray Valley encephalitis virus in southern Australia? In: *Wildlife Diseases of the Pacific Basin and Other Countries.* pp. 36–39. Fruitridge Printing, Sacramento, CA.

Kazmierczak, J. J., E. C. Burgess, and T. E. Amundson. 1988. Suceptibility of the gray wolf (*Canis lupus*) to infection with the Lyme disease agent, *Borrelia burgdorferi.* J. Wildl. Dis. 24:522–527.

Kearney, S. R. and F. F. Gilbert. 1976. Habitat use by white-tailed deer and moose on sympatric range. J. Wildl. Manage. 40:645–657.

Kellogg, F. E., A. K. Prestwood, and R. E. Noble. 1970. Anthrax epizootic in white-tailed deer. J. Wildl. Dis. 6:226–228.

Kelsall, J. P. and W. Prescott. 1971. Moose and deer behavior in snow. Can. Wildl. Serv. Rep. Ser. No. 15.

Kirwin, J. 1988. European scientists track down deadly virus killing North Sea seals. Chron. Higher Educ. 24(5).

Kistner, T. P., G. R. Johnson, and G. A. Rilling. 1977. Naturally occurring neurologic disease in a fallow deer infected with meningeal worms. J. Wildl. Dis. 13:55–58.

Kistner, T. P. and D. Wyse. 1979. Transplacental transmission of *Protostrongylus* sp. in California bighorn sheep (*Ovis canadensis californica*) in Oregon. J. Wildl. Dis. 15(4):561–562.

Knudsen, G.J. 1953. Beaver die-off. Wis. Cons. Bull. 18:20–23.

Kocan, A. A., M. G. Shaw, K. A. Waldrup, and G. J. Kubat. 1982. Distribution of *Parelaphostrongylus tenuis* (Nematoda: Metastrongyloidea) in white-tailed deer from Oklahoma. J. Wildl. Dis. 18(4):457–460.

Krakowka, S., R. J. Higgens, and A. Koestner. 1980. Canine distemper virus: Review of structural and functional modulations in lymphoid tissues. Am. J. Vet. Res. 41:284–292.

Krebs, J. W., M. L. Wilson, and J. E. Childs. 1995. Rabies-Epidemiology, prevention, and future research. J. Mammal. 76:681–694.

Kurtz, H. J., K. Loken, and J. C. Scholtthaeur. 1966. Histopathologic studies on cerebrospinalnematodiasis of moose in Minnesota naturally infected *with Pneumostrongylus tenuis*. Am. J. Vet. Res. 117:548–558.

Labzoffsky, N. A. and J. A. F. Sprent. 1952. Tularemia among beaver and muskrat in Ontario. Can. J. Med. Sci. 30:250–255.

Lankester, M. W. 1974. *Parelophostrongylus tenuis* (Nematoda) and *Fascioloides magna* (Trematoda) in moose of southeastern Manitoba. Can. J. Zool. 52:235–239.

Lankester, M. W. and R. C. Anderson. 1968. Gastropods as intermediate hosts of meningeal worm, *Pneumostrongylus tenuis* Dougherty. Can. J. Zool. 46:373–383.

Lankester, M. W., V. J. Crichton, and H. R. Timmermann. 1976. A protostrongylid nematode (Strongylida: Protostrongylidae) in woodland caribou (*Rangifer tarandus caribou*). Can. J. Zool. 54(5):680–684.

Lawrence, W. H., L. D. Fay, and S. A. Graham. 1956. A report on the beaver die-off in Michigan. J. Wildl. Manage. 20:18.

Lefevre, PC. and W. P. Taylor. 1983. Situation epidemiologique de la fivre catarrhale du mouton (Blue Tongue) au Senegal, Revue d'Elevage et de Medecine Veterinaire des Pays Tropicaux 36(3):241–245. (Engl. Abs.).

Levine, J. F., M. L. Wilson, and A. Spielman. 1985. Mice as reservoirs of the Lyme disease spirochete. Am. J. Trop. Med. Hyg. 35:355–360.

Lloyd, H. G. 1976. Wildlife rabies in Europe and the British situation. Trans. Royal Soc. Trop. Med. Hyg. 70(3):179–187.

Lloyd, H. G., B. Jensen, J. L. vanHaaften, F. J. Wiewold, A. Wandeler, K. Bogel, and A. A. Arata. 1976. Annual turnover of fox populations in Europe. Zentralbl Veterinarmed [B]. 23(7):580–589.

Magnarelli, L. A., J. F. Anderson, W. Burgdorfer, and W. A. Chappell. 1984. Parasitism by *Ixodes dammini* (Acaris: Ixodidae) and antibodies to spirochetes in mammals at Lyme disease foci in Connecticut. U.S.A. J. Med. Entom. 21:52–57.

Mahamooth, T. M. Z. 1943. Rinderpest. Tropic. Agri. 99:20.

Mason, M. J., N. A. Gillett, and B. A. Muggenburg. 1987. Clinical, pathological, and epidemiological aspects of canine parvoviral enteritis in an unvaccinated closed Beagle colony: 1978–1985. J. Am. Anim. Hosp. Assoc. 23(2):183–192.

May, R. M. 1983. Parasitic infections as regulators of animal populations. Am. Sci. 71(1):36–45.

May, R. M. and R. M. Anderson. 1987. Transmission dynamics of HIV infection. Nature 326:137–141.

Maze, R. J. and C. Johnstone. 1986. Gastropod intermediate hosts of the meningeal worm (*Parelaphostrongylus tenuis*) in Pennsylvania: observations on their ecology. Can. J. Zool. 64(1):185–188.

McColl, K. A. and F. M. Spratt. 1981. Parasitic pneumonia in a koala (*Phascolarctos cinereus*) from Victoria, Australia. In: *Wildlife Diseases of the Pacific Basin and Other Countries*. pp. 111. Fruitridge Printing, Sacramento, CA.

McCue, P. M. and T. P. O'Farrell. 1988. Serological survey for selected diseases in the endangered San Joaquin kit fox, *Vulpes macrotis mutica*. J. Wildl. Dis. 24(2):274–281.

McEvedy, C. 1988. The Bubonic Plague. Sci. Am. 258(2);118–123.
McGauchey, C. A. 1961a. The diseases of elephants (Parts I). Ceylon Vet. J. 9:17–21.
McGauchey, C. A. 1961b. The diseases of elephants (Part II). Ceylon Vet. J. 9:41–48.
Meunier, P. C., L. T. Glickman, M. J. G. Appel, and S. J. Shin. 1981. Canine parvovirus in a commercial kennel, epidemiologic and pathologic findings. Cornell Vet. 71(1):96–110.
Miller, M. W., N. T. Hobbs, W. H. Rutherford, and L. L. W. Miller. 1987. Efficacy of injectable Invermectin for treating lungworm infections in mountain sheep. Wildl. Soc. Bull. 15:260–263.
Moegle, H. and F. Knorpp. 1978. The epidemiology of rabies in wildlife, II. Observations on the badger. Zentralblatt f. Veterinarmedizin 225B(5):406–415. (Engl. Abs.).
Moegle, H., F. Knorpp, K. Bogel, A. Arata, K. Dietz, and P. Diethelm. 1974. Epidemiology of wildlife rabies, studies in the southern part of the German Federal Republic. Zentralblatt fur Verterinar-medizin 1974, 21B(Heft 9):647–659.
Morgan, B. B. and P. A. Hawkins. 1949. *Veterinary helminthology*. Burgess Pub., Minneapolis, MN.
Murray, J. D., E. A. Stanley, and D. L. Brown. 1986. On the spatial spread of rabies among foxes. Proc. Roy. Soc. Lond. (Biol.) 229(1255):111–150.
Neiland, K. A. and C. Dukeminier. 1972. A bibliography of the parasites, diseases and disorders of several important wild ruminants of the Northern Hemisphere. Alaska Dept. Fish Game, Wildl. Tech. Bull. 3.
Neitz, W. O. 1965. A check-list and host-list of the zoonoses occuring in mammals and birds in south and southwest Africa. Onderstepoort. J. Vet. Res. 32:189–374.
Nettles, U. F., A. K. Prestwood, R. G. Nichols, and C. J. Whitehead. 1977. Meningeal worm induced neurologic disease in black-tailed deer. J. Wildl. Dis. 13:137–143.
Nikolisch, M. 1965. Rabies -aspects of the history of the disease and its mode of transmission. Blue Book for the Veterinary Professional 10: 7–12.
Nordkvist, M. and K. A. Karlsson. 1962. Epizeotiskt forlopande infection med *Pasteurella multocida* hos ren. Nord Veterinarmed. 14:1–15.
Novak, M. 1972. *The Beaver in Ontario*. Ontario Min. Nat. Resour. 12–15.
Novakowski, N. S., J. G. Cousineau, G. B. Kolenosky, G. S. Wilton, and L. P. E. Choquette. 1963. Parasites and diseases of bison in Canada. II Anthrax epizootic in the Northwest Territories. Trans. N. Am. Wild. Nat. Resour. Conf. 28:233–239.

O'Brien, S. J. and J. F. Evermann. 1988. Interactive influence of infectious disease and genetic diversity in natural populations. Trends Ecol. Evol. 3(10):254–259.
O'Brien, S. J., M. E. Roelke, L. Marker, A. Newman, C. A. Winkler, D. Meltzer, L. Colly, J. F. Evermann, M. Bush, and D. E. Wildt. 1985. Genetic basis for species vulnerability in the cheetah. Science 227:1428–1434.
Obwolo, M. J. 1976. A review of yersiniosis (*Yersinia preutotuberculosis*) infection. Vet. Bull. 46:167–171.
Olsen, A. and A. Woolf. 1978. The development of clinical signs and the population significance of neurologic disease in a captive wapiti herd. J. Wildl. Dis. 14:263–268.
Onderka, D. K. and W. D. Wishart. 1988. Experimental contact transmission of *Pasteurella haemolytica* from clinically normal domestic sheep causing pneumonia in Rocky Mountain bighorn sheep. J. Wildl. Dis. 24:663–667.
Onderka, D. K., S. A. Rawluk, and W. D. Wishart. 1988. Susceptibilty of Rocky Mountain bighorn sheep and domestic sheep to pneumonia induced by bighorn and domestic livestock strains of *Pasteurella haemolytica*. Can. J. Vet. Res. 52:439–444.
Parikh, G. C. 1966. Epizootic hemorrhagic deer disease study. South Dakota Game, Fish, Parks Pittman-Robertson Project W-75-R-8.
Parker, G. R. 1964. Moose disease in Nova Scotia: Gastropod-nematode relationship. Unpubl. M.S. Thesis, Acadia University, Wolfville, Nova Scotia.
Parrish, C. R., P. H. O'Connell, J. F. Evermann, and L. M. Carmichael. 1985. Natural variation of canine parvovirus. Science 230:1046–1048.
Pearsall, W. H. 1954. Biology and landuse in East Africa. New Biol. 17:9.
Percival, A. B. 1918. Game and disease. J. E. Afr. Uganda Nat. Hist. 13:302.
Peterson, M. J. 1991. Wildlife parasitism, science, and management policy. J. Wildl. Manage. 55:782–789.
Peterson, M. J., W. E. Grant, and D. S. Davis. 1991. Bison-brucellosis management simulation of alternative strategies. J. Wildl. Manage. 55:205–213.
Pienaar, U. De V. 1961. A second outbreak of anthrax among game animals in the Kruger National Park, 5th June to 11th Oct. 1960. Kodoe 4:4–16.
Polyanskaya, M. V. 1963. On elaphostrongylosis of reindeer. In: *Helminths of Man, Animals and Plants and Their Control*. Papers presented to academican K. I. Skrjabin on his 85th birthday. pp. 424–425.: Izdatel, Akad. Nank. USSR, Moscow, USSR. [in Russian].

Post, G. 1962. Pasteurellosis of Rocky Mountain bighorn sheep (*Ovis canadensis*). Wildl. Dis. 23:1–14.
Post, G. 1971. The pneumonia complex in bighorn sheep, Trans. N. Am. Wildl. Sheep Conf. 1:98–106.
Potts, M. 1937. Hemorrhagic septicemia in the bighorn sheep of the Rocky Mountain National Park. J. Mammal. 18:105–106.
Prestwood, A. K. and J. F. Smith. 1969. Distribution of meningeal worm (*Pneumostrongylus tenuis*) in deer in the southeastern United States. J. Parasit. 55(4):720–725.
Prestwood, A. K. and T. R. Ridgeway. 1972. Elaeophorosis in white-tailed deer of the southeastern USA: case report and distribution. J. Wildl. Dis. 8:233–236.
Prestwood, A. K., T. P. Kistner, F. E. Kellogg, and F. A. Hayes. 1974. The 1971 outbreak of hemorrhagic disease among white-tailed deer of the southeastern United States. J. Wildl. Dis. 10:217–224.
Price, D. A. and W. T. Hardy. 1954. Isolation of bluetongue virus from Texas sheep-Culicoides shown to be a vector. J. Am. Vet. Med. Assoc. 124:255–258.
Price, E. W. 1953. The fluke situation in American ruminants. J. Parasit. 39:119–134.
Provost, A. 1981. Queries about rinderpest in African wild animals. Wildlife Disease Research and Economic Development, Ottawa, Ontario., IDRC, pp. 19–20.
Rau, M. E. and F. R. Caron. 1979. Parasite-induced susceptibility of moose to hunting. Can. J. Zool. 57(12):2466–2468.
Reidarson, T. H., J. McBain, C. House, D. P. King, J. L. Stott, A. Krafft, J. K. Taubenberger, J. Heyning, and T. P. Lipscomb. 1998. Morbillivirus infection in stranded common dolphins from the Pacific Ocean. J. Wildl. Dis. 34:771–776.
Reilly, J. R. 1970. Tularemia. In: J. W. Davis, L. H. Karstad, and D. O. Trainer (eds.), *Infectious Diseases of Wild Mammals*. pp. 175–199. Iowa State Univ. Press, Ames, IO.
Richards, S. H. 1963. Deer and antelope epizootic in North Dakota Badlands. Proc. North Dakota Acad. Sci. 17:70–71.
Richards, S. H. 1972. Epizootic hemorrhagic disease of deer. North Dakota Outdoors. 34(8):2–4.
Roberts, G. A. 1921. Rinderpest (Peste bouina) in Brazil. J. Am. Vet. Med. Assoc. 13:177.
Robinson, R. M., T. L. Hailey, C. W. Livingston, and J. W. Thomas. 1967. Bluetongue in the desert bighorn sheep. J. Wildl. Manage. 31:165–168.

Robinson, R. M., T. L. Hailey, R. G. Marburger, and L. Weishuhm. 1974. Vaccination trials in desert bighorn sheep against bluetongue virus. J. Wildl. Dis. 10:228–231.

Robinson, R. M., L. P. Jones, T. J. Galvin, and G. M. Harwell. 1978. Elaeophorasis in Sika deer in Texas. J. Wildl. Dis. 14:137–141.

Roneus, O. and M. Nordkvist. 1962. Cerebrospinal and muscular nematodiasis (*Elaphostrongylus rangiferi*) in Swedish reindeer. Acta. Vet. Scand. 3:201–225.

Rosatte, R. C., C. D. MacInnies, R. T. Williams, and O. Williams. 1997. A proactive prevention strategy for raccoon rabies in Ontario, Canada. Wildl. Soc. Bull. 25:110–116.

Rosen, M. N. 1971. Pasteurellosis. In: J. W. Davis, L. H. Karstad, and D. O. Trainer (eds.), *Infectious Diseases of Wild Mammals*. pp. 214–223. Iowa State Univ. Press, Ames, IO.

Roughton, R. D. 1975. An outbreak of a hemorrhagic disease in white-tailed deer in Kentucky. J. Wildl. Dis. 11:177–186.

Samuel, W. M. and M. J. Barker. 1979. The winter tick, *Dermacentor albipictus* (Packard, 1869) on moose, *Alces alces* L., of central Alberta. Proc. N. Am. Moose Conf. Workshop 15:303–348.

Saunders, B. P. 1973. Meningeal worm in white tailed deer in northwestern Ontario and moose population densities. J. Wildl. Manage. 37:327–330.

Schmidt, R. L., C. P. Hibler, T. R. Spraker, and W. H. Rutherford. 1979. An evaluation of drug treatment for lungworm in bighorn sheep. J. Wildl. Manage. 43:461–467.

Schultz, S. R., M. K. Johnson, R. X. Barry, and W. A. Forbes. 1993. White-tailed deer abomasal parasite and fecal egg counts in Louisiana. Wildl. Soc. Bull. 21:256–263.

Schwers, A., J. Barrat, J. Blanco, and M. Maenhoudt. 1983. Prevalence of antibodies against canine parvovirus in foxes *Vulpes vulpes*. Ann. Med. Vet. 127(7):544–546.

Scott, G. R. 1970. Rinderpest. In: J. W. Davis, L. H. Karstad, and D. O. Trainer (eds.). *Infectious Diseases of Wild Mammals*. Iowa State Univ. Press, Ames, IO.

Scott, J. W. 1940. Natural occurrence of tularemia in beaver and its transmission to man. Science 91:263–264.

Scrimenti, R. S. 1970. *Erythema chronicum migrans*. Arch. Dermatol. 102;104–105.

Severinghaus, C. W. and R. W. Darrow. 1976. Failure of elk to survive in the Adirondacks. New York Fish Game J. 23(1):98–99.

Severinghaus, C. W. and L. W. Jackson. 1970. Feasibility of stocking moose in the Adirondacks. New York Fish Game J. 17(1):18–32.
Shope, R. E., L. G. MacNamara, and R. Mangold. 1968. A virus-induced epizootic hemorrhagic disease of the Virginia white-tailed deer. pt. 1. J. Exptl. Med. 111(2):155–1701.
Smith, H. J., R. M. Archibald, and A. H. Corner. 1964. Elaphostrongylosis in maritime moose and deer. Can. Vet. J. 5(11):287–296.
Smith, T. and F. L. Kilborne. 1893. Investigations into the nature, causation and prevention of Texas or southern cattle fever. USDA Bur. Anim. Investigation. Bull. 1.
Spinage, C. A. 1962. Rinderpest and faunal distribution patterns. Afr. Life. 66:55–61.
Spinkler, L. A., R. W. Allen, L. S. Diamond, and J. C. Lotze. 1958. Babesia in a white-tailed deer. J. Protozool. 5 (suppl.):8.
Spraker, T. R., M. W. Miller, E. S. Williams, D. M. Getzy, W. J. Adrian, G. G. Schoonveld, R. A. Spowart, K. I. O'Rourke, J. M. Miller, and P. A. Merz. 1997. Spongiform encephalopathy in free-ranging mule deer (*Odocoileus hemionus*), white-tailed deer (*Odocoileus virginianus*) and Rocky Mountain elk (*Cervus elaphus nelsoni*) in northcentral Colorado. J. Wildl. Dis. 33:1–6.
Stableforth, A. W. and I. A. Galloway. 1959. *Diseases Due to Bacteria, Vol.1:*53–141. Academic Press, New York, NY.
Stair, E. L., R. M. Robinson, and L. P. Jones. 1968. Spontaneous bluetongue disease in white-tailed deer. Can. J. Vet. Sci. 32:382–387.
Steen, M., A. G. Chaband, and C. Rehbinder. 1989. Species of the genus *Elaphostrongylus* parasite of Swedish Cervidae. A description of *E. alces* n.sp. Ann. Parasitol. Hum. Comp. 64:134–142.
Steere, A. C., T. F. Broderick, and S. E. Malawista. 1978. *Erythema chronicum migrans* and Lyme arthritis: epidemiologic evidence for a tick vector. Am. J. Epidemiol. 108:312–321.
Stein, C. D. 1954. The incidence of anthrax livestock during 1953 and the first three quarters of 1954. Proc. U.S. Livestock Assoc. 58:116–122.
Stein, C. D. and M. G. Stoner. 1952. Anthrax in livestock during 1951 and comparative data on the disease from 1945 through 1951. Vet. Med. 47:315–320.
Stevenson-Hamilton. 1911. Game and disease. J. E. Afr. Uganda Nat. Hist. Soc. 13:302.
Stovell, P. L. 1980. Pseudotubercular yersiniosis. In: *CRC Handbook Series in Zoonoses. Section A, Vol. II,* pp. 209–256. CRC Press, Boca Raton, FL.

Swales, W. E. 1935. The life cycle of *Fascioloides magna* (Bassig 1875), the large liverfluke of ruminants, in Canada with observations on the bionomics of the larval stages and the intermediate hosts, pathology of *Fascioloides magna*, and control measures. Can. J. Res. 12(12):177–215.

Telfer, E. S. 1965. Some factors in the ecology of moose and white-tailed deer in Nova Scotia, Northeast Wildlife Conf., (Jan), Harrisburg, PA. (unpubl.).

Theiler, A. 1897. Rinderpest in Sud-Afrika. Schweiz. Arch. Tierheilk. 29:49.

Thomas, J. E. and D. G. Dodds. 1988. Brainworm, *Parelaphostrongylus tenuis* in Moose, *Alces alces*, and White-tailed Deer, *Odocoileus virginianus* of Nova Scotia. Can. Field-Nat. 102:639–642.

Thomas, F. C. and J. Miller. 1971. A comparison of bluetongue virus and EHD virus; electron microscopy and serology. Can. J. Comp. Med. 35:22–27.

Thomas, F. C. and D. O. Trainer. 1970. Bluetongue virus: (1) in pregnant white-tailed deer, (2) a plague reduction neutralization test. J. Wildl. Dis. 6:384–388.

Thomas, N. J., W. J. Foreyt, J. F. Evermann, L. A. Windberg, and F. F. Knowlton. 1984. Seroprevalence of canine parvovirus in wild coyotes from Texas, Utah and Idaho. J. Am. Vet. Med. Assoc. 185:1238–1287.

Thorne, T. 1971. A die-off due to pneumonia in a semi-captive herd of Rocky Mountain bighorn sheep. Trans. N. Am. Wild Sheep Conf. 1:92–97.

Thorne, E. T., N. Kingston, W. R. Jolley, et al. 1982. *Diseases of Wildlife in Wyoming*. 2nd ed. Wyoming Fish and Game Dept., Cheyenne, WY.

Thorne, E. T., J. K. Morton, and G. M. Thomas. 1978. Brucellosis in elk 1. Serologic and bacteriologic survey in Wyoming. J. Wildl. Dis. 14:74–81.

Thorne, E. T., E. S. Williams, T. R. Spraker, W. Helms, and T. Segerstoom. 1988. Bluetongue in free-ranging pronghorn antelope (*Antilocapra americana*) in Wyoming, 1976 and 1984. J. Wild. Dis. 24:113–119.

Thurston, D. R. and R. G. Strout. 1978. Prevalance of meningeal worm *Parelaphostrongylus tenuis*. J. Wild. Dis. 9:376–378.

Titche, A. R., A. K. Prestwood, and C. P. Hibler. 1979. Experimental infections of white-tailed deer with *Elaeophora schneideri*. J. Wildl. Dis. 15:273–280.

Trainer, D. O. 1973. Caribou mortality due to meningeal worm *Parelaphostrongylus tenuis*. J. Wildl. Dis. 9:376–378.

Trainer, D. O. and R. P. Hanson. 1969. Serologic evidence of arbovirus infections in wild ruminants. Am. J. Epidemiol. 90:354–358.

Trainer, D. O. and L. H. Karstad. 1970. Epizootic hemorrhagic disease. In: J. W. Davis, L. H. Karstad, and D. O. Trainer (eds.), *Infectious Diseases of Wild Mammals.* pp. 50–54. Iowa State Univ. Press, Ames, IO.
Uhazy, L. S., J. C. Holmes, and J. G. Stelfox. 1973. Lungworms in the Rocky Mountain bighorn sheep of western Canada. Can. J. Zool. 51(8):817–824.
Ulrich, K. 1940. Ein Fall von Piroplasma beim Rehwild. Tieraerztl. Rundschau 46:331.
Vittoz, R. 1954. Considerations practiques sur le rôle des animaux sauvages dans la transmission des maladis contagieuses et la prophylaxie de celles-ci dans le Sud-Est Asiatique. Bull. Off. Int. Epizoot. 42:206.
Vosdingh, R. A., D. O. Trainer, and B. C. Easterday. 1968. Experimental bluetongue disease in white-tailed deer. Can. J. Vet. Sci. 32:382–387.
Wachendorfer, G. and G. Schwierz. 1980. Epidemiology and control of wildlife rabies-analysis of potential causes for the great reduction of the badger (*Meles meles*) population in Hesse between 1952 and 1977. Deutsche Tierarztliche Wachenschrift 87(7):255–560 (Engl. Abs.).
Wandeler, A., G. Wachendorfer, U. Forester, H. Krekel, W. Schalc, J. Muller, and F. Steck. 1974. Rabies in wild carnivores in central Europe, 1, epidemiological studies. Zentralblatt f. Veterinarmedizin. 21B:765–773.
Wells, E. A., A. D'Alessandro, G. A. Morales, and D. Angel. 1981. Mammalian wildlife diseases are hazards to man and livestock in an area of the Llanos Orientales of Columbia. J. Wildl. Dis. 17(1):153–162.
Wetzel, R. and W. Rieck. 1962. Krankheiten des Wildes. Verlag Paul Parey, Berlin, Germany.
Wetzel, R. and W. Rieck. 1966. *Les Maladies du Sibier.* Librairie Maloine, Paris, France.
Wetzler, T. F. 1970. Pseudotuberculosis. In: J. W. Davis, L. H. Karstad, and D. O. Trainer (eds.). *Infectious Diseases of Wild Mammals.* pp. 224–235. Iowa State Univ. Press, Ames, IO.
Whitlaw, H. A. and M. W. Lankester. 1994. The co-occurence of moose, white-tailed deer and *Parelaphostrongylus tenuis* in Ontario. Can. J. Zool. 72:819–825.
Whitney, E., A. P. Roz, G. A. Rayner, and R. Deibel. 1969. Serologic survey for arbovirus activity in deer sera from nine counties in New York state. Bull. Wildl. Dis. Assoc. 5:392–397.
Whyte, I. J. and S. C. J. Joubert. 1988. Blue wildebeest population trends in the Kruger National Park and the effects of fencing. E. Afr. J. Wildl. Res. 18(3):7–87.

Wildt, D. E., M. Bash, C. P. Goodrowe, A. E. Pusey, J. L. Brown, P. Joslin, and S. J. O'Brien. 1987. Reproductive and genetic consequences of founding isolated lion populations. Nature. 329:328–321.
Witter, J. F. and D. C. O'Meara. 1970. Brucellosis. In: J. W. Davis, L. H. Karstad, and D. O. Trainer (eds.). *Infectious Diseases of Wild Mammals,* pp. 249–255. Iowa State Univ. Press, Ames, IO.
Woolf, A., C. A. Mason, and D. Kradel. 1977. Prevalence and effects of *Parelaphostrongylus tenuis* in a captive Wapiti population. J. Wildl. Dis. 13:149–154.
World Health Organization. 1978. Surveillance and control of rabies, report on a conference. Frankfurt-am-Main, Nov. 15–19, 1977. WHO, Copenhagen, ICP/VPH 001.
Worley, D. E., C. K. Anderson, and K. R. Greer. 1972. Elaeophorosis in moose from Montana. J. Wildl. Dis. 8:242–244.
Worley, D. E. 1975. Observations on epizootiology and distribution of *Elaephora schneideri* in Montana ruminants. J. Wildl. Dis. 11:486–488.
Yuill, T. M. 1986. Diseases as components of mammalian ecosystems: mayhem and subtlety. Can. J. Zool. 65:1061–1066.
Zarnke, R. L. and W. B. Ballard. 1987. Serologic survey for selected microbial pathogens of wolves in Alaska, 1987–1982. J. Wildl. Dis. 23(1):77–85.
Zarnke, R. L. and T. M. Yuill. 1981. Serologic survey for selected microbial agents in mammals from Alberta, 1976. J. Wildl. Dis. 17:453–461.
Zarnke, R. L., C. H. Calisher, and J. Kerschner. 1983. Serologic evidence of arbovirus infections in humans and wild animals in Alaska. J. Wildl. Dis. 19(3):175–179.

Additional Readings

Two important newsletters are produced in North America relating to wildlife disease and parasites. These are SCWDS Briefs, a quarterly newsletter of the Southeastern Cooperative Wildlife Disease Study at the University of Georgia, and the Wildlife Health Centre Newsletter produced by the Canadian Cooperative Wildlife Health Centre at the University of Saskatchewan which is published semiannually.

6 MANAGEMENT SYSTEMS

As it was believed that wildlife stocks originally dwindled in North America because of overexploitation, the first government response was to protect the resource. Protection inevitably took the form of restricting use or harvest. We find legislation designed primarily to perpetuate hunting opportunity for all, reducing the influence of market hunting and "game hogs" on the overall supply of game, or alternatively, to protect completely species whose stocks had been severely depleted. By the time of the American Revolution, twelve of the thirteen colonies had enacted closed seasons on some species, several had prohibited certain destructive equipment and methods such as the infamous duck or punt guns, and some had prohibited the export and sale of deerskins. Upper Canada enacted its first protective legislation for game species in 1829 and its first full-fledged game law in 1839. The proliferation of laws really curtailed only the honest citizen. The need for game law enforcement officers resulted in Massachusetts and New Hampshire developing the first warden (or conservation officer) system in North America in 1850.

The increase in protective legislation and the attendant development of an enforcement system failed to stem the decline in wildlife populations. Human populations continued to grow, and wildlife habitat continued to disappear, and attention turned to the large predators and their control. These animals were competing directly with man for an increasingly limited resource, and mechanisms such as government trapping and bounties were employed to reduce their numbers. The programs were only partially successful. Predator numbers in reality were lowered by the decline in prey species' numbers and the concerted efforts of the agricultural community to remove bears, wolves, and mountain lions that occasionally harassed livestock and seemed threatening to homesteaders. The prime factor leading to predator decline was habitat change associated with agricultural activities. Extensive clearing of forested land created conditions they could not tolerate.

Most of the modern game laws were pretty much in place in the United States by 1880. All states had game laws by that time, and such key regulatory components as licensing of hunters (New York and New Jersey in 1864) and the differentiation between resident and nonresident hunter (New Jersey in 1864), the rest (or closed) day during the week (Maryland in 1872), the banning of market hunting (Maryland in 1872), and the bag limit (Iowa imposed a limit of twenty-five prairie chickens/day in 1878) had been initiated.

Other developments occurring about this same time were designed to maintain some of the fast disappearing wilderness. In 1872 Yellowstone National Park, often considered to be the first national park in the world, was established. (Actually Hot Springs, Arkansas, was the first national park formed in the United States). In 1887, the Rocky Mountain Parks Act gave Canadians their first national park. Whereas the Yellowstone Park Act had prohibited only the "wanton destruction" of wildlife, the Canadian equivalent provided for the "protection and preservation of game, fish, (and) wild birds generally." The inherent strength of the latter statement ensured a sounder legislative base for wildlife protection in the Canadian national parks system. In 1894, Yellowstone Park had to be closed to hunting and thus a precedent was established which had become generally pervasive in North America by the latter part of the twentieth century. However, U.S. legislation has opened at least some of the Alaskan national parks to public hunting.

The national parks system was not the only area of federal jurisdiction where policies differed between Canada and United States. The U.S. Supreme Court had declared in 1842 that wildlife was held in trust for all citizens. This statement set the stage for U.S. federal government intervention and direct involvement in the protection and conservation of wildlife. As trustee of wildlife, the U.S. federal government has successfully met numerous challenges from state governments, and although a doctrine of state ownership of wildlife has developed, U.S. federal wildlife law continues to expand. Its influence is most noticeable in the western states which have extensive federal lands. There the federal government controls the habitat and the state government the regulations. Canada's jurisdiction over wildlife was never clearly stated in the original constitution (the British North America Act). By default, the actual Canadian situation has divided federal and provincial responsibilities based on interpretation and ad hoc arrangements. Provinces may opt in or out of federally funded wildlife programs. Unlike the United States where many states have significant percentages of their land area in federal lands, the Canadian Crown lands are within provincial, not federal control. It is only in the international area and through treaties such as the Migratory Bird Treaty of 1916 and the enabling legislation of the Migratory Bird Convention Act of

1917 that the Canadian federal government obtained responsibility for wildlife management, in this case for waterfowl and other migratory birds.

Despite the great differences in governmental systems and the extent of federal influence, in practice the areas of federal influence do not differ greatly between the two countries. The major difference is the degree of U.S. federal intervention in state activities as a result of extensive federal land holdings. This is still a contentious issue in many western states and was the basis for the "Sagebrush Rebellion" of the 1980s. Some of the southwestern states witnessed citizens' movements designed to bring about the transfer of control from federal to state jurisdiction.

In addition to national parks, habitat protection took the form of bird and game sanctuaries or refuges both public (Lost Mountain Lake: Canada 1887; Pelican Island: United States 1903) and private (Weber's Pond: Wisconsin 1891; Jack Miner's Game Sanctuary: Ontario 1907).

Despite these efforts, wildlife stocks still continued to decline. It took Theodore Roosevelt's idea of "conservation through wise use" and actions precipitated by Roosevelt's doctrine to reverse the trend that prohibition of market hunting, restrictive legislation, and creation of wildlife or wilderness sanctuaries were failing to do. Roosevelt's doctrine was simple. He recognized wildlife as a renewable natural resource just as range and forest land already were recognized. He contended that wildlife stocks would last forever if they were harvested scientifically and not faster than they were being produced. Propagation, stocking, and habitat management were to become other cornerstones of wildlife management as the twentieth century progressed. The biological naivete of the early 1900s which necessitated such statements as Roosevelt's may seem incredible today but it is important to realize we still have too many managers who ignore the biological realities of the populations they are supposedly managing. Even now benign ignorance too often blocks rational wildlife management.

Some Management Principles

What are the mechanisms the modern manager can employ? Manipulation of hunting seasons has been a favorite. Seasons can be opened early or extended in areas where you wish to enlarge the harvest. Shortening a season to reduce harvest often backfires because the increased effective hunting pressure/unit time usually compensates for the decreased number of available days. Males-only seasons (especially for cervids), periodic harvest, and party permits all have been used or suggested as means to maximize recreational opportunity (and license sales), sometimes without regard to the aforementioned biological realities. Economic pressures can be the controlling factors when a management agency derives its income directly or indirectly from license sales.

If we were to have the best of all possible worlds in a management sense, how would harvest be controlled? First, the geographical area to be managed should be divided into ecophysical regions which describe the environmental realities of the area. Factors such as soils, vegetation, physiography, climate, water, human population, and access must be considered and boundaries determined primarily on ecological grounds. The species to be managed will determine whether subunits are necessary, and ideally if the data were available, management should be geared to a discrete population and its critical habitat requirements, be it a white-tailed deer wintering area or a garter snake hibernaculum.

Second, the manager should have sound population estimates for the management area, and because wildlife populations seldom show random distributions, knowledge of the distributional patterns of the species also is needed. Further necessary biological data include natural mortality and natality rates by sex and age class, sex ratios by age class, and factors responsible for any differential rates; basically a complete understanding of the population dynamics of the species.

Third, it is useful to have an idea of the public's demands. How many prospective users are there? What are their expectations? What types of use can be provided? What potential conflicts between users and between users and nonusers exist? This provides the socioeconomic frame for later management decisions.

Finally, a management plan must be developed with options ranging from the most practical to the most desired, offering logical and rational explanations of what the costs and benefits of each approach would be. The desired population levels and attendant use levels, be they harvest or viewing, are to be quantified and mechanisms for control detailed.

Even in our ideal world, there may be legislative restrictions to which one must adhere. What flexibility exists for management decisions? What are the political realities of the jurisdiction, and what mechanisms are available for selling the public and the administration on the value of the management program? What are the funding limitations?

Let's now look at two very different hypothetical examples in a less than ideal world.

Example 1—An Urban Nongame Species

1. recreational viewing
2. aesthetic component of urban areas

Viewing opportunities for this avian species are limited to areas of open grassland interspersed with mature conifers and shrub patches, which are also the breeding and feeding habitat. Habitat requirements are usually met by city parkland and cemeteries but seldom by private homeowners. The species is migratory so viewing opportunity is limited to spring and summer. Normal breeding density is 1 pr/ha with approximately 10 ha of prime habitat still available within the city. Naturalist groups and bird watchers are interested in viewing opportunities, but the species is sensitive to disturbance during the breeding season so productivity is highest (three young/breeding pair) in cemeteries and lowest in public parkland (one young/breeding pair). Demand is 10,000 viewer hours/year. No funds are available for management of this species. Conflict is possible among cemetery owners, bereaved individuals, and resource users. Longevity of the species, natural mortality, and reproductive rate clearly show that park populations are not self-sustaining and that recruitment occurs from the cemetery populations.

Management should be directed to protect the cemetery populations and improve viewing opportunities within the park system. Unless a problem develops, the cemeteries can be left alone. No publicity regarding the abundance of the species there should be made, and a discreet letter to cemetery managers suggesting they contact your agency if bird watchers begin to pose a nuisance should encompass the action (or inaction) requirements. Make parks' officials aware of the critical habitat requirements of the species and ask them to maintain those areas currently supporting breeding populations. See if parks officials and the naturalist groups are willing to fund viewing platforms to at least localize the impact of birdwatchers. Numbers required, siting of the structures, and some mechanism for determining the public response (and the birds' response) should be incorporated into the plan. Modifications are required only if demand and/or supply change drastically.

Example 2—A Big Game Species

1. recreational viewing
2. sport hunting

The species occurs in 75 percent of the management zones (400,000 km^2) in the jurisdiction. Available winter range is critical and potentially limiting habitat. Population is estimated at 60,000 individuals and annual harvest at 20,000. The population has been declining and is currently below the predicted carrying capacity of 85,000 animals. Major mortality factors include poaching, winter starvation in some areas, predators (domestic and wild), and hunting. Nonresident hunters account for 10 percent of the harvest and 30 percent of the revenue from license sales. Forest industries have been reducing critical habitat at a nonreplacement rate of 1000 ha/year. Demands for hunting opportunity and recreational viewing are increasing. The tourist industry demands that no season closures or reductions occur because of the potential economic impact on rural areas. Management direction should allocate hunters and harvest in numbers more representative of regional conditions and population. If adequate management zones do not exist, they must be reorganized and used to distribute hunters more effectively. Impose quota systems to achieve harvest levels which will allow population growth in the underpopulated zones. Hunter opportunity will be reduced by these quotas but some of the demand may be met by primitive weapons seasons (archery or muzzle-loading). Minimum season length should not be reduced below two weeks to avoid concentration of hunters and meet tourism requirements. Consider dog and wild predator control programs for those areas where serious losses to predators occur and explore antipoaching programs such as hot line informer and reward system.

Effective management of critical winter habitat areas must be achieved either through cooperative efforts with the forest industry (by cut restrictions and other forest management guidelines) or through land-use legislation to protect critical wintering areas, or by both of these. License fees for nonresident hunters should be significantly increased (declining numbers of licenses sold are offset by increased cost so revenue remains relatively constant). Area specific licensing for residents achieves the required restrictions on hunter numbers. In areas of high demand, use party permits with one animal per two hunters as the legal limit. Education is required to explain the need for these actions and assure the nonhunting public that managers are aware of their interests. Management programs can increase viewing opportunities. It is also helpful to consider creating nonhunting zones in those areas of highest nonconsumptive recreation activities.

Summary

While these two examples may seem somewhat simplistic or contrived, in reality they represent two extremes in wildlife management. Many nongame management situations are handled adequately by benign neglect or minimal intervention. By contrast, management of popular game species often necessitates comprehensive management at fine levels of detail.

Regulatory Management

Even the most sophisticated model is only an aid. The manager's control over the number of animals being harvested is the ultimate key to effective regulatory practice. Theoretically, the easiest way is to establish a harvest quota and cease the harvest once it is reached. The limitations here are primarily political but also practical. There must be some means of determining the actual harvest level. Those jurisdictions requiring registration within twenty-four to forty-eight hours after harvest can simply monitor the tallies from the registration stations. How then do you effectively close down hunting, trapping, or any form of harvest once the quota is reached? In an uncontrolled licensing system with no limit on the number of licenses sold, inequality of opportunity is likely to occur because of unpredictable season closure. When individuals pay the same fee, they expect the equivalency of opportunity a fixed season provides. At least everyone knows when the season will end in that situation. In Maine, the 1971 white-tailed deer season was curtailed under the emergency powers of the Inland Fish and Game commissioner. A series of severe winters had drastically reduced populations in several management zones (management was actually on the basis of two hunting zones, even though biological data from eight management zones were collected and analyzed). Because Maine has a compulsory registration system for deer, it was easy to compare the actual harvest with safe harvest levels developed from population estimates and indices. The biological advice was to close three management zones because of pending overharvest. The commissioner had to weigh the following factors: (1) potential long-term damage to deer populations; (2) potential long-term and short-term declines in hunting license sales in succeeding years; (3) adverse political response by public, hunting camp operators, and legislators to the use of emergency powers previously only contemplated in years of extreme fire hazards and adverse reponse from others who would support closure to protect the deer and would not buy the economic arguments; and (4) a credibility gap between the managers and public if the severity of the situation became known and nothing was done. Action taken in this sensitive situation was a masterpiece of political tightrope walking. The season was stopped with three days' notice in the northern hunting zone (not the management zones) and shortened slightly in the southern hunting zone. Closure on the basis of long-established hunting boundaries, with sufficient notice for hunters already afield in the north and at least remnant hunting opportunity in the south, did much to temper any adverse response. There were some dissidents who wanted their license fees refunded, and there was a slight decline in license sales in 1972. The public generally accepted the wisdom of the commissioner's action. Such was not the case in a neighboring state.

In Vermont, though severe winter mortality losses suggested compensatory mortality was occurring, entrenched management regulations had long held sway on both deer management and public education policies. As a result of a long-standing bucks-only season, the Vermont range was drastically overpopulated, and the deer populations possibly were psychophysically disrupted. In an effort to balance populations and reduce pressure on the range, a modest antlerless season was initiated based on the premise that these deer would die anyway. Maine showed that their managers were not prepared fully to accept compensatory mortality as fact in their state and warned that overharvest of Maine deer would be a potentially severe depressant to already climatically reduced populations, but Vermont officials reacted differently. Unfortunately, the politics of management more often than not override sound biological judgement. The Vermont officials feared a potentially negative hunter response to their public relations program (which was a justifiable one in light of their ecological realities). They did not explain or perhaps recognize that the Maine and Vermont situations were biologically very different. Even among the Maine management zones, there were real differences. At least one zone bordering on New Hampshire was like Vermont's situation in that it was underharvested but it too had to be closed when the southern hunting zone was curtailed. This is just one small example of the biological and political complexities inherent in management that can result from very restrictive legislation.

Harvest quotas also can be achieved by limiting license sales or hunting opportunity per individual. If license quotas are established based on average success rate per hunter and loss by crippling, adjustment can be made the following year for the slight variations from average values which will occur. After discovering that shortening the deer season did little to limit harvest and the decline of their deer population, Minnesota experimented with a long season. The hunter selected a maximum of three hunting days within the thirty day season, but each day had a maximum number of hunters. This approach effectively distributed hunter pressure throughout the season and provided equality of opportunity.

Hunting opportunity can be allocated by type of hunting. Archery, muzzle-loading, quality or trophy, and wilderness hunting are examples. An allocation system is more subtle but it also serves to distribute hunting pressure. For example, Washington State now requires a hunter to choose a single weapon system per big game species. A hunter may take deer only with bow and arrow if this is the choice but may choose another type of weapon for elk. More hunters can be accommodated because many bow hunters formerly also hunted with rifles. Low success methods like archery or low density hunting, "quality" zones, are means of reducing harvest in sensitive or overharvested areas. Less vulnerable areas are then left to bear the brunt of the general hunting season.

Management problems can result from underharvest, and there are mechanisms for enticing hunters to areas that can handle more of them. Some of the most useful ways to achieve better hunter distribution, involve economic incentives. Lower license fees or increased bag limits are powerful ones. Even within underhunted areas, hunter distribution can be a problem. It has been shown that hunters seldom move distances greater than 1.5 miles (2.4 km) from access roads and the number of hunters generally decreases with distance from an access road or trail. Conversely, studies of animal distribution suggest they may avoid areas close to access roads. More favorable distribution of hunters can be achieved through a better access system. This can be accomplished partially by using either public or private funds to maintain forest roads after cutting. Private funds are user fees paid to the landowner. This must be carefully regulated because some species such as elk, grizzly bear, and moose are vulnerable to disturbance caused by vehicles. Road closure has been used to reduce disturbance and hunting pressure on big game species-especially elk. This has the added benefits of promoting a higher quality hunting experience and reducing danger of hunting accidents in areas with too many access roads. Similarly, seismic grid patterns, which provide travel lanes for both moose and hunter, have necessitated controls on hunter access in places like Alberta and Alaska.

An innovative method of restricting waterfowl harvest was implemented experimentally by the U.S. Fish and Wildlife Service. Each species was allocated a point value according to sex–male mallard, 25 points; female, 40 points. Species in difficulty for which managers wished to reduce harvest had very high point values–female canvasback, 100 points. A hunter could hunt until the accumulated harvest had reached or exceeded the designated point limit. For example, a hunter would have to quit hunting if a female canvasback was shot. The logic behind this regulation is that hunters would concentrate on low point value birds to maximize the number of birds they could get. The system has been tried in at least twelve states, and although successful on this limited basis, it will take some time before general hunting can be controlled by this type of system. The hunter needs only to identify the bird after it is shot for the system to work, but it would be most advantageous to be able to identify the bird on the wing. This system should eventually result in a more knowledgeable hunter. Enforcement is likely to be a problem on a large scale, however, as hunters might be inclined to dispose of high count birds. The interesting outcome of preliminary testing is that the majority of hunters surveyed preferred the point system to the fixed bag limit, possibly because of the potential for increased harvest if the hunter uses the system judiciously.

The effectiveness of regulatory management has been demonstrated many times but perhaps nowhere as dramatically as in Newfoundland. In 1960, biologists determined that moose populations showing winter densities of twelve or more animals per square mile were causing serve damage to balsam fir and white birch regeneration in a central portion of the province. To attract hunters they instituted a long (fifteen weeks) season, a resident license fee at half the normal cost, and a bag limit of three moose of any sex or age (although a separate license was required for each moose taken). With these and other incentives a kill of thirteen moose per square mile was achieved in 1960. This harvest plus that of 1961 resulted in a population decline to less than six moose per square mile in the winter survey; a density level at which browsing pressure was insufficient to cause severe damage to forest regeneration.

One advantage of regulatory, as opposed to habitat management or land-use planning is the immediacy of control it offers. By wise use of liberal or restrictive mechanisms, harvest rates can be adjusted to the desired level. Historically, management agencies have usually been too conservative, probably in response to the massive declines in wildlife numbers in the late 1880s and early 1900s. This has resulted in overpopulations of some species such as white-tailed deer. Bucks-only laws instituted to protect stocks in states like Vermont, New York, and Pennsylvania did the job all too well. Range deterioration and agricultural damage often became problems as it proved impossible to rescind the regulations. To be truly effective, regulatory management must be tied to biological reality. Data on age-specific birth and death rates, breeding ages of males and females, numbers of individuals, sex and age class, and carrying capacity should all be known. A good definition of carrying capacity is "the maximum number of animals of a given species and quality that can survive in a given ecosystem through the least favorable environmental factors with in a state time period." The most limited environmental element will also limit the number of individuals that can live on any area of land. Any management practice that increases the supply or improves the distribution of these elements will tend to increase the carrying capacity of the area up to the point where crowding initiates density-dependent control mechanisms. This concept provides the mechanistic base for habitat management.

The fallacy of state or province wide uniform regulations should become apparent with the realization that animals and hunters are not distributed uniformly over the land area. Regional differences mediated by ecological factors have to exist. They must be acknowledged and regulations formulated accordingly. Characteristics of the hunters must also be known. Ideally managers should apprise hunters of the need for regulatory management changes and obtain their support. More effort is needed to improve the image of hunting and hunters. The best management scheme will fail if the hunting public ignores the program and continues to violate regulations. It is unfortunate that ethics cannot be legislated, for if all hunters were imbued with a land ethic and respect for life and property, antihunting sentiment would not be so strong. We give a small example as a case in point. We approached young hunters as they fired their shot guns from a county road into our neighbor's property. We informed the hunters that they were breaking the law on two counts—firing from a roadway and trespassing—as the property into which they were shooting was posted. They replied, "We didn't know this was private property." At this point we indicated obvious signs. "But we were only shooting tweety-birds," they said. Our disbelief and anger must have been apparent because they decided they had better leave before we told them that this was yet another infraction of the law. Such events are unfortunately all too common. Those hunters who enjoy their sport and take care to obey the law and the rights of others are commonly associated with the far too many gun-carrying miscreants. This issue underlies a considerable amount of the antihunting sentiment and undermines the best efforts of many management agencies. Efforts are underway to counteract the issue of unethical hunter behavior and perhaps the best example is Orion—The Hunters Institute. This private organization based in Montana has taken a lead role in developing programs to educate hunters on hunting ethics.

As we have stated several times already, hunting has often been considered by managers to be a compensatory mortality factor. The idea resulted from a misapplication of Errington's early work on such prolific species as muskrat and bobwhite quail, which were difficult to overharvest. When mortality occurs, species compensate by reducing the age of sexual maturity, increasing survival rates of embryos and young animals, and basically increasing overall productivity per female as the mean age of individuals in the population declines. For species with limited reproductive potential or in situations of extreme local harvest, hunting mortality becomes a prime additive factor; populations are reduced or even extirpated. It is vital to assess the impact on a population of all mortality factors, including hunting. These data indicate whether the population will be able to compensate and to what extent it can do so. Regulations and other management practices then may be framed accordingly.

Bibliography

Bender, L. C. and P. J. Miller. 1999. Effects of elk harvest strategy on bull demographics and herd composition. Wildl. Soc. Bull. 27:1032–1037.

Bergerud, A. T., F. Manuel, and H. Whalen. 1968. The harvest reduction of a moose population in Newfoundland. J. Wildl. Manage. 32:722–728.

Geis, A. D., R. K. Martinson, and D. R. Anderson. 1969. Establishing hunting regulations and allowable harvest of mallards in the United States. J. Wildl. Manage. 33:848–859.

Nelson, L., Jr. and J. B. Low. 1977. Acceptance of the 1970–71 point system season by duck hunters. Wildl. Soc. Bull. 5:52–55.

Rost, G. R. and J. A. Bailey. 1979. Distribution of mule deer and elk in relation to roads. J. Wildl. Manage. 43:634–641.

Sage, R. W., Jr., W. C. Tierson, G. F. Mattfeld, and D.F. Behrend. 1983. White-tailed deer visibility and behavior along forest roads. J. Wildl. Manage. 47:940–953.

Smith, G. W. and J. A. Dubovsky. 1998. The point system and duck-harvest management. Wildl. Soc. Bull. 26:333–341.

Thomas, J. W., J. D. Gill, J. C. Pack, W. M. Healy, and H. R. Sanderson. 1976. Influence of forestland characteristics on spatial distribution of hunters. J. Wildl. Manage. 40:500–506.

Walters, C. J. and P. J. Bandy. 1972. Periodic harvest as a method of increasing big game yields. J. Wildl. Manage. 36:128–134.

7 HABITAT MANAGEMENT

We have made considerable progress in recent decades in the development of management prescriptions for wildlife habitat. Anyone who has compared the first techniques manual of The Wildlife Society with the most recent edition will attest to this progress. Despite considerable improvement in our capability to manipulate the environment for wildlife, we still do so primarily for single species and often secondarily in relation to forest or agricultural management. We are, however, beginning to think and act in terms of ecosystem management, to conceptualize the complex of physical and biological components that provide food, water, cover, and shelter for communities of wildlife species, and to integrate these needs into land-use activities that might otherwise destroy them.

This chapter does not provide management prescriptions but examines issues related to wildlife habitat management and elaborates on mechanisms for ensuring consideration of wildlife needs within the general planning process. It discusses methodologies developed for describing or assessing wildlife habitat and integrating this information into the planning procedure.

Habitat Evaluation and Land-Use Planning

Habitat, literally the place where an animal lives, is often the factor most limiting to wildlife populations. With the increasing demand for land for agriculture, urbanization, water impoundments, and so on, the habitat for many species of wildlife is lost, though a few tolerant and/or adaptable species have their habitats increased. With adequate planning, it is sometimes possible to mitigate the effects of land-use change and in some cases even enhance the natural environment. Large-scale extraction of coal seams, oil shales, or oil sands presents an opportunity for habitat manipulation on a massive scale. European countries have had decades of experience with this, but in North

166 • *The Philosophy and Practice of Wildlife Management*

America the Appalachian rehabilitation of coal-mined areas is the first such major exercise. An example of potential opportunity is development of the Alberta oil sands which could create land forms and water bodies more varied and productive of wildlife than the muskeg found in much of the area now. Similar opportunities sometimes should be exploited by wildlife managers rather than always opposed for the short-term disadvantages involved. As managers, we should maintain a long-term ecological perspective and become actively involved in planning, or events will simply pass wildlife by and the very real opportunities for turning temporary environmental disaster into lasting ecological advantage will be lost.

In the past, wildlife habitat management has tended to be species oriented. Management was practiced individually—for ruffed grouse, perhaps, or white-tailed deer—instead of considering the full ecological consequences of habitat manipulation. Such an approach does have validity when the manager is dealing with critical habitat elements, for example, wintering range for cervids, staging areas for waterfowl, or when endangered or threatened species are involved. Habitat management should otherwise be balanced ecologically to meet the needs of as many different species as possible. Unfortunately, we often do not have data on all the habitat requirements of each species, although much effort has been directed this way in recent years. Thomas (1979) introduced a planning framework for wildlife in forest environments that adopted a more holistic approach.

Although the Pacific Northwest and the managed forests of the Blue Mountains of Oregon and Washington provided the examples, the methodological approach is valid for habitat management in most forested areas of North America. Wildlife management often has to be integrated with forest and range management to be successful. Objective, quantifiable arguments must be presented to justify modification of forestry and range practices to meet wildlife habitat needs.

Thomas's approach is deceptively simple. As a first step, the plant communities and their successional stages are described for an area. Wildlife information is then extracted at four levels. Level 1 is the life form association, that is, the relationship of animal life forms to the vegetative associations for such functions as feeding and reproduction. For example, in the Blue Mountains a number of species feed and reproduce primarily on the ground without specific water or physiographic consideration (elk, dark-eyed junco, and western fence lizard are examples). This type of association describes a life form which includes the three species listed plus others. The advantage of this system is that it reduces the number of faunal units to be considered. In the Blue Mountains case, it meant sixteen life forms represented 378 species (Table 7.1). It also means that forest managers can more readily evaluate the responses of wildlife to habitat types than if they had to consider each species independently.

Level 2 describes the relationship of the individual species to the plant communities and their successional stages for feeding and reproduction (Figures 7.1 and 7.2). This approach is of particular value for endangered, threatened, or indicator (featured) species. The various species are grouped by life form to facilitate comparison with level 1. It becomes apparent as this is done that some forest communities are more productive of wildlife than others.

Level 3 gives a one line summary of key biological information for each species, and at level 4 references are listed for consultation if the information provided in the previous three levels is not sufficient. The adaptability of each species is scored by a versatility index (V) where:

$$V = (C_r + S_r) + (C_f + S_f)$$
C = community
S = successional stage
r = reproduction
f = feeding

Table 7.1—Number of Wildlife Species Oriented to Forested Plant Communities for Feeding and Reproduction (Thomas, 1979)

Life form	Reproduces	Feeds	No. of species[1]	Examples
1	in water	in water	1	bullfrog
2	in water	on ground, in bushes and/or in trees	9	long-toed salamander, western toad, Pacific treefrog,
3	on ground around water	on ground, in bushes trees, and water	45	common garter snake, killdeer, western jumping mouse
4	in cliffs, caves, rimrock, and/or talus	on ground or in air	32	side-blotched lizard, common raven, pika
5	on ground without specific water, cliff, rimrock, or talus association	on ground	48	western fence lizard, dark-eyed junco, elk
6	on ground	in bushes, trees, or air	7	common nighthawk, Lincoln's sparrow, porcupine
7	in bushes	on ground, in water or air	30	American robin, Swainson's thrush, chipping sparrow
8	in bushes	in trees, bushes, or air	6	dusky flycatcher, yellow breasted chat, American goldfinch

Table 7.1 Continued

Life form	Reproduces	Feeds	No. of species[1]	Examples
9	primarily in deciduous trees	in trees, bushes, or air	4	cedar waxwing, northern oriole, house finch
10	primarily in conifers	in trees, bushes, or air	14	golden-crowned kinglet, yellow-rumped warbler, red squirrel
11	in conifers or deciduous trees	in trees, bushes, on ground, or in air	24	goshawk, evening grosbeak, hoary bat
12	on very thick branches	on ground or in water	7	great blue heron, red-tailed hawk, great horned owl
13	in own hole excavated in tree	in trees, bushes, on ground, or in air	13	common flicker, pileated woodpecker, red-breasted nuthatch
14	in hole made by another species or in a natural hole	on ground, in water or air	37	wood duck, American kestrel, northern flying squirrel
15	in burrow underground	on ground or under it	40	rubber boa, burrowing owl, Columbian ground squirrel
16	in burrow underground	in air or water	10	bank swallow, muskrat, river otter
		Total:	327	

[1]Species assignment to life form is based on predominant habitat-use patterns.

170 • *The Philosophy and Practice of Wildlife Management*

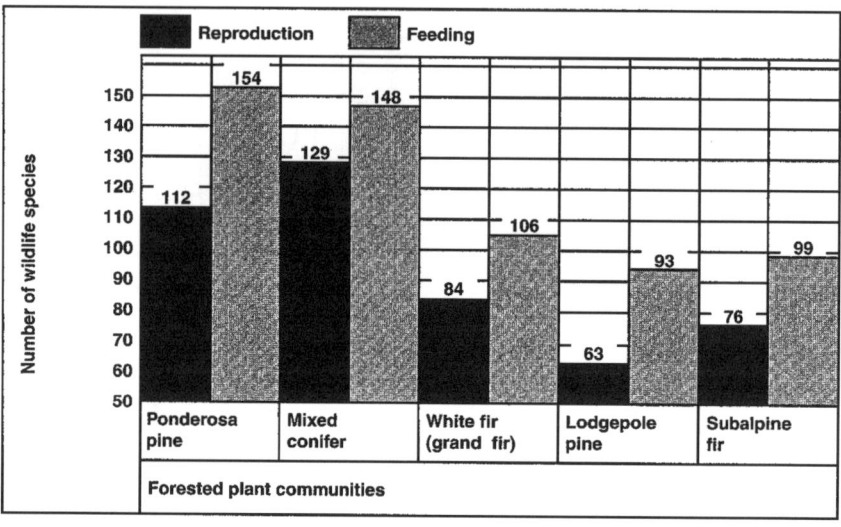

Figure 7.1—Number of wildlife species oriented to forested plant communities for feeding and reproduction (Thomas, 1979)

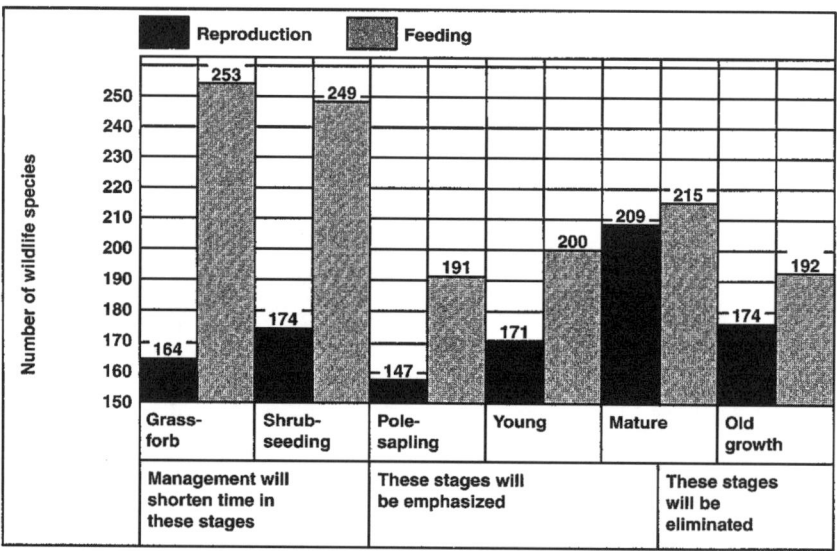

Figure 7.2—Number of wildlife species oriented to forestry successional stages and the potential effect of intensive timber management (Thomas, 1979).

The more specialized wildlife have low V values. Special emphasis is also afforded particular habitat types such as riparian zones, edges, snags, decaying logs, cliffs, talus, and caves.

If featured species management is utilized, actual prescriptions to enhance habitat conditions can be provided. Thomas et al. (1979) gave such details for elk and deer and in the process outlined appropriate silvicultural options. Verner and Boss (1980) developed these concepts into an extensive application for the western Sierra Nevada region of California. The data base was available in computer readable format so that prospective users could plug directly into the species matrices and make predictions on the effects of forest management practices or land-use alterations. Biological information provided by species included: status, distribution/habitat, special habitat requirements, breeding, territory/home range, food habits, other pertinent and key references (Figure 7.3). Habitat was ranked by type as optimum, suitable, or marginal and season(s) of use were given. In effect, it was a valuable planning tool. The approach actually had its background in methodologies proposed to help conduct environmental impact assessments as required by the National Environmental Protection Act (1976).

Attempts to standardize habitat evaluation by federal wildlife personnel resulted in studies by Ellis et al. (1978), who evaluated four methodologies for accuracy and repeatability, and by Baskett et al. (1980), who replaced the original "blue handbook" of the United States Department of the Interior with the "yellow handbook." These early efforts generally have been supplanted by more sophisticated ecosystem classification procedures based on vegetational associations that allow interpretations for wildlife habitat, such as the Ontario Ministry of Natural Resources' Forest Ecosystem Classification for various forest regions of Ontario. In cooperative efforts with a number of provinces,

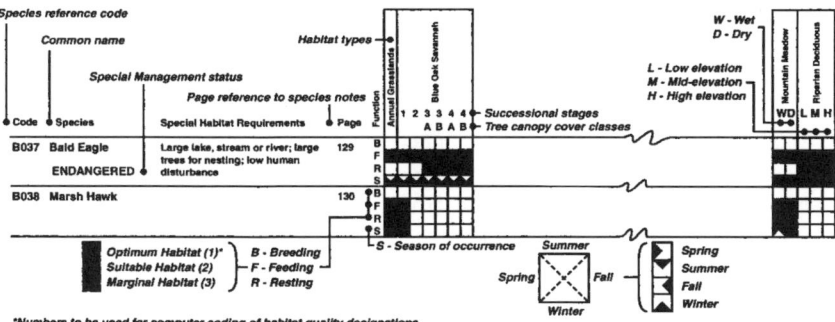

Figure 7.3—The key to elements in the species/habitat matrix given in a bald eagle example (Verner and Boss, 1980)

172 • *The Philosophy and Practice of Wildlife Management*

the Canadian Wildlife Service has used the Golet Wetland Classification system to inventory areas and assign a score for their value to wildlife production. This provides a means of assigning priority to land units for protection, acquisition, or management.

Habitat evaluation procedures are intended to document what is there, what could be there, what would be lost, and the relative value of a given land unit to wildlife populations. In Maine, for example, a system to set priorities for deer wintering areas was used based on number of deer overwintered, size of wintering area, distance to nearest wintering area, and relative contribution to the regional deer population. This allowed managers to decide appropriate forest cutting practices within individual "yards." The outcomes ranged from complete protection to controlled clear-cutting. Inventories of habitat conditions are necessary not only for planning but also for actual manipulative practices, i.e., management.

Increasingly, remote sensing is being used to classify habitat. Aerial and satellite imagery provide the opportunity for mapping and classifying land on a scale not previously possible. Variables commonly used in the classification process include plant associations, land forms, surface topography, soils, aspect, slope, and moisture condition. In Canada, the Canada Committee on Ecological Land Classification provided a focus for wildlife managers attempting to develop uniform methodologies for wildlife habitat classification. No comparable group exists in the United States, but important contributions have been made by using Landsat and aerial imagery. Geographical information systems (GIS) have been an offshoot of remote sensing activities. These systems generally provide a means of overlaying detailed information on a land classification system usually based on wildlife habitats or vegetative communities. The information generated provides both overview and site-specific means of examining the existing data bases. A GIS is a powerful management tool and invaluable for decision making over larger land areas.

Another use has been to analyse data for "metapopulations" and to determine "metahabitats." This approach capitalizes on GIS and comprehensive analysis of the literature to look at species and their needs over broad distributional ranges to assess the key components that constitute essential habitat. Advances in computer technology have made such analyses feasible with personal computers (Hanski and Simberloff, 1997).

Yet with few exceptions, most of the methodologies in use are general in nature and not related solely to game species. This is partly due to the land-management planning policy of the U.S. National Forest Management Act and the Resources Planning Act and explains the forefront activity of the U.S. Forest Service in comprehensive wildlife habitat management. Wildlife becomes an inclusive term representing all species of vertebrates other than fish. Management agencies have been scrambling to fill the void that exists relating to the habitat requirements for many bird, amphibian, reptile, and small mammal species. Examples are the proliferation of nongame programs within state and provincial wildlife agencies and the number of workshops and symposia that have been held recently on nongame habitat management procedures. Management of wildlife habitat is becoming increasingly sophisticated and more indicator species-oriented, sometimes with unfortunate results.

Perhaps of greatest importance to wildlife managers is the question, "How do you preserve wildlife habitat?" Other than direct purchase, there are a number of techniques that have been used. We describe three of them.

Land-Use Planning Controls

These procedures are based on extant legislation that controls or regulates the use of land. Land can be categorized as natural area or wildlife habitat if such classifications exist. Washington State's Environmental Policy Act of 1971 (SEPA), allowed cities and counties to deregulate environmentally sensitive areas. SEPA applied in these areas to some activities which were normally exempt from procedural review. The Massachusetts' Wetlands Protection Act includes wildlife habitat as a function of wetlands to be protected, although the act only protects certain habitats, e.g., riparian, flood plain. Many states have Forest Practices Acts which protect riparian habitat. Official plan designations, zoning bylaws, and transfer of development rights also can be used to protect wildlife habitat. The transfer of development rights has been used to separate the right to develop property within the existing legislative frame from the other rights of land tenure or ownership. The owner is compensated by being given a certain number of development credits which can be transferred and used on another designated area or sold to another party.

With particular reference to wetlands, the 404 Permit Program in the United States, under the Federal Clean Water Act, requires all developers proposing to dredge or fill in wetlands to apply to the Army Corps of Engineers for an approval permit. The granting of permits, in turn, is monitored by the Environmental Protection Agency (EPA) and by conservation groups. The EPA can veto any approval if it feels the land should not be altered from its natural state. The impact of private interest groups on this process has ensured that proposed wetland developments have received considerable public scrutiny when important wildlife habitat is to be impacted. The growing federal initiative to protect wetlands will place increased emphasis on the 404 Permit Program as one of the mechanisms for achieving such protection which in turn has made it the target of development interests.

Habitat conservation plans (*HCPs*) are agreements that allow development to occur that might harm endangered species in return for protection of habitat or mitigation at another location. HCPs were made possible by a 1982 amendment to the Endangered Species Act. Landowners could be given a permit to conduct activities that harm a listed species provided an HCP is devised that "to the maximum extent practicable [will] minimize and mitigate the effects". At this stage there is limited evidence that HCPs have achieved their objective of compensating for development activities that affect endangered species but it is an approach that has promise in achieving compromise solutions in controversial situations.

Purchase of Property Rights

Protection can be given to wildlife habitat by buying the partial rights to property in order to place restrictions on the land use. This is achieved by outright purchase, then placing restrictions or protective covenants on the property title before reselling the land to the public, but often this approach is too expensive. A *conservation easement* is the purchase of partial rights to a property. Under the terms of such an easement there is a servient tenant who retains the right to use and enjoy the property, subject to the rights of the second party the dominant tenant, who places restrictions on the use of the property. Compensation or payment for the easement is usually determined by subtracting the assessed value of the land with the specified development restrictions from the assessed value with no restrictions. An example of this type of easement entails restriction of tree cutting on a given piece of property. The cost of the easement generally reflects the lost commercial value of the unharvested trees. The Nature Conservancy in the United State has effectively used this method in its national wetlands program to protect large watersheds like the Brule River in northern Wisconsin. Many states, including Montana, Oregon, Missouri, and Washington, have enabling legislation for conservation easements. Long-term leasing has also been used to preserve wildlife habitat. The Canadian Wildlife Service entered into many such leases with prairie farmers to protect potholes that were valuable to waterfowl from drainage. The problem with lease arrangements is that they have definite time limits. Although the government, as tenant, retains exclusive possession or control of the property for the fee paid, upon termination of the lease landowners are free to do as they please with the land. Many prairie farmers entered into ten-year lease agreements not to drain potholes which they had never intended to drain. In the face of such attitudes, considerable money can be spent to achieve little. This was also the case for the 1985 and 1990 U.S. Farm Bills, which contained provisions for conservation reserves. The 1990 bill allowed farmers to sell the federal government a conservation easement on their wetlands and toughened the conservation compliance component by requiring farmers to implement plans to prevent soil erosion. Highly erodible agricultural land could be placed under contract with the federal government for a ten-year period and planted to grasses or trees. Wildlife habitat was one recognized objective. The farmer was paid for removing the land from production, but at the end of the ten-year period was free to do whatever he wanted with the land unless the sodbuster provisions were invoked. The end result may be farm subsidy and not conservation but the 1990 provisions may deter this. Nonetheless, evidence that the primary function of the conservation provisions in the 1985 Farm Bill was to reduce commodity production rather than conserve soil or provide wildlife habitat mounted as farmers were allowed to harvest hay from, or graze cattle on, reserve lands during the 1988 drought. Although the 1990 bill appeared to be more environmentally sound than its 1985 counterpart, it too was subject to the same inappropriate agricultural uses.

Management agreements require an agency to perform certain management services providing the landowner does not take certain actions such as posting the land against hunting.

Incentive Programs

The United States Department of Agriculture has a number of programs it administers which serve to enhance or protect wildlife habitat. The Water Bank Act provides payments to landowners who sign an agreement not to drain their lands. It is similar in nature to the leasehold agreement. The National Resources Conservation Service (NRCS) will provide trees and shrubs free or at low cost to prevent soil erosion and provide wildlife habitat, an added benefit. The NRCS also provides wildlife habitat plans for agricultural lands when requested by a farmer. *Preferential tax treatment* can be used as a financial incentive to preserve or produce wildlife habitat. Either property or income tax incentives are offered. In the case of property tax, the landowner may be given an exemption or deferment for maintaining the property in its natural state. Once the property is developed, the exempt status is lost and all deferred taxes have to be paid. Income tax incentives encourage the donation of land to public agencies by allowing it as deduction from income. Another financial incentive that has been proposed in Canada is the use of *loan guarantees* and *reduced interest rates* to encourage environmentally sound farming practices and wildlife habitat preservation especially in the prairie provinces. In such an arrangement, federal and provincial mortgages and loans would include the incentives if the land-use activities for which the money was being borrowed were shown to be sustainable without environmental damage. A *special designation,* although not financial, can provide incentive for a landowner to protect natural habitat. Many of the early bird sanctuaries and nature preserves were private lands whose owners made such designations in order to receive recognition by the government in the form of special certification, publication of the owner's name, or posting of the property. This has worked successfully for many agencies, public and private, in natural resource areas; a good example is the Tree Farm program in the United States.

Declining Wildlife Habitat—Old-Growth Coniferous Forest

Although most forests in North America are the second or third commercial growth on the same site, there are still areas in western Canada, Alaska, and the Pacific Northwest where considerable acreage of old-growth (>180 years) forest remain. Some of these are within wilderness areas or national parks and are thus protected from commercial exploitation, but most of this habitat type outside of such areas was programmed for harvest by the turn of the century. Wildlife species like Vaux's swift, spotted owl, lynx, red-tree vole, Olympic salamander, Oregon slender salamander, and mule deer, which appear partially or wholly dependent on this habitat for survival, will see further population reductions as the amount of old-growth forests, continues to decline. The unique habitat features of old-growth forests, other than the large conifers themselves, are large snags and large dead falls on land and in streams. The forests are structurally complex with multiple vegetational layers and substantial amounts of ground cover. Forest soil building occurs predominantly within old-growth types. The microclimate produced by the vegetation is suitable for fungal growth that promotes decomposition and provides food for many organisms, including a number of small mammals. Crown and ground dwelling lichens are major sources of nitrogen, and lichen has been suggested as a key habitat requirement for woodland caribou populations.

To envision the decline in wildlife fauna possible if further reduction in old-growth occurs, we need only look at one dependent species, the spotted owl. As recently as 1982, projected populations totalled 1365 pairs and the planning goal was 375 pairs; a 75 percent reduction. As the result of an effort to enumerate the species and determine minimum viable population size, the U.S. Forest Service had an objective to maintain 1030 pairs (530 in Oregon and Washington) out of the existing population of perhaps 3000 pairs. Thus a population reduction of about 65 percent was proposed in national forests. The habitat standard set was 1000 acres per breeding pair, although research had indicated that as much as 2000–2500 acres per breeding pair of spotted owls may actually be needed. Unless the habitat standard was increased, it might well have been impossible to save the species except within the national parks and wilderness systems. The proposed spotted owl guidelines in the draft supplement to the final Environmental Impact Statement (EIS) for an amendment to the Pacific Northwest Regional Guide (U.S. Forest Service) would have provided at least 550 spotted owl habitat areas. Although each habitat area (SOHA) might include about 2,200 acres only 1000 acres would be protected from timber sales. Litigation by environmental groups forced reconsideration of the spotted owl for listing as an endangered or threatened species after the U.S. Fish and Wildlife Service (USFWS) had initially declared there was no justification for such status. In 1990, the Fish and Wildlife Service listed the owl as threatened throughout its range. In the meantime, timber sales that were being blocked by court injunctions proceeded under Section 318 of FY 1990 Appropriations Act for the Department of the Interior and Related Agencies, which established measures that the Forest Service must follow to minimize the effect of timber harvest on spotted owls. This compromise agreement allowed 7.7 billion board feet to be sold from Oregon and Washington National Forests. The Forest Service also established advisory boards consisting of environmental, business and community interests in the thirteen National Forests known to be occupied by the northern spotted owl. These boards reviewed and made recommendations related to timber sale designs.

An Interagency Scientific Committee chaired by Jack Ward Thomas and with other scientifically qualified personnel from the Forest Service, Fish and Wildlife Service and the Bureau of Land Management (BLM) was established to: review the biological basis of the Fish and Wildlife Service criteria for review of timber sales and the basis for conference opinions, determine whether current land management strategies of the agencies were reserving options that would allow for long-term conservation of the northern spotted owl, provide recommendations to preserve options until the conservation strategy was completed, define habitat relationships for long-term conservation of northern spotted owls, suggest options to achieve the amount and configuration of habitat needed for long-term conservation of the northern spotted owl throughout its range, and evaluate research, monitoring, and inventory programs to answer critical questions and track the adequacy of management strategies. The Thomas Report, as it is known, recommended that Habitat Conservation Areas (HCAs) replace the SOHAs. The HCA system would allow management for clusters of at least 20 breeding pairs of owls and thus avoided the fragmentation that SOHAs would allow. But the extent of the lands to be "locked up" in the HCAs caused anguished outcries from industry, loggers, truckers, and governments, in the timber dependent communities of the Pacific Northwest. The battle between owls and economics was fully joined. President Bush stated that a "balance" must be found between protection of owl habitat and concern for jobs. Congress was asked to pass legislation to (1) allow the BLM to implement its own owl protection plan rather than the Interagency Scientific Committee's and do so without court challenge being possible, (2) adopt an interim management plan developed by an interagency task force for Forest Service lands and also disallow court challenge, and (3) expand the mandate of the Endangered Species Committee (a politicized vehicle of the president) to allow it to develop a long-term forest management plan for federal lands. In effect, the administration's reaction was to do whatever was possible to ensure the Thomas Report would not be implemented and the power of the Endangered Species Act would be weakened. But events conspired in favor of the owl. In response to a lawsuit brought by the Seattle Audubon Society that the Forest Service had failed to adopt a credible conservation strategy, Judge William L. Dwyer of the Federal District Court ruled against the Forest Service and issued an injunction against further timber sales in spotted owl habitat on National Forest lands until an adequate management plan was developed. Ancillary to this action, a process to designate critical habitat, also forced by Federal District Court action resulted in 6–9 million acres being so designated.

A Scientific Analysis Team was formed to respond to questions raised by Judge Dwyer about the 1992 Final EIS on Management for the Spotted Owl in the National Forests. Again Jack Ward Thomas played a key role as team leader. Judge Dwyer asked the Forest Service to respond to three questions:

1. Did the May 15, 1992 decision by the Endangered Species Committee [the so-called "God-Squad'] to allow cutting of thirteen timber sales by the BLM and judged by the USFWS to cause "jeopardy" for the northern spotted owl require changes in spotted owl viability assessments in the Final EIS?
2. Did the standards and guidelines of the selected alternative in the Final EIS need revision or have the probabilities of maintaining viable populations of the owl assigned in the alternatives changed because of information available since publication of the Final EIS?
3. Would implementation of the selected alternative in the Final EIS lead to the extirpation in the Forest Service planning areas (i.e., National Forests) of any of the thirty-two species identified in the Final EIS as being closely associated with late successional and old-growth forests?

One of the authors was involved in the exercise to answer the last question. The scientific expertise that was brought to bear was impressive and the compilation of existing knowledge regarding faunal and floral associations in old-growth forests of the Pacific Northwest that resulted helped determine the level of viability risk for species. However, it was apparent that distributional data were limited for many species and the best that could be proferred was collective professional judgement on the level of risk involved. What was impressive was that the managers and scientists were prepared to make such professional judgements, a risk that too many wildlife managers have been reluctant to take. We will never have definitive answers to many, if not most, management situations and we must be prepared to make the best educated recommendations in such circumstances. The Report of the Scientific Analysis Team "Viability Assessments and Management Considerations for Species Associated with Late-successional and Old-growth Forests of the Pacific Northwest" (Thomas et al., 1993) should be required reading in any management class because it combines the essential ingredients in modern wildlife management of the impacts of litigation, court action, scientific analysis, combined with political and ecological realities. This is a high octane mix whose volatility is still much in evidence in the old-growth controversy of the Pacific Northwest.

In April 1993, President Clinton convened a Forest Conference in Portland, Oregon. This Conference was a dialogue among the affected parties involved in the old-growth controversy. Clinton asked that a science based forest management plan be developed for the region that would have five goals as its basis:

1. Adherence to the nation's laws.
2. Protection and enhancement of the environment.
3. Provision of a sustainable timber economy.
4. Support of the region's people and communities during the economic transition.
5. Ensurance that federal agencies would work together.

The Federal Forest Management Plan was completed in April 1994 and Tuchman et al. (1996) assessed the first two years of implementation. Because the Plan covered timber harvest rates and protected late-successional and old-growth forests, the short-term economic and social impacts, especially in rural forest resource dependent communities have been high. Rather than dispelling the conflict, there has been ongoing conflict as neither political side accepted the compromise. Congress has attempted to bypass similar processes that could affect east slope Cascades Washington and Oregon forests and National Forests elsewhere, by implementing salvage legislation that could hinder attempts to reestablish healthy forests throughout the USFS system.

A related political controversy has surrounded the Tongass National Forest in southeast Alaska. Here the old-growth hemlock, cedar and spruce provide critical winter habitat for Sitka black-tailed deer. Congressional action has resulted in decreased harvest targets and greater consideration of noncommodity values. Here too, the economic and social well being of the people has been pitted against the temperate rain forest and one wonders if there have been any better outcomes than in the Pacific Northwest.

Habitat availability and quality are but two factors that will decline with a significant reduction in the old-growth forest type. Another major factor is the distributional pattern of the remnant old-growth forests. Although we have no definite knowledge of the best distributional configurations to maintain maximum wildlife values, biogeographical theory clearly predicts that fragmentation or isolation of the remnant old-growth forests will result in local and regional faunal extinctions and an overall decline in species diversity and abundance. Indeed empirical evidence from Mt. Rainier National Park supports the theory as the number of faunal species has declined in less than a century. Therefore, it will be the pattern of cutting and degree of isolation of the remaining patches of old-growth forests that become critical elements in determining the impact such habitat loss has on wildlife. The fragmentation of habitat into isolated units also may result in increased predator and social parasite effectiveness. These realizations have led to landscape level studies and a renewed emphasis on biodiversity as a goal in forest planning and environmental impact analyses. The U.S. Forest Service has adopted an ecosystem approach to forest management in the western coastal states that embraces the maintenance of biological diversity and sustainable outputs of goods and services as was required by the Forest Plan. In fact, ecosystem management, which is what Judge Dwyer prescribed for the Pacific Northwest, is the real focus of the forest management controversy. It has raged within federal management agencies and professional organizations, especially the Society of American Foresters, and Congress. Yet this is precisely what the National Environmental Policy Act of 1969 and the National Forest Management Act of 1976 required. It just took the better part of two decades for forest managers to begin to adjust to those dictates, in many cases dragging their seats firmly behind them. Ecosystem management is the only basis for sustainable forest management, another topic consuming forest managers worldwide. And the criteria for sustainable forestry are not solely ecological; they include social, economic, and environmental considerations that extend well beyond the normal ecosystem functions that might form the basis of ecosystem management. We traverse great unknowns with definitions that many challenge, with knowledge that is grossly limited, with competing demands that pull and tug in sundry directions, with different political agendas and despite this we know deep down that this is the only way we will sustain the forests, the wildlife habitat, and indeed the planet. Yet the continued growth of human populations, the dwindling natural resource base and the increased demand for products from that base make us question whether we will achieve consensus in time to ward off catastrophe. The burning of the Amazonian rain forests for subsistence agriculture, the removal of West African rain forests by commercial harvesting, the gradual warming of the planet by anthropogenic causes, the reluctance to pursue solutions instead of rhetoric, and the basic disagreements that emerge when discussion of solutions does occur, does not bode well. One should be optimistic based on the volume of literature that has emerged recently on adaptive ecosystem management and sustainable management and the network of model forests that has developed globally designed to test the tenets of both concepts, yet the personal disquiet continues. One of us has been a participant in major Canadian initiatives nationally and internationally and witnessed good and partisan intentions clash in the arenas intended to help create solutions. In some cases, fiscal resources were used to sustain traditional approaches or conduct irrelevant studies. Political agendas similar to others in international discussions prevailed where it was more important to be seen as being interested or driving the process rather than creating the solutions. In such circumstances, it is easy to become cynical and accusatory but to what end. Instead one continues to be involved in the hope that the effort might make a difference. Gilbert made the following comments as a panelist at an International Model Forests meeting in Mexico in 1996.

"One of the basic requirements for successful sustainable management of forests is an adequate data base to define what is to be sustained. While we have existing models to assess tree growth and yield and economic outputs related to tree harvesting, the capacity for developing models for other forest based resources such as wildlife, recreation, medicine shrubs and plants and nontraditional forest commodities is limited or nonexistent. The Model Forest Program has addressed this data problem in a number of ways. The most obvious perhaps is research to answer basic questions regarding the ecological relationships that exist within the forest community thus providing the ability to characterize the needs of species other than the commercial trees. The second approach has been to improve decision-making in forest management scenarios through the development of improved decision support systems.

The research to date has demonstrated that our inventory data are not comprehensive enough to provide sufficiently accurate determinations of either existing conditions or future conditions following management prescriptions. Furthermore, even sophisticated decision support systems are dependent on the quality of information being fed into them. These systems will have to be refined by a process of monitoring and assessment tied to the flexibility of adaptive management strategies if we are to move progressively in the direction of sustainable management. Contextually, we will find that there are general principles that can be applied across broad geographic areas but successful sustainable management will require the involvement of people with comprehensive knowledge of and experience with, local conditions. Information transfer will consist of best management practices under existing knowledge conditions. This will continue to be facilitated by the broad based partnerships that have been established as a result of the Model Forests Program. As existing knowledge will continue to accumulate and improve so will best management practices forcing a situation of adaptive management that will move us closer and closer to true sustainability of all forest values. The Model Forest Program appears to be the best national and international effort to provide integration of the accumulated knowledge and its transfer to managers of commercial forests. It has fostered, as examples, the use and development of advanced forest inventory approaches and more sophisticated software for decision support. Also it has ensured that interests other than those traditionally associated with forest management have been heard. Key advantages of the Model Forests are the partnerships that have been formed and the 'working forest' nature of each Forest. The Model Forests transcend a wide variety of forest types, interests, and approaches to use with sustainability as the objective. However, sustainability ultimately must be based on consideration and integration of economic, environmental and social values. The nature of the precise interaction of these values also varies by region and site. Research that will allow us to determine and quantify these values and model their interactions is key to successful sustainable management of forests."

Yet this grand experiment that is the Model Forest Network must continue over commercial and natural rotational time frames and the history of human commitment to such long-term initiatives is not good. Instead there is the possibility that international standards for sustainable forest management based on inadequate existing data bases will supplant the effort to generate those critical data over the life of the commercial forest. There also are pressures to maximize yield from the land base ignoring the lessons of intensive agriculture. There are pressures to lock up land in single uses when truly sustainable forest, or park, or wilderness management suggests that multiple uses should be considered. Politics and human interests often blunt the enthusiasm, good will, and commitment of many in the battle to rationalize our use of the planet's forests. The battle is part of the larger war between aesthetics and exploitation, short-term and long-term economics, sustainability and ecological collapse. The battles must continue but one wonders why there are so few middle grounds sought and when they are achieved as in the Northwest Forest Plan, they continue to be fought over long after the compromise has taken place.

Other examples could have been used to illustrate the same ecological realities of declining and increasingly fragmented wildlife habitat—the loss of bottomland hardwoods in the Mississippi River, the loss of wetland habitat, or the loss of short grass prairie. Each situation suggests that at some stage the remaining habitat will no longer be able to support given wildlife species because of size or distance considerations. When habitats of sufficient size are no longer available and new ones cannot be reached, those species dependent upon them will inevitably become extinct and biodiversity will decline. It is merely a question of time.

Habitat Management—Can It Be Achieved?

Activities posing the greatest threat to wildlife habitat are development (urban, industrial, utilities, mining), agriculture, and forestry. Wildlife management needs to be integrated with these enterprises if we are to maximize the potential of other land uses to produce wildlife habitat or at least to mitigate the impacts. Better communication between government agencies is a first step toward consideration of wildlife values outside of wildlife departments. Interagency panels or committees to explore policies and programs that impinge on wildlife habitat and to whom wildlife managers can express their concerns will also help alleviate the situation. The mere existence of such committees, however, only permits communication, and action is often dependent on the individuals involved. The economic and political components of greatest value often determine what policy or action is ultimately followed. For example, Canadian federal and provincial governments offer incentives to farmers to drain wetland areas, then through other department or ministries sometimes offer incentives to maintain wetlands as wildlife habitat. Such counterproductive actions could be prevented if a coordinated interagency program were developed to provide some means of giving priority to wetlands based on wildlife values. Drainage assistance would then be allowed only for those sites of low value for wildlife and high value for agriculture. The political reality is that agricultural interests have the greatest clout and wildlife habitat suffers as a result. There is hope that attempts to establish a sustainable land management ethic in Canadian agriculture will overcome the decades of countervailing forces.

There often is no meaningful attempt even to set up committee(s) to address the issue(s). The new push toward sustainable agriculture in both countries offers promise for more environmentally sound land management practices in the agriculture sector. Perhaps the abuse of the land base seen this past century will be stopped—we can only hope that new government initiatives will be successful.

Wildlife managers at times have been their own worst enemies. The Ontario Chapter of the Canadian Society of Environmental Biologists (formerly Canadian Society of Wildlife and Fishery Biologists) in the mid-1970s had a committee develop a brief on agricultural land-use effects on wildlife for submission to the provincial government. That committee included representatives from a number of pertinent ministries including Natural Resources (OMNR), Agriculture and Food (OMAF), and Environment, plus the academic community. The brief, when completed, made a number of policy recommendations for implementation by both OMNR and OMAF. The executive, however, refused to transmit it to the government for fear it would be too controversial. Perhaps the effort was ahead of its time politically but by shelving it professional momentum was lost, influence never gained, and action never taken until almost two decades later. Too often the litany of wildlife "professionalism" has been just such inaction. As a profession, it is our responsibility to speak out clearly and strongly on issues which affect wildlife and the environment. To not do so is to abrogate our responsibility. An encouraging sign is the increased advocacy role that professional organizations like The Wildlife Society are taking.

Gravel extraction and other mining activities usually can be turned to advantage by rehabilitation that would enhance wildlife habitat. There are now numerous examples in Europe and North America where agriculture, forestry, and wildlife habitat have been successful land uses after rehabilitation of mining sites. Again, this demonstrates the need for negotiation and involvement in land-use planning by wildlife managers. Adequate habitat inventories are strong weapons in such deliberations because they mean the quantifiable information is available. When it is possible to document the losses occurring and the relative importance of categorical accruals, potent ammunition is added to wildlife habitat advocacy. When this can be translated to losses in animal numbers, economic values can be assigned; and when economic values are involved, wildlife managers are talking in terms that politicians can understand.

The need to preserve habitat to sustain wildlife populations has led Canadian governments (A Wildlife Policy for Canada, 1990) to consider the following actions:

1. Broaden the definition of wildlife to include any species of wild organism.
2. Provide for the conservation of biodiversity in policies and legislation on resources and the environment.
3. Ensure that the concept of wildlife in all policies and legislation includes wildlife habitat.

In relation to wildlife habitat, the following were to be considered by governments and/or aboriginal peoples and NGOs:

1. The maintenance of diversity and the distribution of wildlife habitats.
2. The monitoring and assessment of ecosystems and wildlife populations.
3. The assessment of capabilities of habitats to support wildlife.
4. The assessment of the effects of changing land use, pollution and introduced species on habitats.
5. The determination of gains and losses of habitats.
6. The monitoring of habitat conservation measures.
7. The encouragement of wildlife habitat enhancement activities.
8. The prevention of pollutant discharges that threaten ecosystems.
9. The enforcement of habitat protection laws.
10. The completion and maintenance of comprehensive systems of protected areas representative of ecological types.
11. The adoption of a sound land ethic.
12. The conservation of habitats on public lands.
13. The employment of subsidies, tax credits and other measures to encourage habitat conservation on private lands.
14. The promotion of protected areas on lands managed by aboriginal peoples

The Canadian Wildlife Policy Guidelines (1982) provided a blueprint for the development of meaningful policy; they did not ensure it. But from the 1982 guidelines beginning and continuing through the 1990 policy development, several provinces and NGOs worked to develop parallel policies and strategies appropriate to their particular jurisdictions or within the area of influence of their organizations (e.g. "Wildlife, A New Policy for Nova Scotia, 1987"; "Looking Ahead: A Wildlife Strategy for Ontario, 1991"; "Living with Wildlife, A Strategy for Nova Scotia, 1993"). All were strongly habitat oriented. Still, provincial policies and strategies that have been developed "gather dust on shelves" as one Deputy Minister told us. Policy implementation through action plans requires interpersonal commmunication and mobilization of all interested parties in both bureaucratic and political arenas. In their absence, the wildlife manager works in a vacuum and fails to achieve meaningful results.

As another outgrowth of the wildlife policy guidelines, the Canadian cabinet created Wildlife Habitat Canada (WHC) in 1984 to be an agent of policy implementation, particularly as it relates to habitat acquisition. This represents the only federal agreement involving private agencies such as Ducks Unlimited and the Canadian Wildlife Federation, in addition to provincial jurisdictions. This agency has the major action role in assessing and developing wildlife habitat in Canada. It has had success in establishing conservation partnerships and has initiated a Forest Biodiversity Program and has activities underway in forested, coastal, northern and urban landscapes. In 1996–97, WHC supported forty-six projects across Canada.

The Lakeshore Capacity Study

One illustration of an interagency approach to land-use planning was an effort by the province of Ontario managed through the Ministry of Housing's local planning policy branch. This interministerial study involved the Ministries of Housing, Environment, and Natural Resources. Wildlife was one component in an integrated study designed to measure the capacity of land and lake systems in central Ontario to support cottage development. The wildlife model was to predict quantitatively the impact of proposed development on wildlife and wildlife habitat. Songbirds, loons, raptors, deer, small mammals, mink, reptiles, and amphibians were all investigated as potentially sensitive indicators of change brought about by development of lakeshore areas. The disturbances associated with development include direct habitat loss or change due to clearings, buildings, pathways, and roads, and the indirect effects of noise, wave action (boats), and the presence of humans and their pets. Measures of disturbance were necessary so that cause-effect relationships could be ascertained and the model realistically would predict the effect of different types of development on wildlife populations and habitat. The most apparent effect of development is alteration of vegetation (species composition and structure). By quantifying vegetational change associated with development and measuring differences in songbird and small mammal populations in relation to these environmental changes, a measure of impact was derived. LAKELIFE was the computer information system developed to provide that evaluation. The output predicts the changes in the existing wildlife community that will occur as a result of any particular development proposal. It is a powerful planning tool and when used in an iterative fashion can produce a development with an acceptable level of impact on wildlife communities. The primary inputs are total length of shoreline, length of developed shoreline, number of cottage lots, area of the lake, habitat types (by shoreline segment—usually 50 or 100 m) and average lot size (frontage × 50 m), segments which include deer wintering areas, littoral zone habitat type (by segment), offshore and onshore loon nesting sites (number by segment), number and location of suitable hawk nesting areas, number and location of suitable turtle nesting sites, and total number of streams (by segment). Impact values on small mammals, loons, mink, deer, streams, hawks, songbirds, and five fish species are then derived and used to evaluate whether development should occur and in what segments it should or should not take place. Lot size or location can be altered to reduce the impact. Two basic criteria were used to determine the acceptability of cottage development.

1. Wildlife populations were to be maintained in self-sustaining communities that were as similar as possible to undisturbed shoreline communities, at least on some portion of the lake.
2. No species was to be extirpated from the shoreline community of any lake.

Unfortunately, the provincial government never implemented the findings of the study. Instead it was left to municipal and county planners to capitalize on the results. More comprehensive studies of this type are needed if we are to achieve the necessary predictive capability to demonstrate the effects of habitat alteration and land-use decisions on wildlife populations. The agency controlling the policies for development or exploitation must then regard wildlife as a component to be considered along with water quality, aesthetics, board feet of timber, or agriculture (soils). Often this only occurs when wildlife managers are vociferous enough to ensure that their resource gets equal consideration. Even persuasive wildlife managers can be thwarted by weak ministers, deputy ministers, assistant deputy ministers, game or wildlife commissioners or directors. Advocacy must transcend the organization and must be politically acceptable within the agency before it can be politically successful outside of it.

Other Approaches

In Washington state, a consensus based mechanism has been adopted to avoid increasing the legislative restrictions of the Forest Practices Act. The Timber, Fish, and Wildlife (TFW) agreement facilitates timber management practices which are environmentally sound and protective of wildlife values. Industry, state and federal agencies, tribal groups, conservation organizations, and universities are all represented on the coordinating board of TFW. This approach to habitat management is an appropriate model to emulate as it obviates the confrontational mode of conflict resolution.

The USFWS, through the Habitat Evaluation Procedures Group, has been producing a series of habitat suitability index (HSI) models. The HSIs are designed for use in planning decisions and represent efforts to work with the species' existing biological data bases to provide information on what constitutes optimum habitat. Each model synthesizes the species' habitat use information in such a way that index values between 0.0 (unsuitable habitat) and 1.0 (optimum habitat) can be generated for given environmental factors. Generally only those factors which can be related quantitatively to habitat suitability are selected for the model. Although not perfect, the models are intended to be updated as a result of their application in the field. One major failing of the HISs is the assumption that all relationships between the suitability indices and the environmental factors are linear. Whether this developed as a convention based on the first models produced or whether it is used as a means of simplifying more complex relationships, it is an inappropriate response to "real world" situations. Despite this criticism and problems with applying the models across broad geographical ranges, the HIS is a serious attempt to quantify the important relationships between wildlife and the environments which support it. In addition to the species models, guild and layers of habitat models have also been developed. They allow consideration of broader ecological questions in the planning process.

If we are to be successful in realizing effective habitat management in future years, we must become serious about the preservation of biological diversity and the implementation of sustainable management practices. The preservation of ecosystem diversity and hence the systems' component species and their genetic diversity translates for the wildlife manager into habitat and type diversity allowing communities of species to flourish. Protection of land alone will be insufficient to achieve habitat preservation and sustainable land management. The great majority of the land area will be managed to support the burgeoning human population. It will be how effective we are in managing the residual land base for biodiversity and sustainability from a wildlife perspective and how effective we are in influencing sustainability initiatives on the other lands that will determine the future of wildlife species' habitats. The great challenge then is to implement the policies, plans, and activities that actually will result in sustainability and the maintenance of biodiversity. This means closing the gap between implied commitment and practice on all lands, public or private.

Bibliography

Arcese, P. and A. R. E. Sinclair. 1997. The role of protected areas as ecological baselines. J. Wildl. Manage. 61:587–602.

Avery, M. L. and C. V. Riper III. 1990. Evaluation of wildlife-habitat relationships data base for predicting bird community composition in central California chaparrel and blue oak woodlands. Calif. Fish Game 76:103–117.

Baskett, T. S., D. A. Darrow, D. L. Hallett, M. J. Armbruster, J. A. Ellis, B. F. Sparrowe, and P. A. Korte. 1980. A handbook for terrestrial-habitat evaluation in central Missouri. U.S. Fish Wildl. Serv. Resour. Publ. 133.

Bormann, B. T., M. H. Brookes, E. D. Ford, A. R. Kiester, C. D. Oliver, and J. F. Weigand. 1994. Eastside forest ecosystem health assessment. Vol. 5: A framework for sustainable-ecosystem management. USDA For. Serv. Gen. Tech. Rep., PNW-GTR-331.

Bormann, B. T., P. G. Cunningham, M. H. Brookes, V. W. Manning, and M. W. Collopy. 1994. Adaptive ecosystem management in the Pacific Northwest. USDA For. Serv. Gen. Tech. Rep., PNW-GTR-341.

Bunnell, F. L. and L. L. Kremsater. 1993. Tactics for maintaining biodiversity in forested ecosystems. Proc. IUGB XXI Congr., 1:62–72.

Canadian Forest Service. 1997. *Biodiversity in the Forest. Implementing the Canadian Biodiversity Strategy.* Ottawa, Ontario.

Cannon, R. W., F. L. Knopf, and L. R. Pettinger. 1982. Use of Landsat data to evaluate lesser prairie-chicken habitats in Western Oklahoma. J. Wildl. Manage. 46:915–922.

Clark, K., D. Euler, and E. Armstrong. 1983. Habitat association of breeding birds in cottaged and natural areas of central Ontario. Wilson Bull. 95:77–96.

Cringan, A. T. 1957. History, food habits and range requirements of the woodland caribou of continental North America. Trans. N. Am. Wildl. Conf. 22:487–501.

Crowell, J. B., Jr. 1982. Resource management thrusts and opportunities: fish and wildlife-a fuller dimension to improved resource management. Trans. N. Am. Wildl. Nat. Resour. Conf. 47:17–22.

DeGraff R. M. and N. G. Tilghman. 1980. Workshop proceedings: Management of western forests and grasslands for nongame birds. U.S.D.A. For. Serv. Gen. Tech. Rep. INT-86.

DeGraff, R. M. and W. M. Healy. 1990. Is forest fragmentation a management issue in the Northeast? USDA Gen. Tech. Rep. NE-140.

Dixon, R.S. 1981. Vegetation mapping the barren ground caribou winter range in northern Manitoba using Landsat. Manitoba Surveys and Mapping Branch, Remote Sensing Centre, Dept. Nat. Resour. TR 81-1.

Dodds, D. G. 1994. Toward sustainable forestry in Canada. For. Chron. 70:538–542.

Echelberger, H. E., A. E. Luloff, and F. E. Schmidt. 1991. Northern forest lands: resident attitudes and resource use. USDA For. Serv. NE. For. Exper. Stn., Res. Pap. NE-653.

Edwards, T. C., Jr. and J. M. Scott. 1993. Use of gap analysis as a tool for the management of biodiversity and resource use. Proc. IUGB XXI Congr.,1:73–81.

Edwards, R.Y. and R.W. Ritcey. 1960. Foods of caribou in Wells Gray Park, British Columbia. Can. Field.-Nat. 74:3–7.

Ellis, J. A., J. N. Burroughs, M. J. Armbruster, D. L. Hallett, P. A. Korte, and T. S. Baskett. 1978. Results of testing four methods of habitat evaluation. Report to Proj. Impact Evaluation Team, Div. Ecol. Serv., U.S. Fish Wildl. Serv., Ft. Collins, CO.

Franklin, J. F. and R. H. Waring. 1980. Distinctive features of the northwestern coniferous forest development, structure and function. Proc. Annu. Biol. Colloq. 40:58–86.

Gosz, J. R. 1992. Sustainable forest ecosystem management: interpretations from the sustainable biosphere initiative. School of For., N. Arizona Univ., Flagstaff, 16th W. P. Thompson Mem. Lecture Ser.

Graber, J. W. and R. R. Graber. 1976. Environmental evaluations using birds and their habitats. Biol. Notes No. 97. Illinois Nat. Hist. Survey.

Gutierrez, R. J. and A. B. Carey (eds.). 1985. Ecology and management of the spotted owl in the Pacific northwest. USDA For. Serv. Gen. Tech. Rep. PNW-185.

Hagis, W. and W. Young. 1983. Methods of preserving wildlife habitat. Lands Directorate, Envir. Canada Working Paper No. 25.

Hanski, I. and D. Simberloff. 1997. The metapopulation approach, its history, conceptual domain, and application to conservation. In I. Hanski and M. Gilpin (eds.). Metapopulation Biology: Ecology, Genetics, and Evolution, pp. 5–26. Academic Press, San Diego, CA.

Harding, L. E. and E. McCullum (eds.). 1994. *Biodiversity in British Columbia: Our Changing Environment.* Envir. Can., Can. Wildl. Serv., Ottawa, Ontario.

Haufler, J. B. and L. L. Irwin. 1993. An ecological basis for forest planning for biodiversity and resource use. Proc. IUGB XXI Congr.1:73–81.

Heimer, M. 1975. Bergkamen communal tip: extracts from the explanatory report to the landscape plan. Landscape Planning 2:249–264.

Holthausen, R. S., M. G. Raphael, K. S. McKelvey, E. D. Forsman, E. E. Starkey, and D. E. Seaman. 1995. The contribution of federal and nonfederal habitat to persistence of the northern forest owl on the Olympic Peninsula, Washington: report of the reanalysis team. USDA For. Serv. Gen. Tech. Rep., PNW-GTR-352.

Huff, M. H., S. E. McDonald, and H. Gucinski. 1992. Expanding horizons of forest ecosystem management: Proceedings of the 3rd habitat futures workshop. USDA For. Serv. Gen. Tech. Rep. PNW-GTR-336.

Interagency SEIS Team. 1993. *Forest Ecosystem Management: An Ecological, Economic, and Social assessment. Report of the Forest Ecosystem Management Assessment Team.* Portland, OR.

Isaacson, D. L., D. A. Leckenby, and C. J. Alexander. 1982. The use of large-scale aerial photography for interpreting digital data in an elk habitat-analysis project. J. Appl. Photo. Eng. 8:51–57.

Jensen, M. E. and P.S. Bourgeron (eds.). 1994. Eastside forest ecosystem health assessment. Vol. II: Ecosystem management: principles and applications. USDA For. Serv. Gen. Tech. Rep. PNW-GTR-318.

Kimmins, J. P. 1997. Biodiversity and its relationship to ecosystem health and integrity. For. Chron. 73:229–232.

Kaufmann, M. R., R. T. Graham, D. A. Boyce Jr., W. H. Moir, L. Perry, R. T. Reynolds, R. L. Bassett, P. Mehlhop, C. B. Edminister, W. M. Block, and P. S. Corn. 1994. An ecological basis for ecosystem management. USDA For. Serv. Gen. Tech. Rep. RM-246.

Keating, M. 1989. *Toward a Common Future. A Report on Sustainable Development and Its Implications for Canada.* Env. Can., Ottawa, Ontario.

Larson, F. R. 1992. Downed woody material in southeast Alaska forest stands. USDA For. Serv. Res. Pap., PNW-Rep-452.

Leckenby, D. A., D. L. Isaacson, and S. R. Thomas. 1985. Landsat application to elk habitat management in northeast Oregon. Wildl. Soc. Bull. 13:130–134.

Luckert, M. K. 1997. Towards a tenure policy framework for sustainable forest management in Canada. For. Chron. 73:211–215.

Lynch, J. F. and R. F. Whitcomb. 1978. Effects of the insularization of the eastern deciduous forest on avifaunal diversity and turnover. In: *Classification, Inventory and Analysis of Fish and Wildlife Habitat,* pp. 461–489, U.S. Govt. Printing Off., Washington, D.C.

MacArthur, R. H. and E. O. Wilson. 1967. *The Theory of Island biogeography.* Princeton Univ. Press, Princeton, NJ.

Maki, W. R. and D. C. Olson. 1991. Economic and social impacts of preserving ancient forests in the Pacific Northwest. Am. For. Resour. Alliance, Washington, DC., Tech. Bull. No. 91-07.

Maser, C., J. M. Trappe, and D. C. Ure. 1978. Implications of small mammal mycophagy to the management of western coniferous forests. Trans. N. Am. Wildl. Nat. Resour. Conf. 43:78–88.

Maser, C., J. W. Thomas, I. D. Luman, and R. Anderson. 1979. Wildlife habitats in managed rangelands-the Great Basin of southeastern Oregon manmade habitats. U.S.D.A. For. Serv. Gen. Tech. Rep. PNW-86.

McLain, K. M. and C. P. McLary. 1995. Exploring multiple use and ecosystem management: from policy to operational practices. Proc. FAO/ECE/ILO Int. For. Sem., Prince George, British Columbia.

Natural Resources Canada. 1997. *Safeguarding Our Assets. Securing Our Future.* Ottawa, Ontario.

Niemi, G. J., J. M. Hanowski, A. R. Lima, T. Nicholls, and N. Weiland. 1997. A critical analysis on the use of indicator species in management. J. Wildl. Manage. 61:1240–1252.

Nova Scotia Department of Natural Resources. 1987. *Wildlife. A New Policy for Nova Scotia.* Halifax, Nova Scotia.

Nova Scotia Federation of Agriculture. 1996. *Nova Scotia Environmental Farm Plan.* Truro, Nova Scotia.

Nova Scotia Department of Natural Resources. 1997. *Toward Sustainable Forestry.* Halifax, Nova Scotia.

Oliver, C. D., L. L. Irwin, and W. H. Knapp. 1994. Eastside forest management practices: historical overview, extent of their applications, and their effects on sustainability of ecosystems. USDA For. Serv. Gen. Tech. Rep., PNW-GTR-324.

Ontario Ministry of Natural Resources. 1991. *Looking Ahead: A Wildlife Strategy for Ontario.* Toronto, Ontario.

Quigley, T. M. 1992. Forest health in the Blue Mountains: social and economic perspectives. USDA For. Serv. Gen. Tech. Rep., PNW-GTR-296.

Quigley, T. M., R. W. Haynes, and R. T. Graham (eds.) 1996. Integrated scientific assessment for ecosystem management in the interior Columbia Basin and portions of the Klamath and Great Basins. USDA For. Serv. Gen.Tech. Rep., PNW-GTR-382.

Racey, G. D. and D. L. Euler. 1982. Small mammal and habitat response to shoreline cottage development in central Ontario. Can. J. Zool. 60:865–880.

Racey, G. and D. Euler. 1983. An index of habitat disturbance for lakeshore cottage development. J. Environ. Manage. 16:173–179.

Racey, G. D., T. P. Clark, J. A. McDonnell, and D. L. Euler. 1981. LAKELIFE user's manual. A lake planner's guide to the assessment of impact on wildlife and fish habitat. Wildlife Component Lakeshore Capacity Study Part II. Ontario Min. Nat. Resour. (Nov)

Racey, G. D., T. S. Whitfield, and R. A. Sims. 1989. Northwestern Ontario forest ecosystem interpretations. Ontario Min. Nat. Resour., NWOFTDU Tech. Rep. 46.

Robbins, C. S. 1988. Forest fragmentation and its effects on birds. In: T. E. Johnson (ed), *Managing North Central Forests for Non-timber Values*, pp. 61–65. Soc. Am. For. Pub. No 88-04.

Robinson, S. 1997. Nest gains, nest losses. Nat. Hist. 105:40–47.

Ryder, J. P. and D. A. Boag. 1981. A Canadian paradox-private land, public wildlife: can it be resolved? Can. Field.-Nat. 95:35–38.

Samson, F. B. 1980. Island biogeography and the conservation of non-game birds. Trans. N. Am. Wildl. Nat. Resour. Conf. 45:245–251.

Samson, F. B. and F. L. Knopf. 1993. Managing biodiversity. Wildl. Soc. Bull. 21:509–514.

Schoen, J. W., O. C. Wallmo, and M. D. Kirchhoff. 1981. Wildlife forest relationships: is a re-evaluation of old-growth necessary? Trans. N. Am. Wildl. Nat. Resour. Conf. 46:531–544.

Sims, R. A., W. D. Towill, K. A. Baldwin, and G. M. Wickware. 1989. Field guide to the forest ecosystem classification for northwestern Ontario. Ontario Min. Nat. Resour., NOFTDU.

Society of American Foresters. 1993. *Task Force Report on Sustaining Long-term Forest Health and Productivity*. Bethesda, MD.

Svedarsky, W. D. and R. D. Crawford. 1982. Wildlife values of gravel pits. Symposium proceedings. Univ. Minn. Agric. Exper. Stn. Misc. Publ. 17–1982.

Thomas, J. W. 1979. Wildlife habitats in managed forests. The Blue Mountains of Oregon and Washington. USDA For. Serv. Agric. Handbook. No. 553.

Thomas, J. W. 1985. Toward the managed forest-going places that we've never been. Wildl. Soc. Bull. 13:197–201.

Thomas, J. W., C. Maser, and J. E. Rodiek. 1979. Wildlife habitats in managed rangeland-Great Basin of southeastern Oregon-riparian zones. USDA For. Serv. Gen. Tech. Rep. PNW-80.

Thomas, J. W., H. Black, Jr., R. J. Scherzinger, and R. J. Pederson. 1979. Deer and elk. In: J. W. Thomas (ed.), *Wildlife Habitats in Managed Forests, the Blue Mountains of Oregon and Washington*, pp. 104–127. USDA For. Serv. Agric. Handbook. No. 553.

Thomas, J. W., E. D. Forsman, J. B. Lint, E. C. Meslow, B. R. Noon and J. Verner. 1990. *A Conservation Strategy for the Northern Spotted Owl*. Rep. Interagency Sci. Comm., Portland, OR.

Thomas, J. W., M. G. Raphael, R. G. Anthony, E. D. Forsman, A. G. Gunderson, R. S. Holthausen, B. G. Marcot, G. H. Reeves, J. R. Sedell, and D. M. Solis. 1993. *Viability Assessments and Management Considerations for Species Associated with Late-successional and Old-growth Forests of the Pacific Northwest.* USDA Natl. For. Sys., For. Serv. Res., Washington, D.C.

Thompson, S. and A. Webb (eds.). 1994. *Forest Round Table on Sustainable Development.* Natl. Round Table on the Envir. and Econ., Ottawa, Ontario.

Tuchmann, E. T., K. P. Connaughton, L. E. Freedman, and C. B. Moriwaki. 1996. The Northwest Forest Plan. A report to the President and Congress. USDA For. Serv., PNW Res. Stn.

Verner, J. and A. S. Boss. 1981. California wildlife and their habitats: western Sierra Nevada. USDA For. Serv. Gen. Tech. Rep. PSW-37.

Whitaker, G. A. and F. H. McCuen. 1976. A proposed methodology for assessing the quality of wildlife habitat. Ecol. Model. 2:251–272.

Whitcomb, B. L., R. F. Whitcomb, and D. Bystrak. 1977. III Longterm turnover and effects of selective logging on the avifauna of forest fragments. Am. Birds 31:17–23.

Wickman, B. E. 1992. Forest health in the Blue Mountains: the influence of insects and disease. USDA For. Serv. Gen. Tech. Rep., PNW-GTR-295.

Wildlife Habitat Canada. 1997. *Forest Biodiversity Program.* Ottawa, Ontario.

William, G. L. 1988. An Assessment of HEP (Habitat Evaluation Procedures) applications to Bureau of Reclamation projects. Wildl. Soc. Bull.16:437–447.

Yeo, J. J. and J. M. Peek. 1992. Habitat selection by female Sitka black-tailed deer in logged forests of southeastern Alaska. J. Wildl. Manage. 56:253–261.

Recommended Readings

Bunnell, F. L. 1989. Alchemy and uncertainty: what good are models? USDA For. Serv., Gen-Tech. Rep. PNW-GTR-232. A thorough examination of why wildlife-habitat models often fail and what must be done to make them work.

Canadian Symposia on Remote Sensing. These proceedings contain a number of papers relevant to classification of wildlife habitats using remote sensing techniques.

Committee on Agricultural Land Use and Wildlife Resources. 1970. Land use and wildlife resources. Nat. Acad. Sci., Washington, D.C. Although concentrating on agriculture-wildlife inter-relations, this book not only explores the problem areas but also looks at mechanisms for resolving the points of conflict.

Committee on the Role of Alternative Farming methods in Modern Production Agriculture. 1989. Alternative agriculture Nat. Res. Council., Natl. Academic Press. Washington, D.C. Agriculture is the largest nonpoint source of water pollution, antibiotic, and pesticide residues found in food, Also, soil erosion, soil salinization, and aquifer depletion are continuing problems associated with agriculture. Because of these factors this report was designed to examine alternative production systems which would reduce the adverse environmental effects. The report summarizes the economic and scientific viability of alternative production systems.

Harris, L. D. 1984. *The Fragmented Forest. Island Biogeographic Theory and the Preservation of Biotic Diversity.* Univ. Chicago Press, Chicago, IL. A thorough discussion of the old-growth forest issue with proposals for management of the remnant stands based primarily on the situation in the Pacific Northwest.

Hoover, R. L. and D. L. Wills (eds.). 1984. *Managing Forested Lands for Wildlife.* Colorado Div. Wild. in cooperation with USDA For. Serv., Rocky Mtn. Reg., Denver, CO. One state's attempt to provide information for managing forested lands using silvicultural practices which would improve wildlife habitat.

Ruggiero, L. F., K. B. Aubry, A. B. Carey, and M. H. Huff. 1990. Wildlife and regulation of unmanaged Douglas-Fir Forests. USDA For. Serv. Gen. Tech Res. PNW-GTR. Presentation of community studies in unmanaged Douglas-Fir forests of various ages in the western United States.

8 SPECIES MANAGEMENT

Although the concept of this book has been predicated on a general description of management procedures as they apply to wildlife, this chapter will take a more in-depth look at management at the species level. Several examples representing different groups of wildlife will be given and the management techniques and political problems associated with each outlined in some detail.

Ungulates—Caribou (*Rangifer tarandus*)

Barren-ground and woodland caribou both occur in North America. Most research has been concentrated on the former (by the Canadian Wildlife Service (CWS) in the Northwest Territories and the Alaska Fish and Game Department in Alaska). More recently woodland caribou populations have received increased attention for several reasons. The woodland caribou of the Ungava region of eastern Canada (George River herd) have been increasing at a rapid rate. In contrast, populations elsewhere, in Ontario, Alberta and British Columbia appear to be declining and the Selkirk herd in northern Idaho and Washington and southeastern British Columbia is considered endangered in the United States. In Newfoundland, where woodland caribou have traditionally been a major big game species, studies have assessed the impacts of hydroelectric development, predation, and hunting on the species. Although woodland caribou, and particularly the mountain form, currently offer interesting areas of investigation, this discussion will concentrate on the barren-ground species.

Evidence from aerial surveys in the 1940s and 1950s suggested a substantive and rapid decline in caribou numbers was occurring throughout the Canadian north. The primary cause of the decline was hunting, with harvest and crippling loss equaling or exceeding annual recruitment. However, for sociological, political, and logistical reasons, it was not practical to control the human kill of caribou. Although regulation seemed reasonable in light of the inventory data, inadequate enforcement meant that the traditional disregard of the law by native and white hunters, which resulted in excessive, wasteful killing of the caribou, could not be stopped. Treaty Indians were not required to obey hunting regulations and some native peoples were truly dependent on caribou meat for sustenance and clothing.

As caribou numbers continued to decline rapidly, what was previously a caribou economy had by 1955 reached a point where alternate economic strategies were necessary to prevent starvation of some caribou-dependent people. This occurred despite attempts to curb caribou mortality by hunting (elimination of sport hunting) and predators (estimated annual loss to wolves was 5 percent). Poison baits (alkaloidal strychnine and sometimes 1080 in frozen meat) to kill wolves were placed primarily in caribou wintering areas.

As predator control continued into the 1960s, the rate of decline of the herds slowed and populations appeared to stabilize; by the mid-sixties there were unsubstantiated claims that the caribou herds were growing rapidly. The political pressure generated by these reports caused the Northwest Territories Game Management Service to relax its restrictions on caribou hunting by white residents, and commercial exploitation of the herds followed further liberalization in 1968. A 1967–1968 caribou survey showed, however, that no substantial population increase had occurred since the 1955 census and the liberalized hunting was not justified. This problem of survey accuracy and the incomparability of results plagued caribou managers into the 1980s.

The Kaminuriak herd has received particular attention from biologists and managers. It winters in northern Saskatchewan and Manitoba and summers primarily in the Northwest Territories and transcends three different jurisdictions (four, if the federal government's interest is included). Studies on the biology of these caribou have shown that calf mortality seems to be the primary limiting factor to population growth. Causes of high calf mortality include poor condition of the mother (inadequate milk production), hypothermia (bad weather conditions during the calving period and shortly thereafter), and predation (wolves). A caribou working group (Caribou Management Board) involving representatives of all interested parties (provincial, territorial, and federal governments, plus native peoples) has been working out a management strategy for the Kaminuriak and Beverly herds since 1982. The credibility of government managers who were trying to reduce harvest by use of quotas was severely weakened in 1982 when an improved census showed a population 40 percent above the predicted value. The census findings were duplicated in 1983 showing a population of 100,000–140,000 animals compared to earlier estimates in the 40,000 animal range. Possible reasons for the dramatic increase could relate to the survey technique, immigration into the Kaminuriak herd from another herd, missed calving grounds in previous censuses, or the return of a segment of the Kaminuriak herd which had altered its migration patterns a decade earlier. Whatever the reason(s) for the increase, it left managers in the difficult position of explaining the policy changes necessitated by such a different data base. Nonetheless, a formal agreement was signed in 1983 between the respective governments (Manitoba, Saskatchewan, Northwest Territories, and the federal government represented by the CWS and the Department of Indian and Northern Affairs) and the native peoples. A thirteen-person board with eight native people supplementing five government representatives makes management recommendations for the caribou. This Beverly and Kaminuriak Caribou Management Board and its authorized publication, *Caribou News,* have succeeded in opening discussion on caribou management and establishing mutually agreed upon management strategies. In addition to a detailed long-term management plan, the Board has developed an educational program for use in community schools throughout the caribou range, and set up a scholarship fund for postsecondary students studying caribou management. The Board also recommended that the Northwest Territories government set commercial quotas for the sale of caribou meat as a subsistence food resource. The Board has strongly opposed mining development and other activities which would disturb caribou habitat.

The Alaskan herds also declined by ≥50 percent from the mid-60s to mid-70s with severe winter weather and overhunting the primary factors with predator-caused calf mortality also being important. Insufficient recruitment occurred to replace the adult female losses caused by hunting. Caribou managers in Alaska were unsuccessful, according to Bergerud (1978), because they believed that large herds could not be overharvested; they underestimated predator caused calf mortality, and they thought that caribou populations had to be hunted heavily to prevent overgrazing of the range. Alaska began a program of predator control, opting to reduce calf and adult mortality due to predators rather than substantially reducing harvest. Although popular with Alaskan hunters, this management approach has brought severe criticism from the public in the lower forty-eight states.

Predator control, particularly on a selective basis appears to be increasingly used by big game managers as they realize they are dealing with additive rather than compensatory mortality. The agencies' clientele is the hunting fraternity, and restriction of predator numbers is often more expedient and easier to obtain politically, especially in more rural jurisdictions, than restrictions on harvest or hunter numbers. Therefore, Canadian and U.S. caribou managers both support predator control programs. Furthermore, as Boertje et al. (1996) have shown, there is little doubt that predator control is effective in increasing ungulate populations in the short term.

Because caribou move such great distances and the summer and winter use areas are so large, habitat management is pretty much nonexistent. Forest fires on the winter range have been thought to cause a loss of vital habitat and fire suppression to protect lichen rich mature forests has been supported in Alaska and the Northwest Territories. However, Skoog (1968) questioned the dependency of caribou on lichens, as their diet even in winter is seldom more than 50 percent lichens. A recent study for Coats Island showed that unavailability of lichens in the winter diet contributed to caribou mortality especially of calves during serve winter conditions (Adamczewski et al., 1988). If the species does indeed require lichens in the diet, it must be considered that it takes thirty to fifty years for them to reach reasonable biomass levels. In recent years, there has been a shift in thought toward fire management instead of fire suppression. In support of such management action for caribou, Maikawa and Kershaw (1976) concluded that the occurrence of spruce-stereocaulin woodland in the south-central Northwest Territories was dependent on cyclic burning. Without fire the spruce canopy closes and lichens are replaced by mosses. However, the reburn cycle of about one hundred years in the area studied assured a continuance of suitable caribou winter range. Fire is important in maintaining extensive barren-ground caribou range. Kelsall et al. (1977) caution that this type of statement should be qualified. Fires have been so universal in recent times in areas of major human settlement and mining exploration and development that little if any mature forest remains over large geographical areas. Fire management would be appropriate in these latter areas, with natural wildfire allowed to occur only in remote, noncommercial forests. Furthermore, recent studies by Dan Thomas of the CWS shows lichens making up 75–90 percent of a caribou's diet in winter and the heaviest use of regenerated forest habitat occurring 50–250 years after a fire. These findings have significance because they suggest a longer period of optimal winter range for caribou than predicted in earlier studies.

Another area of controversy has been the effect of northern development, especially of oil and gas resources, on caribou populations. Klein (1971), reporting on studies of Scandinavian reindeer, suggested that highways and railroads would not generally create barriers to movement of caribou. The main problem would be mortality due to collisions with motor vehicles and trains. In North America, though, improved access also would mean increased mortality from hunting—legal and illegal. Johnson and Todd (1977) substantiated these generalizations with a study of mountain caribou in southern British Columbia (the Selkirk herd mentioned earlier), where the animals continued to use a traditional movement route despite mortalities from collisions with vehicles and poaching. It is still possible that increased traffic by trains or motor vehicles on these linear routes ultimately will result in decreased use or abandonment of range. It may also split the population into two units which may be more vulnerable because of smaller unit size. Whitten and Cameron

(1983) showed that caribou avoid moving through the Prudhoe Bay industrial area even though such movements were common prior to 1975.

The effect of pipelines is another concern. In northern Alaska, the trans-Alaska pipeline resulted in some documented avoidance by caribou. It now appears that the Porcupine herd has split into two units as suggested above. In contrast, the Nelchina herd continues to cross the pipeline at preconstruction points and that population has increased substantially. Barriers to traditional migration or movement routes can be caused not only by such transportation corridors but also by impoundments resulting from hydroelectric developments. The areas flooded often have been important calving areas, especially in woodland caribou range.

Preliminary work by Miller and Gunn (1979) suggested that harassment of caribou by low flying aircraft might mean added energetic expenses, particularly to cows and calves, and that calving areas should be avoided by such aircraft. Concern that petroleum exploration in the calving grounds of the Porcupine caribou herd might displace the animals and increase calf mortality prompted a study by Whitten et al. (1992) that confirmed that if the caribou were displaced from the coastal plain to higher elevations calf mortality would likely increase.

A management scenario developing from all this is contingent on better knowledge of caribou movement patterns, productivity, mortality factors, energetics, and population size (see Chapter 4). In varying locales it means predator control, increased enforcement to prevent illegal hunting, curtailment of hunting opportunity, cooperation of native peoples—especially treaty groups—and mitigation of impacts resulting from development activities. Caribou is a species primarily restricted to primitive or wilderness areas and must be managed by a modern approach despite being symbolic of a wildness which no longer exists except in the minds of many North American city dwellers. This means that caribou, as the subsistence base of the Inuit and Indian and a talisman of wilderness to the urbanite, generate their own political realities (another recurrent theme in this book) which must be understood and sometimes addressed by the manager. Decisions made to manage caribou in Alaska and the Northwest Territories are scrutinized in San Francisco, Toronto, and other urban centers. Bad press generated by public outcries far removed from the location of the management activities often dictates what managers can or cannot do. Predator control to enhance caribou populations is such an issue. The input of public controversy is probably best illustrated in our next example, where the best of biological data and management schemes ultimately proved ineffective against a concerted public outcry a continent away.

Bibliography

Adamsczewski, J. Z., C. C. Gates, and R. J. Hudson. 1988. Limiting effects of snow on seasonal habitat use and diets of caribou (*Rangifer tarandus groenlandicus*) on Coats Island, Northwest Territories, Canada. Can. J. Zool. 1986–1996.

Bergerud, A. T. 1978. Caribou. In: J. L. Schmidt and D. L. Gilbert (eds.). *Big Game of North America,* pp. 83–101 Stackpole Books, Harrisburg, PA.

Boertje, R. D., W. C. Gasaway, D. V. Grangaard, and D. G. Kelleyhouse. 1988. Predation on moose and caribou by radio-collared grizzly bears in east central Alaska. Can. J. Zool. 66:2492–2499.

Boertje, R. D., P. Valkenburg, and M. E. McNay. 1996. Increases in moose, caribou, and wolves following wolf control in Alaska. J. Wildl. Manage. 60:474–489.

Cameron, R. D. and K. R. Whitten. 1980. Influence of the Trans-Alaska Pipeline corridor on the local distribution of caribou. In: E. Remers, E. Gaare and S. Skjenneberg (eds.), *Proceedings of the Reindeer/Caribou Symposium,* pp. 475–481, Direktoratet for Viltog Ferskvannskisk, Trondheim, Norway.

Carruthers, D. R., R. D. Jakimchuk, and C. Linkswiler. 1984. Spring and fall movements of Nelchina caribou in relation to the trans-Alaska pipeline. Ren. Resour. Consulting Serv. Ltd., Toronto, Ontario.

Dauphine, T. C., Jr. 1976. Biology of the Kaminuriak population of barren-ground caribou. Part 4: Growth, reproduction and energy reserves. Can. Wildl. Serv. Rep. Ser. No. 38.

Gasaway, W. C., R. 0. Stephenson, J. L. Davis, P. K. Shepherd, and 0. E. Burris. 1983. Interrelationships of wolves, prey and man in interior Alaska. Wildl. Monogr. 84.

Johnson, E. A. and J. S. Rowe. 1975. Fire in the subarctic wintering ground of the Beverly caribou herd. Am. Wildl. Nat. 94:1–14.

Johnson, D. R. and M. C. Todd. 1977. Summer use of highway crossing by mountain caribou. Can. Field-Nat. 91:312–314.

Kelsall, J. P. 1968. *The Migratory Barren-ground Caribou of Canada.* Can. Wildl. Serv., Queen's Printer, Ottawa, Ontario.

Kelsall, J. P., E. S. Telfer, and T. D. Wright. 1977. The effects of fire on the ecology of the boreal forest, with particular reference to the Canadian north: a review and selected bibliography. Can. Wildl. Serv. Occ. Pap. No. 32.

Klein, D. R. 1971. Reaction of reindeer to obstructions and disturbances. Science 173:393–398.

Maikawa, E. and K. A. Kershaw. 1976. Studies on lichen-dominated systems. XIX. The postfire recovery sequence of black spruce-lichen woodland in the Abitou Lake region, N.W.T. Can. J. Bot. 54:2679–2687.

Miller, F. L. 1974. Biology of the Kaminuriak population of barren-ground caribou. Part 2: Dentition as an indicator of age and sex; composition and socialization of the population. Can. Wildl. Serv. Rep. Ser. No. 31.

Miller, D. R. 1976. Biology of the Kaminuriak population of barren-ground caribou. Part 3: Taiga winter range relationships and diet. Can. Wildl. Serv. Rep. Ser. No. 36.

Miller, F. L. and A. Gunn. 1979. Responses of Peary caribou and muskoxen to helicopter harassment. Can. Wildl. Serv. Occ. Pap. No. 40.

Parker, G. R. 1972. Biology of the Kaminuriak population of barren-ground caribou. Part 1: Total numbers, mortality, recruitment, and seasonal distribution. Can. Wildl. Serv. Rep. Ser. No. 20.

Scotter, G. W. 1964. Effects of forest fires on the winter range of barren-ground caribou in northern Saskatchewan. Can. Wildl. Serv. Wildl. Manage. Bull. Ser. 1. No. 18.

Scotter, G. W. 1967. Effects of fire on barren-ground caribou and their forest habitat in northern Canada. Trans. N. Am. Wildl. Nat. Resour. Conf. 32:246–254.

Scotter, G. W. 1971a. Fire, vegetation, soil and barren-ground caribou relations in northern Canada. In: *Proceedings-Fire in the Northern Environment, A Symposium,* pp. 209–230. Pacific Northwest For. Range Exper. Stn., Portland, OR.

Scotter, G. W. 1971b. Wildfires in relation to habitat of the barrenground caribou in the taiga of northern Canada. Proc. Annu. Tall Timbers Fire Ecol. Conf. 10:85–106.

Skoog, R. 0. 1968. Ecology of the caribou (*Rangifer tarandus granti*) in Alaska. Unpubl. PhD Thesis, University of California, Berkeley, Berkeley, CA.

Whitten, K. R. and R. D. Cameron. 1983. Movements of collared caribou, *Rangifer tarandus,* in relation to petroleum development on the Arctic slope of Alaska. Can. Field-Nat. 97:143–146.

Whitten, K. R., G. W. Garner, F. J. Mauer, and R. B. Harris. 1992. Productivity and early calf survival in the Porcupine caribou herd. J. Wildl. Manage. 56:201–212.

The Proceedings of the North American Caribou Workshops (1985–present) contain much useful and current information on caribou often not found anywhere else in the literature.

Marine Mammals—Harp Seal (*Pagophilus groenlandicus*)

The "white-coat", or newborn pup, of the harp seal is one of the most familiar visages of wildlife. The large, sad, brown eyes set in a cuddly countenance and the off-white pelt have made this species an attractive symbol of the anticommercialization forces as well as groups like the Animal Welfare League and Defenders of Wildlife. The Canadian government reacted to growing public pressure against the east coast seal hunt in the 1970s by funding considerable research on the harp seal to ensure appropriate quotas were set to maintain populations at a level sustaining an annual harvest of 180,000 animals. To understand the management strategies some background is necessary. There are three separate and distinct breeding stocks of harp seals in the North Atlantic. The largest stock occurs off the east coast of Canada within the Gulf of St. Lawrence and on the Front (off the coast of Labrador). The adult females spend January–March in this area and give birth before returning to the Canadian Arctic and the waters off west Greenland for summer breeding. All three stocks have been exploited commercially for centuries, but concern mounted because harvests of the Gulf and Front stocks during the 1950s and 1960s were responsible for major declines in harp seal numbers. In 1961, the establishment of opening and closing dates was the first regulatory management step to be taken. In 1965, adult females on the whelping patches were protected; Norway stopped sealing in the Gulf and Canada imposed a quota of 50,000 seals on Canadian sealers operating there. By 1970, the stock was only 33–50 percent of the 1951 level. Pressure from humane groups such as the International Fund for Animal Welfare was growing. In 1971, the Canadian government acted by imposing quotas for both the Gulf and Front and establishing an independent Committee on Seals and Sealing (COSS) composed of scientists, veterinarians, and executive members of Canadian and international humane societies. COSS was to examine the economic, sociological, ecological, and humanitarian aspects of the seal hunt and recommend to the government any changes in regulation which might be needed.

The quotas established were done so under the International Commission for the Northwest Atlantic Fisheries (ICNAF) Harp Seal and Hood Seal Protocol, which since 1961 had assumed management responsibilities for the international (Canada, Denmark, and Norway) hunt. The 1971 quotas were 200,000 harp seals to be taken by sealing ships and 45,000 by landsmen. In 1972, the allowable take was sharply reduced to 150,000 animals (120,000 by vessels, 30,000 by landsmen) and remained at this level through 1975 while scientific studies on the actual status of the population were under way. The population model developed projected a population of $\geq$ 1.2 million animals/year which would produce about 320,000 pups in 1977. The model assumed a carrying capacity of about 3.7 million seals with a Maximum Sustained Yield (MSY) of 1.6 million animals. The Sustained Yield (SY) for the 1977 population was projected at 190,000 animals, with a total allowable catch of 170,000 animals suggested to permit the stock to reach the MSY in ten to fifteen years. This was considerably higher than the 1976 quota of 127,000.

Within a year, through scientific agreement that natural mortality values used in the model were too high and should be set at 11 percent, the SY estimates ranged from 227,000 to 245,000 seals. Pup production estimates were 310,000 to 350,000. The Minister of Fisheries set a management strategy for 1978 that would restrict harvest to a maximum 75 percent of SY, thus allowing continued population growth. A quota of 180,000 seals was set for 1978. Although conflicting views from the scientific community continued, there was no dispute that the population should progress toward the MSY level with the target allowable catch value.

Following 1978, Canada took a more independent role in determining seal management and accepted ICNAF recommendations only if they were deemed satisfactory. This stance occurred in conjunction with a political move which extended jurisdictional responsibility to 200 miles offshore instead of the old twelve mile zone.

COSS continued to make recommendations which were generally adopted to upgrade the humaneness of the sealing operations. Although alternatives were examined, the regulation bat or club was found to fulfill the requirements of humane slaughter. Emphasis was placed on educational programs for the sealers and biological studies, particularly of the behavior and energetics of the harp seal.

It appeared that significant progress was being made toward scientific management of harp seals that would make the harvest acceptable to the public. Canadian public opinion began to shift in favor of the hunt in 1980. Then in 1981, ice formed offshore of Prince Edward Island and landsmen there had direct access to the breeding stock for the first time in many years. Although quotas were not exceeded, the gore of the slaughter and the unprofessional attitude of the amateur sealers were viewed nationally and internationally on television. Public opinion was drastically influenced. This event fed the antiseal hunt lobby, giving it added momentum and heightening the European response which eventually led the European Economic Council in 1983 to ban importing of seal skins or products. The 1983 quota was meaningless, for very few sealers bothered to harvest. The primary economic market had disappeared; the commercial nature of the hunt and the economic factors which had previously driven it became apparent to all. The political reality was that all the biological knowledge and reasonable management criteria that had been developed for this species were meaningless when the consuming public deemed the exercise to be cruel and inhumane.

Even without a commercial market, harp seals will still be killed, just as grey seals are, when their activities, directly or indirectly (such as feeding, net fouling) affect fishermen's work. However, secondary markets have been secured and the harp seal is again the focus of a major antiharvest effort by the International Fund for Animals fueled in part by a video provided by sealers who admitted that some of the events had been staged. We doubt that the sealing controversy will be settled in the near future as we are again hearing disputes over population estimates and harvest rates from those involved in the earlier debate.

Bibliography

Herscovici, A. 1985. *Second Nature: The Animal-rights Controversy*. CBC Enterprises, Toronto, Ontario.

Lavigne, D. M. 1978. The harp seal controversy reconsidered. Queen's Quart. 85:377–388.

Lett, P. F., R. K. Mohn, and D. F. Gray. 1979. Density-dependent processes and management strategy for the Northwest Atlantic harp seal population. ICNAF Select. Pap. 5:61–80.

Malouf, A. H. 1986. Seals and sealing in Canada. Report of the Royal Commission (3 Vols.), Can. Govt. Publ., Ottawa, Ontario.

Mercer, M. C. 1977. The seal hunt. Inform. Branch, Dept. Fish Environ., Ottawa, Ontario.

Ronald, K., J. Selley, and P. Healey. 1982. Seals. In: J. A. Chapman and G. A. Feldhamer (eds), *Wild Mammals of North America,* pp. 769–827. John Hopkins Univ. Press, Baltimore, MD.

Furbearers — Beaver (*Castor canadensis*)

The beaver is the primary North American furbearing species in terms of economic importance. Its fur value and the economic impact of flooding and tree damage represent many millions of dollars each year in North America. In fact, its fur value led to extirpation throughout much of North America by 1900. Stringent or total protection of the species was law almost universally by 1915. The focus of early management during this century was to restore populations throughout much of the former range. Beaver were live-trapped from remnant populations and transplanted successfully to many new locations. Beaver now occur over virtually all their historical range. Only two jurisdictions within the beaver's distributional range, Rhode Island and Delaware, do not have trapping seasons for the species.

The resurgence of beaver populations was not without its problems. Many southern states found that considerable economic losses were being sus-

Table 8.1 — Beaver Land Capability Classification System (Slough and Sadleir 1977)

	Class		No. of beaver colony sites per shoreline mile	
No.	Description		Lakes	Streams
1	No biophysical limitations affect beaver production		3+	6+
2	Slight limitations		2–<3	4–<6
3	Moderate limitations		1–<2	2–<4
4	Severe limitations		<1	<2
5	Limitations preclude beaver production		0	0

Table 8.2 — Descriptions of Beaver Land Capabilities Classification System Showing Limiting and Special Subclasses (Slough and Sadleir, 1977)

	Symbol	Subclass description
		Limiting subclasses
Lakes	S	Shoreline configuration allows buildup of waves
	O	Outlet not regulated by beaver dam(s)[a]
Streams	W	Width restricts damming[a]
	G	Gradient restricts damming[a]
Both	F	Absence of major food and construction species (aspen, willow, and alder)
		Special subclasses
	H	Human disturbance of shoreline (e.g., roads, railways, land clearing)
	T	Natural topography limiting as above
	D	Lake depth limiting. Freezes to bottom in winter

[a]These factors result in water level instability. Limitations imposed by stream gradient (i.e. flow rate).

tained by timber growers and agricultural interests. While such problems are not limited to the south, the fur value of southern beaver (Georgia, Alabama, Mississippi) is low, and there is often little incentive to trap the animals. Alternate forms of population control are needed. Some extreme methods have included consideration of alligator releases to reduce beaver populations by predation. Dispersal of alligators and/or beaver likely would limit the effectiveness of such an approach. Generally the most that can be accomplished without trapping is control of the water level by drains in the impoundments created by beaver. This is useful where such activities would otherwise flood roads. Fences have been used successfully to prevent clogging of culverts and dams dynamited to eliminate flooding-after the beaver have been removed by trapping.

Trapping in North America is either by registered trapline, harvest quota, license quota, or on an area restriction basis. The trapper using a registered trapline usually has a harvest quota established by the management agency. Registered traplines are on Crown or other public land; the area is limited geographically and may be censused by the agency to ensure the quota is appropriate. Censusing consists of flying over the area just before freeze-up to count the food piles (representing active lodges) and multiplying by the number of beaver/colony (usually about 4.0). This method overlooks den or bank beaver and requires an accurate determination of actual number of animals/active lodge.

Much of northern North America is too remote for economic census. In such cases, the land capability system for beaver developed by Slough and Sadleir (1977) or some related methodology may have real applicability. By using remotely sensed images of the area and comparing habitat characteristics to beaver requirements, a measure of the land's capability to produce beaver can be obtained. The regression model developed by the above authors resulted in a classification system with subclasses representing limitations and special considerations (Tables 8.1–8.2). The equations for estimating beaver numbers are:

$$\hat{Y}(\text{lakes}) = -3.84 - 0.781(P_L) \\ + 1.43E{-}3(A_L) + 0.555(A_L^{1/2}) \\ - 5.10E{-}4(R_L^2) + 1.24(W_L) \\ + 1.79(TA_L^{1/2}) + 6.32(N_L)$$

and

$$\hat{Y}(\text{stream sections}) = 74.2 + 24.1(L_s) \\ - 0.554(L_s^2) - 98.5(L_s^{1/2}) \\ + 56.2(\log_{10} L_s) - 2.43E{-}4(W_s^2) \\ + 4.42(G_s^{-1}) + 0.954(TA_s) + 0.600(NS_s^2)$$

where:

P_L = perimeter
A_L = area
R_L = area:perimeter
W_L = water level stability index
TA_L = length of aspen shoreline
N_L = length of nonproductive brush shoreline
L_s = length
W_s = width
G_s = gradient index
NS_s = length of nonproductive brush and swamp shoreline

Similar models can be used for all wildlife species when we have enough valid information on the habitat requirements of the species. One would expect that the Habitat Suitability Index for beaver could be used to predict populations in occupied habitat based on quality of habitat. But as was indicated in Chapter 7, the Indices have not been validated in all cases and when the beaver Index was tested in Kansas there was no correlation (Robel et al. 1993) between the Index values and ground counts of colonies. Easter-Pilcher (1990) and Broschart et al. (1989) provide two other examples of more simplified methods for predicting beaver population density.

Most of the larger Canadian provinces have some form of registered trapline or fur block (Saskatchewan) system whereby specific trappers are licensed to trap specific areas. The vast majority of North American jurisdictions sell only resident trappers' licenses that permit them to trap on their own property, any private property for which they can obtain permission, and sometimes public property. The competition on unregulated public lands in such jurisdictions can be extreme. Ontario has a management system whereby the number of trappers on private lands are limited by zone. To obtain a license, a new trapper must operate in a zone where vacancies exist or go on a waiting list for the trapping area of preference if the quota is filled. This system provides more control over trappers and harvest and is aimed at eliminating the possibility of overexploitation that became evident when fur prices soared in the late 1970s.

Closed seasons exist except in those areas where beaver are considered an economic liability. The open season is generally sometime between November and March, coinciding with primeness of the pelt. Often a royalty or stamp fee is paid to the management agency, and the pelt is stamped. This allows a record to be kept of the number of pelts, individual trapper performance, and marketing activities through fur dealers and auction houses. The actual trade in fur can be estimated from mail surveys, inspection of fur buyers' records, shipping permit records, and pelt tagging records.

Many jurisdictions now require novices to take a trapper education course that deals with the proper sets to use for given furbearers, how to use traps safely and effectively, landowner trapper relationships, skinning and casing procedures, and more recently, how to trap and kill an animal as humanely as possible. Some trappers' associations hold workshops to keep their members informed of the latest developments in legislation, public opinion, trapping methods, and trap development. It should be evident that management effectiveness depends to a large extent on the cooperation of the trappers. Many associations have developed their own trapper education programs when jurisdictions do not have compulsory ones. Trappers are vulnerable to public criticism because of the commercial nature of their profession and the perceived cruelty of devices used to catch animals (see Chapter 9). Associations often provide an effective political buffer to this antipathy just by being organized and using the management agency as their political lever.

Two diseases are particularly relevant when discussing furbearers–rabies and tularemia. While rabies seldom occurs in beaver, tularemia, caused by the bacterial agent *Francisella tularensis,* can be an important disease of beaver (see Chapter 5).The risk of infection is high for trappers and wildlife managers who may be handling contaminated live animals, carcasses, or fur.

Bibliography

Arner, D. H., C. Mason, and C. J. Perkins. 1981. Practicality of reducing a beaver population through the release of alligators. In: J. A. Chapman and D. Pursley (eds.), *Worldwide Furbearer Conference Proceedings,* Vol. III, pp.1799–1805. Frostburg, MD.

Boettger, R. W. and M. Smart. 1968. Beaver flowages converted from liabilities to assets. Maine Fish Game. 10(3):5–7.

Broschart, M. R., C. A. Johnston, and R. J. Naiman. 1989. Predicting beaver colony density in boreal landscapes. J. Wildl. Manage. 53:929–934.

Easter-Pilcher, A. 1990. Cache size as an index to beaver colony size in northwestern Montana. Wildl. Soc. Bull. 18:110–113.

Novak, M. 1987. Beaver. In: M. Novak, J. A. Baker, M. E. Obbard and B. Mallory (eds.), *Wild Furbearer Management and Conservation in North America,* pp. 283–312 Ontario Min. Nat. Resour. Toronto, Ontario.

Robel, R. J., L. B. Fox, and K. E. Kemp. 1993. Relationship between Habitat Suitability Index values and ground counts of beaver colonies in Kansas. Wildl. Soc. Bull. 21:415–421.

Slough, B. G. and R. M. F. S. Sadleir. 1977. A land capability system for beaver (*Castor canadensis Kuhl*). Can. J. Zool. 55:132–1335.

Waterfowl—Black Duck (*Anas rubripes*)

For the Atlantic flyway, the black duck is the most important bird in the hunter's bag. The species apparently has been declining in numbers, however, and by 1982 had shown a 60 percent decline from 1955. Recent winter survey information suggests that the decline has stopped. The reliability of the major indicators of population trends has been questioned even though Christmas bird counts tended to confirm the decline trend (70 percent decline from 1949–50 to 1982–1983) and the stabilization after more limited harvests beginning in 1983. What was responsible for this massive decline has been the subject of considerable speculation. Although there appears to be suitable unoccupied breeding and winter habitat available, there have been massive die-offs from starvation in New Jersey. This suggests that wintering habitat may be declining in some areas because of loss of salt marshes. Biologists have been able to increase brood production by liming dystrophic lakes. The many lakes which appear suitable for black duck breeding purposes may simply be unproductive bodies of water, and all suitable breeding habitat may be saturated. If this is so, the acid precipitation problem is likely to make more lakes unsuitable and breeding habitat may be a key limiting factor.

Toxic chemicals such as DDT and PCBs have been found in high concentrations in black duck eggs. DDT and its metabolites decreased in importance as PCBs increased. Both groups of chemicals can, and do, cause eggshell thinning. Measurements of black duck eggshell thickness in 1964 averaged significantly less than eggs collected prior to 1940. Eggshell thickness has increased with the decline in use of DDT, but is still below the pre-1940 value. Thin eggshells can lead to cracking during incubation. The hen will then remove any defective eggs or abandon the nest. If environmental contaminants are a major cause of decline, one would expect to find decreased productivity per female, but a study of brood size and production from 1956–1981 failed to reveal any meaningful changes.

Genetic swamping by mallards has been supported by a number of biologists as a cause of the decline. Black ducks and mallards readily hybridize, and wing samples had shown 13 percent hybrids by 1980. The rate of hybridization is estimated at 3–5 percent per year. The release of mallards by states and private organizations in the Atlantic flyway has contributed to this problem. Black duck characters are replaced by mallard characters so that the hybrids tend to be more mallard-like, and over time this genetic mixing alone could doom the black duck, especially when studies have indicated that male mallards dominate male black ducks when both compete for the same female. Lower productivity of black ducks compared to mallards was supposed to aggravate the situation but work by Longcore et al. (1998) suggests that productivity is similar for the two species. Despite consideration of these other factors, it appears that overhunting was most responsible for the black duck decline. Managers often have looked at hunting as a compensatory mortality factor and following this concept overhunting is simply not likely, whether it be for deer or ducks. In waterfowl, this concept was based on mallard data and in reality was a relatively recent change in philosophy from the additive approach. It is believed that that if there is the potential for substantial natural mortality, hunting losses can be compensatory. Hunting is known to cause 50–60 percent of the annual mortality in black ducks. With liberal regulations in the mid 1950s large kills were made which may have exceeded productivity despite the high duck population. By 1959, when the black duck population was obviously reduced, the more restrictive regulations were insufficient to halt the decline. The smaller kills were still proportionately too large to allow a population increase. In 1983–1984, restrictive regulations were imposed which ultimately showed a reversal in the decline. The conservative approach to harvest regulations continued into the 1990s but there have been only modest gains in black duck numbers. Francis et al. (1998) present data on the effect of the restrictive harvest regulations and it has been suggested that managers focus their attention on the additive hunting mortality component of the black duck population equation. Not suprisingly there has been similar controversy over hunting and its effect on mallard populations, some of it generated by a paper by Smith and Reynolds (1992).

What should have been the management reaction in the black duck situation? Season closure is an ideal candidate. Yet the black duck is the major bag species on the flyway, and management agencies derive their revenue primarily from the sale of licenses. Many waterfowl hunters are after black duck, and a season closure would result in loud complaints from hunters and significant declines in revenues to state agencies. Not only was there no closure during the early stages of the controversy, but in 1970 the U.S. Fish and Wildlife Service actually liberalized black duck hunting regulations.

Beginning in 1972 and continuing each year until 1975, minor restrictions were imposed, but seasons were still more liberal than those in 1968. By 1982, states like Maine and Massachusetts were considering season restrictions or closure because of alarming statistics showing large short-term declines in black duck numbers and a link between hunting and the decline. Although waterfowl management is ultimately a federal responsibility, states are empowered to impose more restrictive regulations than the federal ones. Maine biologists recommended unilaterally closing the season in that state in 1982. But the politicians decided it would be unfair to Maine hunters unless the other states were also closed. Some states, including Maine, adopted their own restrictions to reduce harvest, but the Atlantic Waterfowl Council put off consideration of any flyway restrictions until 1983. At the same time, the Humane Society of the United States, the Maine Audubon Society, and a private citizen filed a lawsuit to block the 1982–1983 black duck season. It was rejected. In the winter of 1983, the inventory of black ducks was the lowest ever recorded. A 1983 management objective was to achieve a 25 percent reduction in state kill for those states harvesting 5000 or more black ducks annually. This would achieve an overall kill reduction of 12 percent in the

Atlantic Flyway when Canada was included. This level of reduction was adopted not because modelling indicated it would reduce kill sufficiently to allow the population to stabilize and recuperate, but because it was the minimum measurable reduction which biologists thought could be made. It also was the maximum reduction state fish and game directors in the flyway would accept. And as we have seen, although populations have subsequently stabilized, they have not rebounded.

Are we then managing the resource for the resource's sake or for political goals which may ultimately deplete the resource? In truth we do both, and this is the real conundrum of wildlife management. When the two conflict we may fail to respond as quickly as we should because the political costs are too high, but by failing to respond quickly, the long-term economic and biological costs may be higher than the immediate ones (see Chapter 6, the Maine deer example).

We are learning more about the black duck as a result of the controversy but we still are unsure of the proper management responses as the roles of the various factors are still unclear because the data are often conflicting. Nonetheless, it appears that populations have rebounded as a result of regulations and other factors over the past five years. Winter counts for 1999 showed about 265,000 birds. Genetic swamping and habitat loss are still viewed as possible primary factors in the general decline from historic highs.

Bibliography

Blandin, W. W. 1982. Population characteristics and simulation modelling of black ducks. Unpubl. PhD. Thesis, Clark University, Worcester, MA.

Conroy, M. J., G. R. Costanzo, and D. B. Stotts. 1989. Winter survival of female American black ducks on the Atlantic coast. J. Wildl. Manage. 53:99–109.

Conroy, M. J., J. R. Goldsberry, J. E. Hines, and D. B. Stotts. 1988. Evolution of aerial transect surveys for wintering American black ducks. J. Wildl. Manage. 52:694–703.

Francis, C. M., J. R. Sauer, and J. R. Serie. 1998. Effect of restrictive harvest regulations on survival and recovery rates of American black ducks. J. Wildl. Manage. 62:1544–1557.

Grandy, J. W. 1983. The North American black duck (*Anas rubripes*): a case study of 28 years of failure in American wildlife management. Intl. J. Study Anim. Problems 4(4):1–35.

Feierabend, J. S. 1984. The black duck: an international resource on trial in the United States. Wildl. Soc. Bull. 12:128–134.

Heusmann, H. W. 1982. The black duck situation (and what to do about it). Mass. Wildl. (May–June):14–19.

Krementz, D. G., M. J. Conroy, J. E. Hines., and H. F. Percival. 1988. The effects of hunting on survival rates of American Black ducks. J. Wildl. Manage. 52:214–226.

Longcore, J.R., D.A. Clugston, and D.G. McAuley. 1998. Brood sizes of sympatric American black ducks and mallards in Maine. J. Wildl. Manage. 62:142–151.

Longcore, J. R., D. G. McAuley, and C. Frazer. 1991. Survival of postfledging female American black ducks. J. Wildl. Manage. 55:573–580.

Martinson, R. K., A. S. Geis, and R. I. Smith. 1968. Black duck harvest and population dynamics in eastern Canada and the Atlantic flyway. In: P. Barske (ed.), *The Black Duck Evaluation, Management and Research: A Symposium,* pp. 21–52. Atlantic Waterfowl Council and Wildl. Manage. Instit., Washington, D.C.

Nichols, J. D. 1991. Science, population ecology, and the management of the American black duck. J. Wildl. Manage. 55:790–799.

Nichols, J. D. and F. A. Johnson. 1989. Evaluation and experimentation with duck management strategies. Trans. N. Am. Wildl. Nat. Resour. Conf. 54:566–593.

Nichols, J. D., M. J. Conroy, D. R. Anderson, and K. P. Burnham. 1984. Compensatory mortality in waterfowl populations: a review of the evidence and implications for research and management. Trans. N. Am. Wildl. Nat. Resour. Conf. 48:241–256.

Rusch, D. H., C. D. Ankney, H. Boyd., J. R. Longcore, J. K. Kingelman, and V. D. Stotts. 1989. Population ecology and harvest of the American black duck; a review. Wildl. Soc. Bull. 17:379–406.

Sedinger, J. S. and E. A. Rexstad. 1994. Do restrictive harvest regulations result in higher survival rates in mallards? A comment. J. Wildl. Manage. 58:571–577.

Smith, G. W. and R. E. Reynolds. 1992. Hunting and mallard survival, 1979–88. J. Wildl. Manage. 56:306–316.

Smith, G. W. and R. E. Reynolds. 1994. Hunting and mallard survival: a reply. J. Wildl. Manage. 58:578–581.

Spencer, H. E. 1982. Black ducks-a state of concern. Maine Dept. Inland Fish Game, (Apr).

U.S. Fish and Wildlile Service. 1980. Important resources problem strategy paper: black duck-coastal mid-Atlantic (IRP No. 504). Habitat preservation. Reg. 5., Newton Corner, MA.

U.S. Fish Wildlife Service. 1983. Public information package concerning black ducks. Washington, D.C.

Upland Game Birds—Ruffed Grouse (*Bonasa umbellus*)

The ruffed grouse is the most ubiquitous of the North American grouse. There are excellent descriptions of the species and its management in Johnsgard (1973), Bump et al. (1947) and Atwater and Schnell (1989). The ruffed grouse is one of the species that many wildlife managers consider to be immune to hunting pressure. Fischer and Keith (1974), reporting on a central Alberta population where they banded 1132 birds, concluded that although fall hunting increased total annual mortality in certain cohorts, (territorial males), it had no measurable effect on spring population levels. Their study echoed the findings of numerous other researchers and has helped support the concept that hunting is a compensatory mortality factor for this species. One important aspect of Fischer and Keith's study and another by Gullion (1970) was the finding that hunting kill decreased with increasing distance between banding site and the nearest road or access trail. This means that in large continuous tracts of habitat with limited road access hunters are unlikely to influence the population sufficiently to offset the normally large reserve of juveniles. However, in much of eastern North America ruffed grouse habitat is no longer continuous, and those small patches open to public hunting may receive such high hunting pressure that local populations can be affected. What allows them to rebuild is the decreasing return effect. As the population is effectively reduced by hunting, the return to hunters per unit effort decreases until it reaches the point that they abandon the area. With decreased hunting pressure, the remnant population or immigrating birds can build the numbers back up. It is apparent that small tracts (100–200 acres) can have the local grouse populations wiped out if hunting pressure is severe enough.

The drumming log is an important component of ruffed grouse habitat. Forest management practices can have a substantial influence on ruffed grouse by affecting this component of the male's territory. The drumming stage itself is needed but it will not be used by the male if stem density and canopy coverage in the shrub layer are insufficient. When the shrub layer is removed within an area of 50 m^2 centered on a previously active drumming site, it renders that site unacceptable as a display location. Thus forest plantations with little or no shrub layer and clean forest practices (removal of downed material) mean the loss of potential drumming stages. When birds do select these sites, they become more vulnerable to avian predators. There is a critical balance between sufficient shrubby material to screen the bird from potential predation and sufficient openness to allow the grouse to see at least 6 m in all directions. During periods of population depression, ideal habitat is the reserve.

The most consistently used male territories contain a number of acceptable display locations and are near the middle of an aggregation of territories (lek concept of territorial behavior in grouse). Because of the nearly ideal habitat conditions, they are used by long-lived birds. This reservoir of breeding animals serves as the focus for repopulating more marginal habitats which, because of biological or physical constraints, are less preferred or less safe and are "home" to more transitory populations.

Ruffed grouse show seasonal shifts in habitat use. Deciduous types are preferred in spring and summer, but as temperatures drop the animals make increasing use of conifer cover. During periods of snow cover the birds are likely to be found in conifer stands. The birds still move to the deciduous trees to feed on buds in the winter and must do so on a regular basis because they carry very small fat reserves. The lower critical temperature for the species measured at -6.0 °C in February and 0.3 °C in March near midlatitude of the eastern distributional range in Massachusetts indicated that the animal must either ingest considerable energy or conserve it in some fashion. Plasma glucose levels increase during the winter in grouse while plasma lipid levels remain stable. The increased glycogen levels in liver and pectoral muscles suggest that shivering thermogenesis could be enhanced and that the animal responds to energy needs glycolytically, not lipolytically. Because the birds use snow burrows (which reduce radiative and convective heat loss) and have excellent plumage insulation, the lower critical temperature is probably seldom reached except when the birds are feeding. During feeding, they are able to select the most nutritious and highest energy buds, and as the feeding bouts are short (fifteen to twenty-five minutes) and infrequent (two per day) the energy costs of feeding are kept as low as possible. It is important from a management perspective to have sufficient male aspen or other budding trees (white birch, black cherry) available to provide the birds with good winter forage.

With this species, habitat management assumes greater value than regulatory management except in those situations where heavy hunting pressure impacts small localized populations. The only political question is usually associated with hunter access and landowner complaints. Grouse hunters usually use shotguns, so landowners are not as concerned about safety problems. The ruffed grouse has shown considerable resilience to man's activities and has been relatively easy to manage. Many early management efforts such as small forest clearings and seeding log landings and roads to clover were more effective in making the birds available to hunters than increasing populations. The ruffed grouse is thus an example of an uncontroversial species for which considerable biological and management knowledge exists. It can be difficult to hunt because it flies well, is generally in brushy areas, and becomes more flighty as hunting pressure increases. It also is widely distributed, yet can be easily studied and so serves as an exception to most game species for which management action is more critical.

Bibliography

Atwater, S. and J. Schnell (eds). 1989. *Ruffed Grouse.* Stackpole Books, Harrisburg, PA.
Boag, D. A. 1976. The effect of shrub removal on occupancy of ruffed grouse drumming sites. J. Wildl. Manage. 40:105–110.
Boag, D. A. and K. M. Sumanik. 1969. Characteristics of drumming sites selected by ruffed grouse in Alberta. J. Wildl. Manage. 33:621–629.
Bump, G., R. W. Darrow, F. C. Edminster, and W. F. Crissey. 1947. *The Ruffed Grouse. Life History. Propagation. Management.* N.Y. State Conserv. Dept.
Dorney, R. S. and C. Kabat. 1960. Relation of weather, parasitic disease, and hunting to Wisconsin ruffed grouse populations. Wisconsin Conserv. Dept. Tech. Bull. 20.
Fischer, C. A. and L. B. Keith. 1974. Population responses of central Alberta ruffed grouse to hunting. J. Wildl. Manage. 38:585–600.
Gullion, G. W. 1970. Factors influencing ruffed grouse populations. Trans. N. Am. Wildl. Nat. Resour. Conf. 35:93–105.
Gullion, G. W. and W. H. Marshall. 1968. Survival of ruffed grouse in a boreal forest. Living Bird. 7:117–167.
Johnsgard, P. A. 1973. *Grouse and Quails of North America.* University of Nebraska Press, Lincoln, NE.
Pietz, P. J. and J. R. Tester. 1982. Habitat selection by sympatric spruce and ruffed grouse in north central Minnesota. J. Wildl. Manage. 46:391–403.
Rasmussen, G. and R. Brander. 1973. Standard metabolic rate and lower critical temperature for the ruffed grouse. Wilson Bull. 85:223–229.
Robel, R. J. 1972. Possible function of the lek in regulating tetraonid populations. Proc. Intl. Ornith. Congr. 15:121–133.
Svoboda, F. J. and G. W. Gullion. 1972. Preferential use of aspen by ruffed grouse in northern Minnesota. J. Wildl. Manage. 36:1166–1180.
Thomas, V. G., H. G. Lumsden, and D. H. Price. 1975. Aspects of the winter metabolism of ruffed grouse (*Bonasa umbellus*) with special reference to energy reserves. Can. J. Zool. 53:434–440.
Thompson, F. R. and E. K. Fritzell. 1989. Habitat use, home range, and survival of territorial male ruffed grouse. J. Wildl. Manage. 53:15–23.

Raptors—Peregrine Falcon (*Falco peregrinus*)

The peregrine falcon showed tremendous declines in populations throughout the Northern Hemisphere beginning in the 1940s and continuing into the 1970s. Pesticides, particularly DDT, were responsible for eggshell thinning and hatching failure. The species disappeared from much of its former range but like the bald eagle, the osprey (*Pandion haliaetus*), and other raptors affected by environmental contamination by persistent pesticides, populations were expected to increase with the reduced use of DDT. This has happened even though relatively recent studies have still found DDT metabolite levels in eggs in Oregon above those known to cause hatching failure.

Action was precipitated by a conference held in 1965 in Madison, Wisconsin. Raptor experts from North America, Europe, and Great Britain met solely to discuss the plight of the peregrine falcon. In 1970 the Canadian Federal-Provincial Wildlife Conference (an annual meeting of senior personnel from provincial, federal, and territorial wildlife agencies) authorized the CWS to initiate a captive breeding program for the peregrine (jurisdictionally within the provinces' area of responsibility). The function of the program was to raise sufficient birds to allow reintroductions in their former range. At about the same time as the CWS project, Cornell University was also involved in a peregrine breeding program. Cornell birds were the first to be released into the wild, in 1974. As the number of releases increased it became apparent that avian predators such as great horned owls (*Bubo virginianus*) and golden eagles (*Aquila chrysaetos*) could cause havoc by preying on the young peregrines. Control measures are needed or releases have to be made in areas where these other avian predators are not a threat.

Releases are generally made at historical eyrie sites where human disturbance is minimal. In natural areas, eyries are usually found on cliffs with ledges for perching, roosting sites, plucking, and feeding areas. High-rise buildings in urban areas previously served as nesting sites for peregrines and successful releases have been made in such cities as Montreal, Baltimore, New York, Spokane, and Edmonton where the birds feed primarily on pigeons and English sparrows. The ledges of the buildings meet all the biological requirements of the species just as well as natural rock ledges. The release of young birds is achieved by a process known as hacking. The birds are kept in an artificial nest box and fed until they are ready to fledge. The food source is gradually eliminated as the birds learn to fend for themselves. The management hope is that the birds will successfully migrate and return to the release site in future years to breed. This has occurred in Edmonton and Baltimore, among other locations. Citizens and visitors in both cities have had the rare opportunity of observing, through a closed circuit video system, the whole process of incubation, hatching, and rearing the young on a ledge of one of their high rises. The public relations value has been superb and much of the populace now considers the birds to be "their peregrines."

The lack of disturbance sought at release sites has not been the norm for much of the remaining wild population. Egg collectors, falconers seeking new birds, private individuals with their own captive breeding programs, and naturalists observing or photographing raptors are all sources of disturbance which tend to deplete natural recruitment and the wild breeding population. Raptors are still persecuted by hunters and ranchers. Trapping, poisoning, and shooting, although illegal in most jurisdictions, take a considerable toll on species such as the peregrine.

Legislation has been passed in the United States to make raptor propagation permits available under the Migratory Bird Treaty Act. The permit allows propagators and certain other people to purchase, sell, or barter captive-bred raptors for scientific, educational and falconry purposes if corresponding state regulations are passed. The legislation toughens the restrictions on endangered or threatened raptors listed under the Endangered Species Act although it allows continued trade in these species if the animals were in captivity prior to 1978. An intended primary function of the legislation is to reduce pressure on wild raptor populations. Without increased enforcement, the more restrictive legislation by itself is unlikely to have a significant impact on the illicit trade in falcons.

Management efforts for the peregrine falcon have been similar to those for other endangered nongame species. Captive breeding, reintroduction, and protection of existing eyries from disturbance coupled with a public education program to prevent unwanted mortality and to sensitize the public to their precarious status are the prime components. The peregrine falcon management effort has been so successful that it has been removed from the endangered species list and it certainly can be used as the model for other raptor species management. Nonetheless, the whole program would have been doomed to failure if the connection between persistent hydrocarbons such as DDT and the disastrous decline in population level had not been made and government action taken to reduce or eliminate use of the culprit chemicals in the environment. As managers, we must continue to use wildlife species as sentinals or early warning signals of declining environmental quality and must alert the public and governments to the consequences of continued contamination, despoilation, or incorrect resource extraction procedures which may be accounting for the problem. If this necessitates linking the potential consequences to human populations, we should do it. If the data are adequate and the public is convinced it is in its collective interest to seek a ban on the use of certain chemicals to save certain habitat types, to reduce sulphuric acid in the atmosphere, to reduce massive clear-cutting operations, to stop draining wetlands (as it inevitably is in the long run), then the chance of political success increases. We should capitalize on those symptoms of our declining natural heritage to point out the long-term ecological consequences of a man controlled environment. Our dominance must be made to reflect the need for sustaining the diversity of life forms we have been fortunate enough to inherit. Future generations may well look back on this generation as the one which met or failed its responsibility to maintain an ecologically sound human interface with the life support system of this planet. If we fail, they will reap the disastrous consequences.

Bibliography

Fyfe, R. W. and R. K. Olendortf. 1976. Minimizing the dangers of nesting studies to raptors and other sensitive species. Can. Wildl. Serv. Occ. Pap. No. 23.
Henny, C. J. and M. W. Nelson. 1981. Decline and present status of breeding peregrine falcons in Oregon. Murrelet 62:43–53.
Hickey, J. J. and D. W. Anderson. 1968. Chlorinated hydrocarbons and egg shell changes in raptorial and fish-eating birds. Science 162:271–273.
Peakall, D. B. 1976. The peregrine falcon (*Falco peregrinus*) and pesticides. Can. Field-Nat. 90:301–307.
Peakall, D. B., T. J. Cade, C. M. White, and J. R. Haugh. 1975. Organochlorine residues in Alaskan peregrines. Pestic. Monit. J. 8:255–260.
Ratcliffe, D. A. 1967. Decrease in eggshell weight in certain birds of prey. Nature 215:208–210.

Black-footed Ferret (*Mustela nigripes*)

The most endangered mammal species in North America probably is the black-footed ferret. Once distributed over much of the western plains of Canada and the United States, the species was considered extinct in 1947 when the only known population in South Dakota disappeared and captive breeding proved unsuccessful. Because the ferret is dependent on prairie dog colonies for both prey and shelter, its fate was inextricably linked to that of the prairie dog. As the prairie became cropland and eradication programs were conducted on the prairie dog colonies both species' futures were impacted early in this century. Poisoning of prairie dogs was widespread and successful and the loss of prey base in addition to secondary poisoning resulted in precipitous declines of black-footed ferret numbers.

In 1981, a population of black-footed ferrets was discovered near Meeteetse, Wyoming; that event initiated a saga which exemplifies the best and worst of endangered species management in the United States. In the west, there are strong public vs. private lands and federal vs. state jurisdictions conflicts. The Sagebrush Rebellion, as one example, in reality combined aspects of both. The Endangered Species Act clearly identifies the USFWS as the responsible agency for terrestrial wildlife but there was only one known population of ferrets and it occurred within the state boundaries of Wyoming. That state thus considered it part of its resident wildlife and claimed responsibility for the species' management. A similar situation existed with the California condor in the state of California. Additionally, the critical prairie dog colonies were primarily on private land, albeit much of that belonging to a cooperative rancher, but the private landowners were more wary of the "Feds" than they were of the state people and thus were less likely to approve of federal intervention although clearly the weight of the law would support such action. What happened was the USFWS rapidly delegated its responsibility to the Wyoming Game and Fish Department. The Department then set up a Black-footed Ferret Advisory Team (BFAT) consisting of a private rancher, representatives from the Department, the Bureau of Land Management, the Forest Service, the USFWS, the Wyoming State Lands Board, and the University of Wyoming. Another complicating factor was an ongoing research effort by Idaho State University and a consultant firm, Biota, funded in part by a major conservation agency, the Wildlife Preservation Trust. This ultimately resulted in representation of wildlife conservation agencies on BFAT by the National Wildlife Federation. The stage was set for interaction, positive and negative, between federal and state management agencies, private landowners, conservation interests, and researchers.

BFAT moved cautiously but a coordinated research program involving federal, state, and university biologists was soon underway. Censusing and radio-collaring of ferrets, predator competition studies, prey base analysis, and discussion of recovery procedures all proceeded as part of BFAT's agenda of study. While the highly intrusive field research could be justified on the basis of the paucity of knowledge on the species, early concerns that the Meeteetse population was vulnerable to catastrophic events and that captive breeding should have a high and immediate priority were too easily set aside. BFAT did agree that at some point ferrets should be brought into captivity but, although planning for this event was conducted, there was no sense of urgency in the advisory team's actions. A workshop was held in 1984 in Laramie to present information on the ferret's biology, the basic tenets of conservation biology, and considerations important to a captive breeding program. Still, there was an atmosphere that it was necessary to explore all the potential pitfalls before committing to captive breeding. Much emphasis was placed on the failure of the Patuxent captive breeding effort using South Dakota ferrets, even though that population exhibited severe genetic problems which likely precluded successful breeding.

In 1985, disaster struck. Die-offs in the prairie dog colonies from sylvatic plague caused concern about loss of the prey base for the ferrets. Prairie dog burrows were treated with Sevin (active ingredient carbaryl) to kill the fleas which transmit the bacteria. BFAT agreed to plans to capture three pairs of ferrets for a captive breeding program to be conducted by Wyoming Game and Fish at the Sybille Research Unit. During all these events, some of the researchers at Meeteetse raised concerns that the ferret population seemed to be in decline and the program to capture six ferrets may be too little, too late. BFAT and Wyoming Game and Fish were considering whether the wild population was large enough to capture additional ferrets for another facility, the Front Royal Wildlife Research and Conservation Center in Virginia. The interest was to protect the Meeteetse population by not removing too many animals. One of the six ferrets captured for Sybille died of canine distemper on 21 October and the mark-recapture count at Meeteetse resulted in a population estimate of 31 animals down from 130 animals in the 1984 census. It was evident that distemper was in the wild population and that ferret numbers were declining rapidly. The initial capture group succumbed to the disease but a decision was made to capture as many of the remaining wild ferrets as possible and this time take preventative measures such as isolation to stop the spread of the disease from infected animals. The worst case scenario had occurred. The pleas of researchers that captive breeding should have been the first priority of BFAT now had the weight of a prophecy of doom come true.

By 1986, all that remained of a once thriving wild population, the only known population of black-footed ferrets, were 18 animals all in captivity. A captive breeding advisory committee was formed and now the research was geared to maximizing production from the captive remnants of the Meeteetse population. Fortunately, there is now a substantial captive population housed at several captive breeding sites. Furthermore, reintroduction of the species has occurred at sites in four western states, although the reintroductions have not all been successful as sylvatic plague eliminated the ferrets reintroduced to the original Wyoming site.

At the time of crisis, full cooperation of the interest groups was finally achieved. The politics of management prior to the crisis may have ensured its occurrence. The rivalry between agencies and conservation interests, the reluctance of the USFWS to apply its jurisdictional rights and employ the full power of the Endangered Species Act may have been unavoidable. There is a lingering suspicion that other colonies of ferrets exist in the wild in Wyoming but they are on private land. As is evidenced by the proposed revisions to the Endangered Species Act which would provide greater protection to private land owners there is considerable support for the ascendancy of land owner rights in situations such as this. There is suspicion that if any government agency had exerted its power over the ranchers at Meeteetse, rather than cooperating with land owners and in essence operating at their largesse as was the case, any other ferret populations that might exist in Wyoming were doomed. The last thing that the independent Wyoming ranchers would tolerate would be the threat of government intervention on their land because of an endangered species. What is even more disturbing is the knowledge that the habitat for this species is so easily defined. The species exists only in large complexes of prairie dog colonies. Protect the prairie dogs and you protect black-footed ferret habitat. Ranchers do not like prairie dogs. It is felt that the prairie dogs compete with cattle for limited forage and there is a risk that cattle or horses will damage their limbs if they accidently step in a burrow opening, and farm equipment may be broken if a burrow system collapses under its weight. There is little evidence to support competition, indeed some studies show that forage for cattle may actually be increased by the presence of prairie dogs. The arguments appear economically based but in final analysis they are not. At the time, the USFWS was unwilling to invoke the habitat protection component of the Endangered Species Act possibly because it would have resulted in a challenge to the Act and a potential weakening of it. This may occur anyway as the Act is reauthorized (see Chapter 10). It was left to the northern spotted owl to see how great an economic challenge was necessary to alter consideration of the protection provided by the Act for endangered species.

Having witnessed the black-footed ferret saga, there are two relevant conclusions. The first is that wildlife professionals have vested political interests that may mitigate against effective and appropriate wildlife management activities. The second is that in times of crisis, it is possible to see the best of wildlife managers and management as personal, or agency, interests are set aside for full consideration of the resource's needs. The pattern is not unique to wildlife management. The current debates over global warming, the ozone layer's depletion, and acid precipitation are mechanisms which effectively delay meaningful or appropriate actions until the crises are at hand. Wildlife managers are no better, or no worse, than human beings in other walks of life. Perhaps this is the fatal flaw in our own species which may someday make us endangered.

Bibliography

Clark, T. W. 1989. Conservation biology of the black-footed ferret, *Mustela nigripes*. Wildlife Preservation Trust. Spec. Sci. Rep. 3.

Dobson, A. and A. Lyles. 2000. Black-footed ferret recovery. Science 288: 985–988.

Forrest, S. C., T. W. Clark, L. Richardson, and T. M. Campbell, III. 1985. Black-footed ferret habitat: some management and reintroduction considerations. Bureau Land Management (Wyoming) Wildl. Tech. Bull. 2.

Seal, U. S., E. T. Thorne, M. A. Bogan, and S. H. Anderson (eds). 1989. *Conservation Biology and the Black-footed Ferret.* Yale Univ. Press, New Haven CT.

Ubico, S. R., G. O. Maupin, K. A. Fagerstone, and R. G. McLean. 1988. A plague epizootic in the white-tailed prairie dogs (*Cynomys leucurus*) of Meeteetse, Wyoming. J. Wildl. Dis. 24:399–405.

9 SOME SPECIALIZED AREAS OF MANAGEMENT

In past years, the art and science of wildlife management have been primarily concerned with five major principles. The first was protection, complete or partial, normally applied through regulations and their enforcement. The second was predator control. Third, if neither protection nor predator control was adequate to increase populations, perhaps establishing an inviolate area for the wildlife to reproduce in would do the trick. Fourthly, if none of these approaches worked, maybe raising animals in captivity and releasing them into the wild would do it. This last activity was referred to by sportsmen as stocking. It was and still is very popular indeed in some jurisdictions. The fifth and final idea to come of age was the concept of habitat improvement (Chapter 7). As we have seen, however, management is daily becoming more complex, and there are a myriad of specialized wildlife areas and concerns the manager must be aware of. In this chapter we present a few we feel are among the more important.

Protected Areas

The original concept behind establishing protected areas was to provide sanctuary (refuge or asylum) for one or more species to reproduce in without harassment from human activities. Many people also felt that progeny from these protected areas would venture forth and repopulate extensive ranges outside the protected zone. This latter hope, as we now understand things, depends upon the home range and dispersal of the species as well as upon the size of the protected area. We have noted (Chapter 1) that the concept of setting aside areas for wildlife protection is not new; both Henry VIII and James I did so. It should be recalled also that although these areas were set aside, they were protected from everyone except these same two gentlemen. Private parks for deer and exotic birds were present prior to 1900, principally on lands of the wealthy in Europe and North America. In this century, man has designed an amazing variety of areas, protected to some degree, for wild things, plants as well as animals.

Sanctuaries are public or private tracts of land which in their true form protect animals and plants within their boundaries. Certain elements of systems within a sanctuary may be removed to maintain either habitat conditions or animal populations at a particular level. Refuge is another term for sanctuary. These protected areas may be under federal, state, provincial, county, or municipal jurisdiction in North America. Some may be privately owned if trespass legislation or more specific legislation is present.

Parks are usually sanctuaries for wildlife except that fishing is normally allowed. Parks also exist within the several levels of jurisdiction in North America. There are federal, state, provincial, county, and municipal park systems in the United States and Canada.

Designated wilderness areas on federal lands in the United States also provide protection to many wildlife species. These areas are roadless and thus less vulnerable to public use. Restrictions on motorized vehicles and land-use activities ensure preservation of natural conditions.

Management areas inherently admit to human intervention although they may provide some protection for specific purposes (nesting, brood areas) or for certain species. Federal and/or provincial and state law with attendant regulations applies in all protected areas, however, and specific legislation for various protected areas often is present in separate statutes.

The largest refuge (sanctuary) system in North America and probably in existence anywhere is the U.S. Wildlife Refuge System run by the U.S. Fish and Wildlife Service which encompasses some 37.5 million hectares (90 million acres). These areas have been established primarily to protect either waterfowl, other migratory birds, or larger mammals but also have been created since passage of the Endangered Species Act to protect threatened or endangered species. The refuge concept in North America began in 1893 with George Bird Grinnel and Theodore Roosevelt, and the first refuge was Pelican Island, Florida, established in 1903. The acquisition authority for the waterfowl refuge system was provided in 1929 under the auspices of the Migratory Bird Conservation Act. The Migratory Bird Hunting Stamp Act of 1934 assured a steady source of funding but also assured that the primary emphasis would be on waterfowl and that there would be pressure for hunting within the refuges. Amendments to the Conservation Act now allow hunting in up to 40 percent of the total area of any refuge if compatible with the major purpose(s) for which the area was established. Management on the refuge areas may be restricted to protection or may involve the production of grain crops for food, water level control, plant-water interspersion maintenance, reseeding, reforesting, and game harvesting. In the latter instance, muskrats and other furbearers have sometimes been trapped in the Montezuma Wildlife Refuge (New York) and the Horicon Marsh Refuge (Wisconsin) to aid in maintaining near optimum ratios of cover and open water for waterfowl. The overall administration and management of the U.S. Wildlife Refuge System, and its units have been considered by Giles (1978) and Drabelle (1985).

In Canada, the first sanctuary was established in 1887 by Edgar Davidney who was at the time lieutenant governor of the Northwest Territories. This was the Lost Mountain Lake area 130 kilometers northwest of Regina. Today, there are some 82 federal migratory bird sanctuaries in Canada totaling about 11.4 million hectares (about 28 million acres). Many of these areas serve to protect mammals, reptiles, and amphibians as well as birds.

Hot Springs (Arkansas) was the first national park (1872) in the United States and Banff (originally Rocky Mountains National Park) the first in Canada (1885). National park policy throughout the world today generally protects wildlife within park boundaries. Wildlife, by definition, now usually also includes vertebrates other than birds and mammals, some invertebrates, and plant life. Wildlife management within national parks always has been somewhat controversial; for if the park represents natural ecological conditions, apart from man's intervention, how do managers account for the establishment of the park as an inviolate area when it is itself an intervention? Inviolate areas are subject to population increases of many game species because neither direct nor indirect action by humans reduces their numbers. Some species actually learn to avoid harassment by not venturing outside. Such is the case with elk and white-tailed deer which during open hunting seasons, may move to areas where they are not disturbed—often parks or sanctuaries. Park managers and interpreters also require roads, open areas for public use, buildings, and trails for access as well as areas for camping, administration, recreation, and interpretation. Roads and trails, in particular, often provide ecotones and open grazing areas which not only allow the public to view animals feeding but also increase the carrying capacity of the protected area. Transmission lines may provide additional interspersion especially valuable in producing food plants for grazing and browsing wildlife and for songbird nesting habitat. Faced with such changes, managers often have had to act to reduce game populations either through live trapping and removal or by shooting excess animals in order to help reduce both damage to vegetation and increased erosion. Elk were removed from Yellowstone for stocking elsewhere and deer were killed to reduce the population density in Acadia National Park in Maine. A control program has been developed for the mountain goat, a species introduced to Olympia National Park in Washington. The Yellowstone elk problem has been controversial for many years, but it serves as an excellent example of the problem ungulates may present when protection is provided. Harvests by hunters outside the park have removed thousands but have brought the wrath of members of the public who despise the "slaughter." Harvests by park authorities have also accounted for many animals, but even when the carcasses are distributed and wisely used, the public complains and costs are high. The expense of livetrapping and translocating is even higher. Should the matter of the elk be left to its natural conclusion, allowing range deterioration and perhaps a repeat of the massive die-off of 1914? Such a complex question must be carefully studied by park wildlife staff and state wildlife managers to maintain both the integrity of the park and sound, socially acceptable species management within budgets provided. The reintroduction of wolves to Yellowstone was in part designed to provide a natural control to the elk population. What happens inside a park also affects land areas outside. Houston (1982) and Chase (1986) provided contrasting views of the elk situation in Yellowstone.

Management of wildlife within parks is most often restricted to manipulation of their habitat. Roadside, open area, and power line management to encourage the growth of palatable, nutritious plant species while discouraging others through hand labor (cutting them out), herbicides, seeding or planting native varieties is often possible. In other situations, winter feeding may be necessary (and responsible) on occasion to get animal populations of marginally high densities over brief periods of food shortage. In still other cases, the removal of animals by livetrapping or culling will be necessary.

Wildlife managers in parks can also play a major role in advising on road and trail placement, service area locations, campground layouts, and various day-use or seasonal facilities to the benefit of wildlife and for better wildlife viewing. Advice on the use of salt on highways, fertilization, erosion control, speed limits, the location of garbage dumps, and many other concerns may involve the wildlife manager and one must also cope with nuisance wildlife that can pose problems.

Perhaps no nuisance has reached such proportions as the bear-human interface. Black bears are attracted to dumps and people are attracted to black bears. The presence of grizzly bears in areas of even low human activity can be a major safety problem. The dump concern may not confront wildlife managers often in the future, however, because park dump locations should be determined at the planning stage so that humans are not allowed access to them. The grizzly bear situation is not so easy. These bears require extensive range areas and protection. Bear country can never be considered high density use areas for humans. Issues associated with the increased opportunity for bear-human interaction resulting from increased human use of parks and accommodation by bears to the presence of humans in a nonhunting environment now consume considerable attention, particularly following an attack on someone. Other nuisance wildlife such as skunks, porcupines, or rodents are usually thought of as part of the natural environment for park visitors to accept on the animals' terms. Only the beaver that floods the road may have to be moved.

Management areas may be within federal, provincial, or state jurisdiction and may go by different names. They allow for multiple use activity by humans within the bounds of the necessary management for one or more species in each area. Even hunting may be allowed. Canada has forty-four national or cooperative (joint federal-provincial management) wildlife areas. With less management involved, the waterfowl production areas and coordination areas administered by the USFWS also qualify as management areas, if for no more than habitat maintenance and seasonal protection. States and provinces also have wildlife management areas. Some are managed specifically for wildlife and others are management areas in name only and are used almost entirely for other purposes. Such is the case with the Tobeatic Wildlife Management Area of Nova Scotia, which is primarily for forest management. Land use and motor vehicle access are both controlled by industry.

Private sanctuaries exist by virtue of land trespass legislation in many jurisdictions. The landowners take advantage of protection offered under statute, suitably mark the property, and refrain from killing the animals. Such areas are common in parts of North America but few are true private sanctuaries, known to the public and recognized as such. One very successful private sanctuary is located at Kingsville, Ontario. The story of the Jack Miner Bird Sanctuary began in 1904 when Miner, a pioneer conservationist, woodsman, and sportsman first lured four Canada geese to his home pond and clipped their wings. With this beginning of four live decoys, he eventually wintered thousands of Canada geese and wild ducks. Jack Miner's work was continued by his son Manly and the Kingsville sanctuary is still open to visitors, human and avian, from October to April each year.

Other early examples of private sanctuaries for wildfowl are those along Maryland's eastern shore. In the 1930s W. P. Chrysler, J. J. Raskob, and others, planted food and released propagated birds. They created highly successful protected areas even though limited hunting took place on some of them. Winter feeding became necessary and upwards of 4000 bushels of corn were fed one winter on one property alone. These areas also provided an element of protection for quail and pheasant. An interesting variation in protected area management is described from Connecticut (Bishop, 1949). It began in 1940 and was one of the earliest controlled hunting experiments in the eastern United States. Permits were required of hunters, bag limits established, areas divided into coverts for hunting, pheasants stocked, and wild birds (or stocked escapees) totally protected in blocks of 75 hectares (180 acres) to 177 hectares (424 acres) within the management unit. These protected areas were known as seed-stock refuges. Since then, a number of states have passed legislation allowing landowners to establish managed hunting areas with some degree of protection included for various species (for example, the New York State Fish and Wildlife Management Act). In North America today, refuges or sanctuaries for wildlife are very successful and of most consequence for two groups of animals; waterfowl and endangered species. The waterfowl (migratory bird) refuge system and the international migratory bird legislation allowing waterfowl to be managed on a flyway basis have been of paramount importance for ducks and geese. Protection of breeding and nesting areas, brood areas, and staging areas hold the highest priority in current and future management. For many endangered species, protected areas hold out the only hope for their continued survival; even protected areas may not be all that is required for such severely endangered species as the California condor, although the captive propagation and release program has had some early success.

The size and most particularly, the actual boundaries of most protected areas are usually determined by socioeconomic and political factors rather than biological or ecological ones. Nevertheless there are a number of factors which under ideal conditions, the wildlife manager should consider prior to establishing a protected area. In relation to the land, the carrying capacity for the target species at the time of establishing the reserve should be assessed. In relation to the target species involved, the populations, breeding behavior, productivity, and mobility are all factors that should be related to the size of the area and location of boundaries of the area. Adjacent land areas and land use should also be assessed relative to their contribution to protected area populations and vice versa. If both populations and the land area can be managed, the manager should attempt to maintain the estimated or theoretical optimum population sizes for survival of each protected species (Schonewald-Cox, 1983).

We should not leave the subject of protected areas without mentioning wilderness, which may or may not be protected. Sometimes extensive, remote wilderness areas are protected only by legislation restricting certain types of access (e. g., float planes). At other times, smaller wilderness areas are established as inviolate refuges or sanctuaries. In the United States, wilderness areas are generally roadless, and machines are not allowed unless traditional use areas have been established. Grazing is often permitted. Since World War II, planetwide concern has been expressed for the retention of wilderness. Of the International Union for the Conservation of Nature's three main objectives in its World Conservation Strategy, two may eventually be possible only if wilderness is preserved: the maintenance of essential ecological processes and the preservation of genetic diversity. For the wildlife manager, these objectives are important in dealing with wild animals, plants, and ecosystems on an everyday basis. Attempts to deal with them on an extended basis have been only partially successful. The International Biological Program (1964–74) was a cooperative project among the International Council of Scientific Unions. As one of its mandates, it attempted to describe, list, and reserve through legislation, representative terrestrial communities worldwide. This program did result in an increase in terrestrial community preservation. In the future, it seems likely that it will become far easier to maintain several small areas under wilderness conditions than to protect a few large ones. Those species able to maintain genetic diversity within small ecosystem units will obviously be favored most for survival. Wilderness is increasingly threatened and "once our master, then our enemy and now, finally, it has become our pensioner" (Gaston, 1982). The wildlife manager will be called on regularly, either to protect wilderness, or to destroy it, through management.

Bibliography

Bishop, J. S. 1949. Seed-stock refuge investigation. , Connecticut State Board of Fisheries and Game. (PR-4R).

Chase, A. 1986. *Playing God in Yellowstone: The Destruction of America's First National Park*. Atlantic Monthly Press, Boston, MA.

Drabelle, D. 1985. The national wildlife refuge system. In: *Audubon Wildlife Report 1985.*, pp. 151–179. Natl. Audubon Soc., New York, NY.

Gaston, T. 1982. Wilderness: the passing of a dream. Northern Perspectives 10(4):9–11.

Giles, R. H., Jr. 1978. *Wildlife Management*. W.H. Freeman and Company, San Francisco, CA.

Houston, D. B. 1982. *The Northern Yellowstone Elk: Ecology and Management*. MacMillan Publ. Co., New York, NY.

Leopold, A. S., S. A. Cain, C. Cottam, I. N. Gabrielson, and T. Kimball. 1963. Wildlife Management in the National Parks. The Leopold Committee Report. Rep. from American Forests (Apr).

Levi, H. W. 1952. Evaluation of wildlife importations. The Scientific Monthly LXXIV No. 6:315–322.

Miner, J. 1977. *Wild Goose Jack*. Paper Jacks, Markham, Ontario.

Schonewald-Cox, C. M. 1983. Guidelines to management: A beginning attempt. In: Schonewald-Cox, Chambers, MacBryde, and Thomas (eds.), *Genetics and Conservation*. Benjamin/Cummings, Menlo Park, CA.

Trefethan, J. B. 1961. *Crusade for Wildlife*. The Telegraph Press, Harrisburg, PA.

Exotic Species

Mankind has always been interested in that which is unusual or exotic in nature, and animals foreign to one's native land are unusual—at least until they become easily recognizable and eventually common in a new home. For many people, the sight of an unusual foreign bird or mammal provides an aesthetic experience somewhat greater than a similar experience involving an indigenous animal. Similarly, the sportsman often gains more pleasure from hunting an introduced species than from seeking out a local animal as a target. Then too, there are the examples of immigrants to a new area who long for the presence of a part of the natural system to which they once belonged. They may want an animal for food or perhaps to make them feel more comfortable in their new homeland. As Teer (1979) has noted, "From his earliest beginnings, man has taken plants and animals with him wherever he went."

Because we are prone to like the unusual as well as the animals we are used to, we have hundreds of examples of exotic introductions into most parts of the world visited and settled by humans from distant homelands. Sometimes when a landmass was observed not to have many animals using it, its new human inhabitants got carried away. In New Zealand, of some 45 species of mammals introduced about 25 became established. King (1984) chronicles the impact of introduced mammalian predators in New Zealand on the native fauna. Because most avian niches were already filled, however, only around 24 of 130 species of introduced birds were successful. Usually, the number of exotics introduced has been less than the New Zealand example, but most jurisdictions in the industrialized world have dealt with new animals. Only a small percentage of the introductions have been successful in the past, and today's approach to the use of exotics is more cautious than it was years ago.

There are several questions the manager must be concerned with when dealing with a potential exotic introduction into the wild. Will the newcomer create a disruption in the system? Will diseases or parasites brought in be damaging to native species, or will diseases and parasites already present affect the exotic? Are all habitat requirements present and is a trophic niche available? Is the source stock viable and is the reproductive potential of the exotic capable of insuring success?

In the case of system disruptions, we cannot always be sure what will happen, but we can take precautions. We know, for instance, that niche segregation among indigenous animals functions to reduce competition except under certain conditions which may alter the habitat either on a seasonal basis (winter; the dry season) or during protracted habitat stresses (long-term droughts). In the case of red deer (*Cervus elaphus*) introduced into New Zealand in 1851, the animals were confronted with lush vegetation and no other mammals (except bats), so there was no niche segregation. Other species were soon introduced, but over the next seventy years the red deer increased with essentially no controls and became an agricultural pest of great proportions. Their feeding not only hurt the farmer but it modified the range to the detriment of other species and to man. We cannot blame the red deer, for the New Zealand story is one of man's failures, and more recently, successes. Harvests did not begin until the late 1920s, and soon after that hunters (cullers) were paid to kill red deer to reduce the populations. These cullings became commercial, and now meat and antlers produced on game ranches are significant components of the New Zealand gross national product. The introduction of red deer into New Zealand was certainly successful, but it was also detrimental as were the successful introductions of the cavity nesting starling (*Sturnus vulgaris*) into the United States in 1872 and the burrowing European rabbit (*Oryctolagus cuniculus*) into Australia. The last two are competitive species with high reproductive rates and diverse feeding habits. In these and many other cases, we might have been able to predict what would happen, but the effects of introducing the grey squirrel (*Sciurus carolinensis*) to England and the snowshoe hare (*Lepus americanus*) to Newfoundland might not have been predictable even with today's knowledge and techniques. The grey squirrel competed successfully with the native red squirrel (*Sciurus vulgaris*) and became a forest pest of considerable proportions. The snowshoe hare increased following release, allowing high lynx densities to increase, which in turn reduced the Arctic hare (*Lepus arcticus*) numbers to low levels in restricted habitat.

Usually a quarantine period prior to release will ensure that diseases and parasites of birds are not introduced, but for mammals susceptible to organisms such as rabies with long and variable incubation of six to twelve months, a quarantine alone might not be adequate. In the United States, only the offspring of big game mammals held in captivity may be released into the wild. All industrialized countries have restrictions on the importation of exotics. Usually it is the health and agriculture agencies that are involved, but in North America, there may be both federal and provincial or state legislation preventing importation for any purposes, except by permit. The United States has had the State-Federal Cooperative Foreign Game Program in operation since 1950. This program has three objectives:

1. To provide a reservoir of sound ecological and life history information to individuals and government agencies against which to evaluate foreign species suggested for trial acclimatization
2. To discourage unwise introductions by making these facts available to all concerned
3. To provide an alternate course of action by meeting the recognized need for filling vacant or drastically understocked habitats by a scientifically conceived and executed program of research and trial introductions (Bump, 1968)

One of the keys to success or failure based on the objectives listed is understanding why, in objective (3), for example, the habitats are understocked. Why are indigenous species not present? For highly developed jurisdictions such as in Europe and North America, land-use changes have so altered our native habitats that many species have already been lost or their numbers severely reduced. Replacing these habitats we have new ones, habitats that prevent natural succession and keep lands in crop production year after year. In forestry, we also eliminate stands and often prevent natural succession by intensive management using fire, herbicides, reforestation, and the several mechanized and hand-labor tools of the forester. In essence, we have created habitats dependent upon human management of an intensive nature where exotics are sometimes better suited than are our native species. We find in eastern Washington that three of the four upland game birds present in non-forested habitats are pheasant, gray partridge, and chukar partridge (*Alectoris chukar*), all exotics.

Both release site habitats and source sites require intensive study. It may be that we do not have to match them exactly to allow for success, but cover and food types for the requirements and behavior of the animal must be present, whether identical or not.

In addition to available habitat with an available niche, we also have to concern ourselves with the population from which we will take our animals for release. For a plastic species that may differ throughout a native range in ability to withstand various weather conditions, we should obviously obtain stock from areas with weather similar to that of the release site. Both mean minimum and maximum temperatures may be important, as are snowfall, ice, precipitation, and other weather factors. The animals must be from viable stock relative to survival and should reflect as well optimum reproductive rates for the species.

Animals released into new habitats may exhibit any of several different population patterns. They may increase to reasonable numbers compatible with their habitat, at or below carrying capacity; they may increase to a point above carrying capacity and then decline rapidly (boom-bust); or they may exist in low numbers for an indefinite period. The causes that eventually trigger success or failure may be obvious or unknown to us. Each instance is different and requires its own approach by managers.

Usually it is wise to afford newly released animals as much protection as possible from both man and opportunistic predators. It is also wise to monitor movements of released animals, for sometimes dispersal from release sites is rapid and wide ranging, yet we cannot expect one species from one source area to do the same thing in two different release sites. During ruffed grouse releases in Newfoundland, we had birds staying in the release site area and establishing a population in some cases and dispersing rapidly and widely from preselected release sites in others.

We also need to keep in mind the breeding behavior of the animals we are dealing with. If a critical number at one site is a factor, such as in sharptails (*Pedioecetes phasianellus*), then both the sex ratio and the number of birds available in one small area will be important. With birds like ruffed grouse which do not require a social gathering place for more than two individuals, the number of birds released may be less critical than the sex ratio, which should favor females. The question of "how many" depends greatly on the behavior of the species, its dispersal, its courtship and breeding behavior, its reproductive potential, and its overall plasticity. One additional factor of importance is cost. Thousands of gray partridge were released in upstate New York and thousands of pheasants were released in many different states and in Canada, in Ontario and Nova Scotia, but the entire present population of Newfoundland moose resulted from no more than six animals. The cost-benefit ratios in these cases are considerably different. Exotics are imported into North America for zoos, sportsmen and nonsportsmen to fill vacant or understocked niches and for commercial hunting programs. In this latter circumstance, large mammals have been successfully introduced into several states from Europe, Asia, and Africa. As far as it is possible, the wildlife manager must, however, accept the responsibility of maintaining the integrity of the native ecosystems. Where feasible, native species should be used before exotics.

Reintroductions are often successful if the original factors responsible for the decline and loss of the population are now absent or under control and if the habitat is still present. Translocations or transplantations are also often successful. Turkey (*Meleagris gallapavo*) have been successfully reintroduced into areas of New York, Idaho, Washington, and New England and beaver have been successful in many jurisdictions. Sea otter (*Enhydra lutris*) translocations on North America's west coast from Alaska to Oregon have been more successful in the north than in the south.

Sometimes exotics may be used for "put-and-take" management purposes. In this approach, animals are released in the knowledge that they will survive in the wild at very low rates, if at all, and that a great many will be shot. The purpose is to provide targets for the hunter, and for economic feasibility, as high a proportion of animals as possible should be taken. Pheasants have often been used (as in Michigan, Ontario, Washington, and Maine) on a put-and-take basis although the practice is viewed today with less favor by the public and managers than it was in the past.

In releasing exotics into the wild or in releasing native species into the wild in areas away from their normal range, the manager may employ variations of two basic approaches. In the violent release approach, the animals are released with no opportunity to become familiar with their new surroundings. Early introductions were usually of this type. In the gentle release approach, animals are held in enclosures on the release site for a period of time prior to release, and in some cases, only the young born to these animals are released. This has been the preferred approach for release of endangered species, whether wolves or black-footed ferret. The behavior of the animal and its known tolerance to variable habitat types are two of the parameters to consider in deciding on a release approach.

Outside of wildlife management but impinging on management in the countries of origin, exotics are used in medical research and to help sustain a gigantic pet industry. Colombia has been and remains a leading supplier of primates used for the study of malaria, cancer, and other human maladies. The pet industry uses many tropical and subtropical species of birds, reptiles, and some mammals. Of course captives also reproduce and help maintain stocks. The trade in exotic animal products is sometimes controlled in the market place. Products of animals not listed in the Convention for International Trade in Endangered Species' appendices may be purchased, and there is also a limited domestic market for all products in the countries of origin.

Bibliography

Bergerud, A. T. 1967. The distribution and abundance of arctic hares in Newfoundland. Can. Field-Nat. 81(4):242–248.

Bump, G. 1968. Exotics and the role of the State-Federal Foreign Game Investigation Program. In: *Symposium: Introduction of Exotic Animals: Ecological and Socioeconomic Considerations.* Caesar Kleberg Research Program in Wildlife Ecology, College Station, TX.

Jameson, R. J., K. W. Kenyon, A. M. Johnson, and H. M. Wight. 1982. History and status of translocated sea otter populations in North America. Wildl. Soc. Bull. 10(2):100–107.

King, C. 1984. *Immigrant Killers. Introduced Predators and the Conservation of Birds in New Zealand.* Oxford Univ. Press, New York, NY.

Levi, H. W. 1952. Evaluation of wildlife importations. Scientific Monthly LXXIV No. 6: 315–322.

Parker, R. L. 1968. Quarantine and health problems associated with introductions of animals. In: *Symposium: Introduction of Exotic Animals: Ecological and Socioeconomic Considerations.* Caesar Kleberg Research Program in Wildlife Ecology, College Station, TX.

Pimlott, D. 1953. Newfoundland moose. Trans. N. Am. Wildl. Conf. 18:563–581.

Schonewald-Cox, C. M. 1983. Guidelines to management: A beginning attempt. In: Schonewald-Cox, Chambers, MacBryde, and Thomas (eds.). *Genetics and Conservation*. Benjamin/Cummings, Menlo Park, CA.

Teer, J. G. 1979. Introduction of exotic animals. In: R. D. Teague and E. Decker (eds.), *Wildlife Conservation, Principles and Practices*. pp. 172–177. The Wildlife Society, Washington, D.C.

Vale, R. V. 1936. *Wings, Fur and Shot*. Stackpole Books, Harrisburg, PA.

Shooting Preserves and Put-and-Take

Shooting preserves have been around for a long time. In fact Leopold (1933) suggests that Henry VIII may have used one when he protected pheasants, herons, and partridges near Westminster Palace. Kozicky and Madson (1979) indicate that shooting preserves "have been around for about 70 years and are as American as quail and cornpone." Shooting preserves are far less common in Canada than in the United States, although legislation providing for them occurs in several provinces. Private pay-to-hunt preserves are successful in Ontario and Nova Scotia, for example. A shooting preserve definition is given by Kozicky and Madson (1979) as "an area owned or leased for the purpose of releasing pen-reared game birds for hunting over a period of five or six months." The benefits that are seen to exist are:

1. Legislation provides a longer season and thus an extended period of recreation.
2. Quality, uncrowded hunting is provided near high density human population centers.
3. The use of preserves by a proportion of the hunting public relieves pressure on public lands.
4. Preserves provide a place to hunt in areas where public access for the purpose of hunting may be denied on private lands.

A few released birds may escape to areas outside and survive to reproduce. Some knowledge and expertise concerning the biology and management of the species used are also gained.

As human populations continue to increase, as lands become less available for wildlife and to the hunting public, and as we ease further away from the extensive management approach in North America toward intensive management, shooting preserves may play an increasingly important role in wildlife management. This probably will hold true even though a high proportion of the public opposing hunting sees the shooting preserve as one of the worst examples of blood sports.

A large number of birds and mammals are popularly advertised as available on preserves or "resorts" as they are referred to in Texas. Various species of Asian and North American ungulates, wild boar (*Sus scrofa*), quail (*Colinus virginianus*), turkey, mallard ducks, gray partridge, and ring-necked pheasant are among the scores of animals raised under captive or semi-captive conditions and hunted on private preserves for profit. The pheasant, however, is the most important target wherever preserves exist within its range. Upland game birds used must flush readily and fly high and fast, providing a good target for the hunter. Recovery rates of released birds must also be high to ensure a reasonable profit. Birds raised in captivity also must lend themselves well to captive conditions and offer excellent survival rates in captivity while retaining their wildness.

Hunters who do not favor shooting preserves may disagree with the put-and-take concept or simply feel that artificially propagated birds are too tame. Others may be displeased as a matter of class distinction or income level, because they are not able to afford artificial hunting and do not approve of it because of the class or cost factor. Feelings sometimes run high, which shows that all hunters are not of similar mind when it comes to all hunting practices.

The shooting preserve concept is a natural extension of two management principles, stocking and protected areas. Stocking of both exotic and native species has a long history (Chapter 1). It has been a mainstay in fishery management and during the 1920s and 1930s was a popular element in approaches to wildlife management. *Forest and Stream* ran a column called "Practical Game Breeding," written by G.C. Corsan for years, and although game breeding was in the title of the column, stocking and area protection were often discussed with mention of private sanctuaries, including Jack Miner's, occurring commonly. Since the stocking of exotics, reintroduction, and translocations are mentioned generally elsewhere, we are concerned here only with put-and-take stocking, which is used most efficiently on preserves. The obvious reason is that the releases of birds are coincident with known hunting pressure. Preserve management combines the skills of habitat management, propagation, and business acumen. Cover types and feeding and roosting areas are maintained specifically for upland game birds in such areas. Habitats are as close to the optimum for the birds as possible yet natural for hunters and dogs. Although some escapees or even wild stock may be present from time to time, the majority of birds harvested are released from the pens forming the game farm portion of the preserve. In some situations, the preserve and game farm may operate independently, with birds purchased and released at regular intervals. In the case of pheasants where hunting may be restricted to males, depending upon the legislation, many more males than females are released. Females are occasionally retained for sale as meat birds, if the law allows. For species such as gray partridge, both sexes are released and harvested. The use of one upland game species over another depends upon a number of factors including breakeven costs, hunter choice, habitat available on the preserve, and the availability of birds. Since it is the recovery rate that is important on preserves and survival of pen-reared birds in the wild is low in most cases, the viability of the stock is not important here although it is of obvious importance in other forms of stocking. Put-and-take stocking is used on public lands in some instances, but public lands are even now inadequate to provide hunting opportunities close to many metropolitan areas. The put-and-take stocking efforts can never be as efficient on such lands as they are on preserves because of costs involved and because neither hunting pressure nor hunter distribution can be strictly controlled.

Besides wildlife agencies, the North American Game Breeders and Shooting Preserve Association of Goose Lake, Iowa, is the most up-to-date source of regular information for both game breeding and shooting preserves. Wildlife managers should be familiar with the legislation in the jurisdiction where they work and the major problems of game breeding for species most commonly used. They must also understand the basic concepts of preserve management.

Bibliography

Byers, S. M. and G. V. Burger. 1979. Evaluation of three partridge species for put and take hunting. Wildl. Soc. Bull. 7(1):17–20.

Hawley, A. (ed.). 1993. *Commercialization of Wildlife in North America.* Krieger Publ., Malabar, FL.

Kozicky, E. L. and J. B. Madson. 1979. The shooting preserve concept. In: *Wildlife Conservation, Principles and Practices.* pp. 156–160. The Wildlife Society, Washington, D.C.

Leopold, A. 1933. *Game Management.* Charles Scribner's Sons, New York, NY.

Ratti, J. T. and G. W. Workman. 1976. Hunter characteristics and attitudes relating to Utah shooting preserves. Wildl. Soc. Bull. 4(1):21–25.

Schreiner, C., III. 1968. Uses of exotic animals in a commercial hunting program. In: *Symposium: Introduction of Exotic Animals: Ecological and Socio-economic Considerations.* Caesar Kleberg Research Program in Wildlife Ecology, College Station, TX.

Migratory Animals

Migration has always fascinated biologists. The "why" question has been answered for most species but the "how" still remains a question for many.

Among the mammals, the barren-ground caribou in North America and the reindeer of Russia (both *Rangifer tarandus*) move for food and calving purposes. In savannah-land areas of Africa, ungulates move for food as periods of rainfall and drought alter the availability of grasses and forbs. In mountainous regions, ungulates may undertake altitudinal movements for food and comfort. Migration in birds is a subject which has kept the interest of naturalists and researchers high for hundreds of years. Some species, like the crow (*Corvus corax*), may move short distances while others, like most warblers (family Parulidae), fly thousands of kilometers. Where man has not disrupted the process, migration allows mammals and birds to escape periods of cold or drought, to produce their young at the most propitious times in relation to weather and climate, and to have food for them. Like hibernation, migration has evolved over millions of years and provided selective advantages that have allowed for the successful existence of hundreds of species.

Regular (or nearly so) movements of animals, whether fish, reptiles, amphibians, birds, or mammals provide certain unique problems to the wildlife manager for only a part of the year. Perhaps this will be summer or the dry season. In winter or during the rains, another manager may be concerned with the species. As long as the animals remain within the boundaries of one jurisdiction, management usually can be coordinated rather well. With animals of limited mobility, there is little chance that political boundaries will be crossed because they move short distances but this is not true of most migratory species. In North America, coordinated, cooperative management was made possible through the migratory bird treaties and implementing legislation in Canada, the United States, and Mexico. For certain marine mammals like whales, some cooperative management has also been possible, although not always effective. For many species, however, cooperative arrangements without the support of legislation have had to develop and the problems have often been complex. Such is the case today, for example, with the Kaminuriak caribou herd in Canada.

Recently, international concern has been apparent in relation to migratory species. Beginning about 1962, a series of intergovernment conferences on wetlands considered the topic from many points of view and for many purposes, one of which was for waterfowl habitat. In 1971, the Convention on Wetlands of International Importance Especially as Waterfowl Habitat was adopted at a conference in Ramsar, Iran. Parties to this intergovernmental treaty accept two major obligations:

1. To recognize the international importance of at least one wetland within its territory by placing it on the convention list

2. To formulate and implement planning to promote the conservation of wetlands so listed and wise use in general of wetlands within their territory

Prior to this convention only the regulatory management of waterfowl and other migratory birds was possible under international legislation in North America. The protection of habitats and reservation of lands was accomplished on a national, state, or provincial basis in Canada, the United States, and Mexico. Canada became a party to the Ramsar Convention in 1981 and by 1989 had listed eighteen wetlands with the convention because of their importance primarily to migratory birds. Some are staging areas while some are stopover or breeding, nesting, and brood areas. Under the basic criterion of the convention, one percent or more of a regional population of a species must use the area at some time during the year. Although protection could, of course, be provided under national or provincial legislation, the listing under an international convention provides a sounder long-term guarantee of protection and opens the door to further international cooperation in management.

In 1979, the Convention on the Conservation of Migratory Species of Wild Animals was held in Bonn, West Germany. This was an umbrella convention to provide agreements covering individual migratory species or groups of species throughout their twelve-month range(s). International conventions can serve to restrict national autonomy in the management of migratory species, and it may be for this reason, as stocks of fish in particular became the focus for international conflict in resource management, that North American jurisdictions have not rushed to become party to the Bonn convention. Nevertheless we can look to the increased effectiveness of international conventions in restricting national domination over resources in the future, for this is really the only way in which resources of the "commons" and migratory animals can be properly managed on a world wide basis.

The IUCN supported by the United Nations Environmental Program was responsible for authoring the draft agreement for the Bonn conference that was signed by twenty-two nations, many of them underdeveloped countries. Some of the problems related to international conventions of this nature are noted by Johnson (1980), and it is significant that he notes lobbying by non-government organizations (NGOs) as being one of them. Many antihunting and other preservationist groups are among these NGOs. They argue primarily from moralistic and emotional positions and as Johnson notes " . . . nearly 70 lobbyist groups were represented at Bonn and some of these carry a great deal of political clout in their home countries. Only a few management groups attended or were represented." Here we see once again the potential for decision making by nonprofessionals. Johnson further urges that "Wildlife managers must increase their involvement in these matters."

Bibliography

Johnson, M. 1980. New treaty arouses management issues. Wildl. Soc. Bull. (2):152–156.
Vontobel, R., (ed.). 1982. *Man and Wildlife in a Shared Environment*. Can. Wildl. Serv., Ottawa, Ontario.

Urban Wildlife

There is a growing realization that urban centers can be made more natural and contain at least some wildlife components which will allow the human residents to understand more readily the ecological ties which bind us all. While this realization has come almost too late for some inner city areas, there have been welcome successes in many urban cores—from the hacking of peregrines on downtown office buildings in Montreal, Baltimore, and Edmonton to architectural designs which include roof and ledge gardens for avian habitat.

The city generally has been antithetical to nature and to man's psyche. Puri (1974) pointed out that the stresses and tensions in people living in cities cause numerous mental and psychosomatic disorders. Many of these problems of urban living cannot be treated successfully with conventional medicine. Instead, relaxation is one of the major sources of relief. How often have you found the solace of an urban park to be sufficient to ease the tension of noise, pollution, and the masses of humanity and concrete which defile our earthly base? How often did you think that it was not just the vegetation—the trees, shrubs, flowers—that was achieving this relaxation but the animals also—the squirrels, pigeons, robins, and sparrows? Geist (1975) echoed this sentiment when he stated that "the management and support of wildlife in urban areas is in the final analysis an exercise in preventive medicine." The late Constantinas Doxiadis (1975) looked at the problem of declining natural areas at the global level and felt there was only one solution and that required an "overset system" of wildlife and human settlements. In a global ecology context, there should be a balance between agriculture, industry, cities (settlements), and natural areas; there also should be intergradation so that natural areas literally invade urban complexes. Nature and wildlife must be part of the cities of the future if mankind is to avoid the mistakes of past civilizations which sought to remove or overpower nature and in the process created wastelands.

Perhaps the best mechanism for ensuring consideration of wildlife in urban areas is by specifically planning for it. Subject to management are open space and building design, the only two factors necessary to ensure both aesthetically pleasing human environments and wildlife habitat. Wildlife professionals too often avoid the urban scene for more rural surroundings. Wildlife management is first and foremost dealing with people, and over 70 percent of North America's people are found in cities. We have a responsibility for contributing to the urban planning process and lobbying effectively for natural ecological values; there is considerable public approval for such action. We have failed so far to capitalize on much of that latent political support.

What are some of the elements of wildlife management in an urban context? At the planning level, it means advocating use of open space areas as wildlife habitat or designing corridors of habitat that link natural areas together. Often existing linear systems can be used to this end—railway and utility rights of way, streams, and even highway verges and medians. These corridors will tie together the urban ecological preserves and parks. Most cities already have natural or seminatural areas that are undeveloped (hazard lands) or have land uses of little impact (e.g., cemeteries). By linking these and providing access for viewing and educational purposes, Doxiadis' dream might be realized in every city! These areas need protection, however. If narrow belts of wildlife habitat or other sensitive places are adjacent to high-density residential sections they are doomed to abuse and possible destruction. Adjunct land uses should be carefully controlled; low-density residential, hospital or convalescent homes, housing for the elderly, and light industry are appropriate.

Architectural designs which minimize or eliminate nesting or loafing areas for species like starlings, English sparrows, and pigeons can be incorporated into new construction. Landscaping with native shrubs and trees of high wildlife values and reducing the need for mowing (i.e., grassed areas) will encourage wildlife. There are now countless prescriptions for landowners on how to develop backyard wildlife habitat. The USFWS Biological Services Program also has a document outlining methodologies for planning for wildlife in urban and suburban areas. The Urban Wildlife Research Center, Inc., founded in 1973, (now the National Institute for Urban Wildlife) serves as a storehouse of information for wildlife planning in cities. It also conducts research relating to "(1) effects of man's activities on wildlife and wildlife habitat; (2) environmental and wildlife planning and management for new residential, commercial and industrial developments; and (3) human-wildlife relationships in existing urban or other developed areas."

With the burgeoning interest in urban wildlife and its management, there is a growing need for wildlife managers committed to working within and improving metropolitan environments. Unfortunately, too many "wildlifers" are still attracted to the profession by the call of the wild and are trying to escape the city rather than improve it. Until our profession makes the necessary moves to ensure a wildlife voice is heard in urban planning circles, we will continue to lose, not by malicious acts of planners but by the lack of consideration of alternate ways of doing things which are compatible with or enhance wildlife values.

A very important consideration is that most of the human population lives in cities, many of whose inhabitants have lost touch with their ecological roots; the educational system is not adequately geared to redress that deficiency. More direct understanding of nature and how our planet functions should be incorporated into elementary curricula. Too often field trips to natural areas or interpretive centers are considered luxury items and are dropped when school budgets are cut. Many inner city schools are within districts which do not have a strong tax base, so ecological field exercises become too expensive to be kept in the program. We should show that such experience at the elementary school level is essential to healthy mental development and enrichment as well as to understanding the complexity of life.

Project-Wild has been developed by the Western Association of Fish and Wildlife Agencies and the Western Regional Environmental Education Council with the support of agencies such as the National Wildlife Federation and Defenders of Wildlife. It has been adapted for use in several states and provinces. This educational program uses wildlife examples pertinent to a child's understanding in diverse subject areas like mathematics and physics and is suitable in classes from kindergarten to high school level. It is hoped that by using examples from nature in courses other than biology or environmental studies the students who do not take these courses will at least get some exposure to ecological principles and to wildlife as real components of their "world." Project-Wild is introduced to teachers through workshops conducted by the local wildlife and education agencies. In this way, it is expected to receive greater attention and use than if curricular guides were simply provided without proper instruction in their use. Many of the problems associated with the management of wildlife species in cities have been addressed and adequately summarized in recent symposia. The reader should review these for more in-depth analysis of the particular attributes of this specialized wildlife management area (see Noyes and Progulske, 1974; McKeating, 1975; Euler et al., 1975; USDA, 1977; Adams and Leedy, 1987).

Bibliography

Adams. L. W. and L. E. Dore. 1989. *Wildlife Reserves and Corridors in the Urban Environment. A Guide to Ecological Landscape Planning and Resource Conservation.* Natl. Inst. for Urban Wildl., Columbia, MD.

Adams. L. W. and D. L. Leedy (eds.). 1987. *Integrating Man and Nature in the Metropolitan Environment.* Natl. Inst. for Urban Wildl., Columbia, MD.

Doxiadis, C. A. 1975. Wildlife and human settlements. In: D. Euler, F. Gilbert, and G. McKeating (eds.), *Proceedings of the Symposium—Wildlife in Urban Canada.* pp. 2–22. Off. Cont. Ed., Univ. Guelph, Guelph, Ontario.

Euler, D., F. Gilbert, and G. McKeating (eds.). 1975. *Proceedings of the symposium—wildlife in urban Canada.* Off. Cont. Ed., Univ. Guelph, Guelph, Ontario.

Geis, A. D. 1975. Urban planning and urban wildlife; a case study of a planned city near Washington, D.C. In: D. Euler, F. Gilbert, and G. McKeating (eds.), *Proceedings of the Symposium—Wildlife in Urban Canada,* pp. 79–84. Off. Cont. Ed., Univ Guelph, Guelph, Ontario.

Geist, V. 1975. Wildlife and people in an urban environment—the biology of cohabitation. In: D. Euler, F. Gilbert, and G. McKeating (eds.), *Proceedings of the Symposium—Wildlife in Urban Canada,* pp. 36–47. Off. Cont. Ed., Univ. Guelph, Guelph, Ontario.

Kelcey, J. G. 1975. Opportunities for wildlife habitats on road verges in a new city. Urban Ecol. 1:271–284.

Leedy, D. L. and L. W. Adams. 1984. *A Guide to Urban Wildlife Management.* Natl. Inst. for Urban Wildl., Columbia, MD.

McKeating, G. B. 1975. *Nature and Urban Man.* Can. Nat. Fed. Spec. Publ. No. 4. Ottawa, Canada.

McKeating, G. B. and W. A. Creighton. 1975. *Backyard Habitat.* Ontario Min. Nat. Resour., Toronto, Ontario.

Noyes, J. H. and D. R. Progulske. 1974. *Wildlife in an urbanizing environment.* Univ. Mass., Planning and Resource Development Ser. No. 28, Amherst, MA.

Puri, G. S. 1974. The health hazard of stress in the urban-industrial environment of developed countries. Intl. J. Ecol. Environ. Sci. 1:53–60.

Thomas, J. W., R. O. Brush, and R. M. DeGraaf. 1973. Invite wildlife to your backyard. Natl. Wildl. Mag. (Apr–May).

U.S. Department of Agriculture, Forest Service. 1977. *Children, Nature, and the Urban Environment.* Gen. Tech. Rep. NG30.

Depredations

Many wildlife species project contrasting images depending on which segment of the public is doing the viewing. Wolves are symbolic of wilderness to many, but to a livestock rancher they are vermin. Red-winged blackbirds are pretty birds to some, but to agriculturists they can be economically devastating if a farmer happens to grow corn, sunflowers, or any number of crops close to a summer or winter roost. These are the classic confrontations that, exclusive of the regulatory management of game species, have occupied a disproportionate share of attention in wildlife management. This has occurred because, as with game species, economic values can be assigned. Damage to food production, be the commodity sheep (coyote), honey (bears), fruit (deer, small mammals and birds), or grains (small mammals and birds), totals in the billions of dollars and has focused attention on means of controlling populations of depredating wildlife. The USFWS, until 1986 when the Department of Agriculture assumed the responsibility, operated a predator control section that had concentrated primarily on the coyote. Prior to the 1960s, the leg hold trap, M-44 ("coyote-getter"), shooting, and toxicants such as 1080 were used to reduce coyote numbers-generally without significant success according to the carnivore catches from survey traplines. Some jurisdictions used bounties but they, too, usually have proven ineffective. Bounties are frequently paid for animals that would have been taken anyway. Biologically, it is difficult to find any cases where population levels of carnivores have been reduced as a result of bounties.

Two major reports on predator control have been submitted to the U. S. federal government. The first of these (Leopold et al. 1964) contained the conclusion that "All native animals are resources of inherent interest and value to people . . . [therefore] Basic government policy should be one of husbandry of all forms of wildlife." However, in situations where a species was causing significant damage to other resources or crops or where it endangered public health or safety (e.g., rabies), local population control was an acceptable and essential management tool. The authors cautioned that control had to be limited to the troublesome species, or preferably individuals, and to the locality where the damage or danger existed.

A second report (Cain et al., 1972) authorized through the Advisory Board on Wildlife Management (as was the Leopold report) recommended cessation of toxicant use. This was one of fifteen points intended to improve environmental safeguards associated with predator management. The committee contained five wildlife ecologists, one plant ecologist, and one political scientist and was criticized strongly for not having representation from the livestock industries, especially when a month later an executive order was signed into law which restricted the use of toxicants on federal lands and in federal programs of mammal and bird damage control. In addition to the political repercussions from stockmen who tried to have predator control turned over to the states, this situation stimulated efforts to find alternative methods of coyote management; hence, new research was funded. The stockmen failed in their attempt to have more sympathetic management agencies (state agricultural departments) take over from the federal authority. If they had succeeded, it would have been antithetical to the concept that wildlife is the property of all the people and, therefore, should be managed by agencies responsible to a plurality of public values. A report prepared by the Mammalian and Avian Pest Management Committee on coyote predation in Ontario addressed this particular issue by suggesting that county predator control committees be formed in areas designated by a provincial predation control coordinator. The coordinator, an employee of the Fish and Wildlife Division of the Ministry of Natural Resources, was to work closely with the director of the Livestock Branch of the Ministry of Agriculture and Food. The Predator Control Committee would be given responsibility for directing management actions to prevent stock predation. The advantage of the system was that farmers and control agents were to operate through a local committee that should be more responsive and, therefore, more effective than a bureaucracy operating out of regional or headquarter offices. The committee representation was from the Ministry of Natural Resources (1 conservation officer), Ministry of Agriculture and Food (1 agriculture representative), municipal government (1), and producers. The committee, chaired by the agriculture representative, was to designate the location, intensity, and method of predator control, educate the producers, and coordinate local sportsmen and farmers to reduce the need for future control efforts. The scheme was never adopted by the provincial government; instead it continued to provide compensation to producers for documented livestock losses to wolves or coyotes.

A simulation model of coyote population control showed that the primary effect of killing coyotes was to stimulate density dependent changes in birth and natural mortality rates and that coyotes could thus sustain their populations except at the very highest levels of control. If 75 percent of a population were eliminated each year, it could be exterminated in slightly over fifty years. However, populations reduced by intensive control measures could recover to their precontrol densities in as few as three years once the control measures ceased. In effect, this supports the need for control efforts to be targeted at the problem individuals rather than the population as a whole, unless the control measures can be selective against pregnant females or females and their litters (a knowledge of the location of natal dens is required).

For the moment, management efforts are awaiting results of the research generated by the removal of 1080 and M-44s as control agents. Whether aversive conditioning (lithium chloride sheep collars), sterilization (diethyl stilbestrol), or repellents ever prove effective is a moot question. Perhaps better husbandry and more selective removal methods are all that are actually needed. The toxicant 1080 has been approved for such selective use.

Blackbird control poses another set of interesting management dilemmas. Some of the aversive conditioning agents like Avitrol (4-aminopyridine) have side effects which include nontarget species poisoning and distress vocalizations of the target species (which results in the aversive conditioning). The nontarget problem can be minimized or even eliminated by not treating the headlands and first rows of the crop, but the visual and auditory stimuli can be unpleasant or misinterpreted by an uninformed observer. This may not be a problem in agricultural areas. When the chemical was used in an urban situation (Toronto), the city folk did not take kindly to birds dropping out of the sky and flopping around on the lawns of a hospital. The use of Avitrol is only cost effective in certain situations, and movement of birds to different foraging sites may necessitate repeated applications. Therefore, this chemical agent and others such as methiocarb (3,5 dimethyl-4-(methylthio) phenol methyl-carbamate) did not offer the growers any real panacea.

Until recently, the USFWS had conducted a program of population reduction at winter roost sites. A detergent (Turgitol) was used to break down the oil on feathers and destroy their insulating value when they were wetted. This resulted in hypothermia and death of the birds so treated. The approach had value in reducing disease and depredation problems associated with the roosts, but had little if any impact on summer populations and crop depredation near summer roost sites. Because Turgitol is effective at temperatures below 10 °C (50 °F), a postbreeding, summer roost population test program was proposed for a site where intensive studies of population, biology, and depredation had been conducted. Despite stringent environmental precautions and proposed water monitoring, pre- and post-spray, the provincial government did not give approval until after it had been reelected with a substantial majority. The lesson here is that potentially controversial wildlife management procedures probably should not be proposed or conducted close to election time. The proposal to kill a local population of blackbirds concerned an area where farmers (a small but vocal voting group) strongly supported the action. In terms of political realities, however, the proposal caused fear that the press might make an issue of it and thus generate a negative influence on urban voters. While all governments and politicians are not so sensitive, it is a truism that most are, especially if there is a precedent which resulted in bad press. Canadians, in particular, seem especially prone to such a reaction possibly because of the harp seal issue and the strength of the antivivisectionist movement in that country.

The Jack H. Berryman Institute at Utah State University has become a major center for research on depredation. Its faculty and graduate students have been successful in carrying out a mission to develop integrated approaches to resolving human/wildlife conflicts such as those mentioned in this section.

Bibliography

Cain, S. A., J. A. Kadlec, D .L. Allen, R. A. Cooley, M. H. Hornocker, A. S. Leopold, and F. H. Wagner. 1972. *Predator Control—1971 report to the Council on Environmental Quality and Department of the Interior by the Advisory Committee on Predator Control.* Univ. Michigan Press VIII, Ann Arbor, MI.

Connolly, G. E. and W. M. Longhurst. 1975. *The Effects of Control on Coyote Populations: A Simulation Model.* Dir. Agric. Sci. Univ. Calif. Berkeley. Bull. 1872.

Dolbeer, R. A. 1980. *Blackbirds and Corn in Ohio.* U.S. Fish Wildl. Serv. Resour. Publ. 136.

Gilbert, F. F. 1995. Historical perspectives on wolf management in North America with special reference to humane treatment in capture methods. In: *Ecology and Conservation of Wolves in a Changing World,* pp. 13–18. Occas. Publ. Ser. No. 35. Can.Circumpolar Inst., Edmonton, Alberta.

Leopold, A. S., S. A. Cain, C. M. Cottam, I. N. Gabrielson, and T. L. Kimball. 1964. Predator and rodent control in the United States. Trans. N. Am. Wildl. Nat. Resour. Conf. 29:27–49.

Linhart, S. B. and W. B. Robinson. 1972. Some relative carnivore densities in areas under sustained coyote control. J. Mammal. 53:880–884.

Somers, J. D., F. F. Gilbert, D. E. Joyner, R. J. Brooks, and R. G. Gartshore. 1981. Use of 4-aminopyridine in cornfields under high foraging stress. J. Wildl. Manage. 45:702–709.

Somers, J. D., R. G. Gartshore, F. F. Gilbert, and R. J. Brooks. 1981. Movements and habitat use by depredating red-winged blackbirds in Simcoe County, Ontario. Can. J. Zool. 51:2206–2214.

Stickley, A. R., Jr., D. L. Otis, and D. T. Palmer. 1979. Evaluation and results of a survey of blackbird and mammal damage to mature field corn over a large (three-state) area. In: *Vertebrate Pest Control and Management,* pp. 169–177. Am. Soc. Testing and Materials. Spec.Tech. Publ. 680. Philadelphia, PA.

Tyler, B. M. J. and L. W. Kannenberg. 1980. Blackbird damage to ripening field corn in Ontario. Can. J. Zool. 58:469–472.

Windberg, L. A. and F. F. Knowlton. 1990. Relative vulnerability of coyotes to some capture procedures. Wildl. Soc. Bull. 18:282–290.

Wywialowski, A. P. 1994. Agricultural producers' perceptions of wildlife-caused losses. Wildl. Soc. Bull. 22:370–382.

Some Specialized Areas of Management • 251

Humane Trapping

One last emotional issue in wildlife management we will look at is the trapping of furbearing animals. Many humane organizations have focused their attention on the steel-jawed leg-hold trap, targeted because of the damage it causes to the animal, its general nonselectivity and its common use by trappers. In reality, much of their activity is overtly or covertly designed to eliminate commercial trapping altogether. As managers of wildlife, we have been remiss in supporting wildlife trapping without qualification. In Ohio, for example, blatantly false impressions were generated to defeat a proposed trapping ban. Instead of speaking out against the ludicrous claims that such diseases as rabies and bubonic plague will increase in prevalence without the use of the leg-hold trap, managers tacitly approved of the approach. This type of biased approach does little to enhance the profession's credibility even though it may help achieve policy or political objectives. Excellent biological arguments can be made for population management of such herbivores as beaver and muskrat and even for carnivores, which can be culled by means of careful management. Unfortunately, the management expertise needed to exact the proper age and sex quotas and the means selectively to trap individuals from populations are often lacking. As an example, only relatively recently have regulations been changed to protect adult, female fishers which become increasingly vulnerable to trapping as the season progresses. With a January closing, harvest in Ontario was primarily restricted to juveniles and adult males. Without using live traps, which have the ancillary benefit of allowing the release of unwanted individuals, it is unlikely that the selectivity required for intensive management can be achieved—unless tremendous progress is made with pheromones or other chemical attractants or repellents.

Bearing this in mind, it is critical that traps and trap sets be made as selective as possible and that they either dispatch the animal humanely or hold the animal with minimal discomfort. Trapper education can be used to demonstrate the proper traps and sets for given species, but standards are necessary to ensure that traps meet accepted humane criteria. Recent improvements to leg holding devices include soft hold traps which have rubber jaws, leg snares, and triggers which can be set to different tripping pressures (and thus have a weight threshold to increase selectivity).

In Canada, the Federal Provincial Committee for Humane Trapping (FPCHT) submitted a final report in 1981 which contained a number of recommendations intended to maximize the "humaneness" in holding or killing furbearers. This government steering committee, created by the Federal-Provincial Wildlife Conference in 1973, grew out of the efforts of Canadian humane societies to focus public attention on the capture of furbearing animals. Along with sensationalist activities, like "They take so long to die"—a film dramatizing the trapping of wild animals, objective efforts such as those by the Humane Trap Development Committee (HTDC) of the Canadian Federation of Humane Societies were undertaken. The HTDC made a performance evaluation of quick-kill traps from both mechanical and biological perspectives. Ultimately the FPCHT settled on a similar tack when a scientific and technical subcommittee became involved in determining policy and direction for the parent group.

A number of scientific studies were undertaken to determine the following:

1. The energy thresholds, by species, required to effect a humane death (humane by standard definition means rapid loss of consciousness followed by irreversible subsidence to death—for killing traps—but for the purpose of laboratory testing three minutes to loss of consciousness was used)
2. The mechanical characteristics, including the energy output of traps
3. The interface between traps and target species

The end products of this research were that killing traps should strike the animal in the head/neck region of the body, that threshold graphs could be produced by species showing the impact and clamping forces (and combinations thereof) necessary to meet the humane criterion and that traps existed or could be modified to meet the mechanical criteria. The major deficiency of the FPCHT efforts was that virtually no field testing was undertaken to corroborate the research findings.

Examination of leg hold traps in "drowning" sets led to recommendations that they only be used for muskrat and mink and not for beaver. Similarly, snare studies showed that these devices did not meet the humane criterion established for red squirrels or canids. Needless to say, the recommendations were controversial for several reasons, not the least of which was that implementation would revolutionize how fur trapping was conducted in Canada. Some provinces such as British Columbia and Ontario adopted some of the recommendations relatively quickly because they were under the greatest pressure from anti-trapping groups. But Canada has moved rapidly toward more uniform trapping regulations that restrict the use of leg-hold traps, reduce trap check times, and the establishment of standards for killing and restraining trap devices. Trappers generally opposed restrictions on the use of snares and the steel-jawed leg-hold traps because these were the cheapest, and to them, most effective traps.

The FPCHT findings provided the basis for the first humane standard for killing traps established through the Canadian Government Standards Board. The Canadian government, in response to pressure from certain CITES nations which suggested sanctions against fur taken by inhumane means, proposed adoption of humane trapping by the international community. The action precipitated significant activity by the American National Standards Institute to establish a U.S. Technical Advisory Group (TAG) which would examine the issue within that country. This was an important step because many of the Canadian recommendations, which related primarily to killing devices, had little applicability in the United States. Many jurisdictions in the United States have banned the use of killing traps because of the danger to pets and fears for human safety in high human density areas. It was hoped that the TAG would be able to overcome the extremely polarized positions of the trapping (trappers, managers) and anti-trapping (humane societies, animal lovers) groups. A sufficient number of nations, including Canada and the United States, voted to establish an International Standards Organization (ISO) Technical Committee on humane animal traps. Work began in 1986 to pursue meaningful international standards for killing and restraining trap devices. In addition, Canada established the Fur Institute of Canada (FIC) in 1983 to continue the work initiated by the FPCHT. The FIC embarked on a very ambitious research and public relations effort designed to offset the emotional approach of the many anti-trapping groups. But while considerable progress was made in research and the various working groups of the ISO process helped delineate options for international standards, anti-trapping forces caused the effort to fail. The irony was that the very groups that decried the inhumaneness of trapping became the obstacle to implementing standards that could have improved the situation for millions of animals. In 1998, all that was left of the prodigious effort to make a meaningful step worldwide toward truly humane trapping devices was a testing standard for traps. Undeterred, Canadian interests pursued, with some success, the development of new Canadian standards for humane traps that provided incentives to produce new humane devices within a reasonable time frame in areas where no traps currently met the standards.

Coincident with the ISO effort and a function of its impending failure, a Quadrilateral Task Force made up of European Union (EU), United States, Russian, and Canadian delegates was established to try to develop a compromise that would prevent a trade embargo by the EU and a trade challenge to GATT by the United States. Although this body managed to reach a shaky consensus, the United States only became signatory to the agreement at the last moment. Despite a reluctance to pursue this option the United States has shown its commitment by providing significant new funding for humane trap research. Meanwhile Canadian jurisdictions are moving to end the use of the leghold trap for the few remaining species for which it was still allowed.

As one of the authors was an active participant in the U.S., Canadian and international efforts, he was provided with an unique perspective of how vested interests can circumvent the best intentions of people of good will to achieve difficult objectives. For example, so-called animal welfare interests in the United States, that in reality represented anti-trapping interests, did not become involved in the standards process until it became apparent that the better part of a decade of effort might actually result in international and, subsequently, national, standards. The spectre of humane standards that would allow trapping and possibly even use of the hated leghold trap to continue was enough to mobilise them into action. The pseudoscience, half-truths, character defamation, and political lobbying that were all part of the process to discredit the work of wildlife managers, wildlife biologists, scientists, and others involved in the standards development were impressive in their magnitude and disregard for fact. The saddest aspect of the whole thing was that the biggest losers were the animals that continue to be trapped for fur, damage control, health, and many other reasons.

What might have seemed like a simplistic question of "To trap or not to trap?" becomes in reality a more sophisticated association of "How to trap? With what to trap? Who should trap? What should be trapped? and Why should it be trapped?" Wildlife managers were forced into the scientific and political arenas to become effective as professionals. But even the ability to communicate honestly and at the proper level of technicality for the audience,

Some Specialized Areas of Management • 255

while a key asset for any manager, may not be enough to overcome the political force of a committed adversary. Protecting and managing the resource requires educating, informing, and consulting the public interested in that resource and only with the support of the public will the force of logic prevail in the trapping controversy. Decisions cannot and probably should not be made solely by agency personnel, for to do so risks jeopardizing the scientific basis of management by exposing it to the emotional arena of politics. That is the responsibility of elected officials, not civil servants. As managers we must countenance, respect, and perhaps respond to those interests that may be different from our own but are nonetheless elements representing the general tax-paying public. A manager in reality only makes recommendations to government regarding action to be taken. If there is scientifically documented rationale for the recommendations and the impacts on the resource and on the public have been adequately demonstrated, governments must be prepared to act. It is more likely to act favorably if the management agency has made its case well to the public. In Canada, the governments did just that and although slower to respond U.S. governments are beginning to do the same by supporting research and considering the possibility of developing national humane trap standards. Therein lies the secret to successful wildlife management. Scientific knowledge may be the underpinning, but the capability of "selling" a management scheme to politicians (and their electors), no matter how long it may take, is equally crucial to success.

Wildlife managers must be aware that there are methods being used to circumvent this legitimate management agency role and this process. Such was the case in a 1983 referendum in Maine to prohibit moose hunting. If the referendum had been successful, the moose management program would have been damaged and a precedent established to manage by public fiat rather than on the advice of professional scientists. If the public had been uninformed and the biological case for the moose season had not been made adequately, wildlife management in Maine would have suffered a serious setback. So in reality, the voters sustained the management decision because the management agency had been open and truthful with the public. In a democratic system, we, as wildlife managers, must be prepared to defend sound management decisions in any forum, including the political one. Similar challenges are becoming more commonplace and Colorado and California have seen management decisions made by public ballot. The American system allows for referenda to be placed on the ballot, which usurps the normal representative governmental process. If managers ignore this reality and operate outside the public arena, they must be prepared to accept the consequences. In California, voters decided to close the mountain lion season despite managers' opinions that the population was growing and could support a harvest. They have voted as well to ban the use of leg hold traps. Decisions by voter initiative thus are becoming commonplace in California and it is possible that the voter may decide the fate of hunting in that state by this process in the near future. In the case of the mountain lion, recent human mortalities caused by mountain lions may place it once again in an open hunting season. In the case of leg hold traps, arguments that endangered species are now at risk because their predators cannot be controlled by trapping are being used to try to overturn the referendum action.

One perspective on furbearer trapping and the biological, social, and political issues surrounding that controversy in the United States is provided by Andelt et al. (1999). These authors conclude that wildlife managers have focused on biology to the exclusion of social science and politics, a position we have gone at some length to take in this book. In the case of the trapping issue, they offer sensible action for managers that are consistent with the development of standards and being responsive to the legitimate concerns of all segments of the public.

Bibliography

Andelt, W. F., R. L. Phillips, R. H. Schmidt, and R. B. Gill. 1999. Trapping furbearers: an overview of the biological and social issues surrounding a public policy controversy. Wildl. Soc. Bull. 27:53–64.

Federal-Provincial Wildlife Conference. 1981. Report of the Federal Provincial Committee for Humane Trapping. Ottawa, Ontario.

Gilbert, F. F. 1991. Trapping-an animal rights issue or a legitimate wildlife management technique. The move to international standards. Trans. N. Am. Wildl. Nat. Resour. Conf. 56:400–408.

Goodrich, J. W. 1979. Political assault on wildlife management: is there a defense? Trans. N. Am. Wild. Nat. Resour. Conf. 44:327–336.

Lautenschlager, R. S. and R. T. Bowyer. 1985. Wildlife management by referendum: when professionals fail to communicate. Wildl. Soc. Bull. 13:546–570.

Linhardt, S. B., G. J. Dasch, and F. J. Turkowski. 1981. The steel leg-hold trap: techniques for reducing foot injury and increasing selectivity. In: F. A. Chapman and D. Pursley (eds.), *Worldwide Furbearer Conference Proceedings, Vol. III*, pp. 1560–1578. Frostburg, MD.

Novak, M. 1981. The foot-snare and the leg-hold traps: a comparison. In: J. A. Chapman and D. Pursley (eds.), *Worldwide Furbearer Conference Proceedings, Vol. III*, pp. 1671–1685. Frostburg, MD.

Strickland, M. A. and C. W. Douglas. 1981. The status of fisher in North America and its management in southern Ontario. In: J. A. Chapman and D. Pursley (eds.), *Worldwide Furbearer Conference Proceedings, Vol. II*. pp. 1443–1458. Frostburg, MD.

10 ENDANGERED SPECIES—Some Management Strategies

Almost every day on television, or in the press, we are told about a spotted or striped cat, a white or black rhino, a seal, a whale, or even a salamander that is reported to be threatened or endangered. We are constantly asked to support organizations whose job it is to conserve or protect animals and sometimes plants from extinction. The message is often dramatic, and we are led to believe that the extinction of mammals and birds is happening hourly, daily, or weekly and that the rate of extinction is increasing rapidly. If the only contact we have with our wildlife is through the popular media, we may get the impression that the pending extinction of animals in our time is the most important wildlife issue there is. And this is the impression we are supposed to get with the somewhat exaggerated facts. However, we certainly know that the endangered species issue is paramount in the minds of many and that something is being done about it both nationally and internationally. Just what is an endangered species? Perhaps we should briefly review something about extinction and then ask ourselves the question once again.

Fossil records tell us of thousands of extinct species of animals and plants, perhaps as many or more than we have living today. Bird remains, fragile as they are, are not preserved well over periods of time, and yet in 1952, some 787 extinct species were already known from their fossil records. Hundreds of mammals have lived and become extinct over the past ± 150 million years, and there are many amphibians, reptiles, fishes, and invertebrates, all of which first preceded and then coexisted with birds and mammals that are also extinct. Endangerment and extinction are certainly not new although the circumstances surrounding them may be. Quaternary geologists and paleozoologists are gradually joining together the patterns of mammalian distribution dating from about 70,000 to 10,000 B.P. At times in the past, whole populations died simultaneously, and species were snuffed out like so many candles. With the stage swept bare, it appears that new participants made the scene in a brand new play of evolutionary history. New species presided on the land masses and in the waters of the earth again. Millions of years intervened and millions more transpired before the newer cast of players would exit, but exit they did. To be sure, some have survived that shared the earth's habitats from ages past, particularly with quaternary species. Harrington (1977) suggests that mountain goats lived in British Columbia 70,000–23,000 years ago with the now extinct muskox (*Symbos cavifrons*), the ground sloth (*Megalonyx* sp.), mastodons, and mammoths. But why have so many species of wildlife been so abundant and then suddenly become extinct? This question has been a lively one for decades.

Uniformitarian geological theory as espoused by Hutton and Lyell in the eighteenth and nineteenth centuries has been accepted by most scientists up to the present. Cataclysmic theories developed to explain the most abrupt changes in flora and fauna in the fossil record were shunned in most professional circles. In recent years, however, approaches involving a combination of the theory of uniformity and of the acceptance of catastrophic events have been gaining credibility. This is because of findings that there have been at least four mass extinctions in the geologic past. Newell (1963) reviewed the theories dealing with changes in atmospheric oxygen, disease, cosmic radiation, and other factors and argued that sea level changes were probably at the root of the mystery. More recently, meteorite impact has been suggested as the causative factor for the catastrophic theory with some pretty convincing corroborative findings from the geologic record. Whatever the cause of the major changes, beginning with the earliest dated Cretaceous-Tertiary occurrence, extinction has happened on a grand scale a number of times in the past.

In more recent geological and ancient historical periods, we have a few documented causes for species extinction along with several reasonable theories to explain the loss of others. All of these conditions may also have occurred in the past, and most will probably recur. The original distribution of marsupials in the Americas suggests that climatic changes and the resulting alterations of habitat might have been factors in both distributional change and species decline. Overspecialization could have caused the demise of the Irish elk (*Magalacerus giganteus*). A combination of commercial and subsistence harvesting of birds and the gathering of eggs for human food probably eliminated the great auk (*Pinguinus impennus*), and perhaps genetic impoverishment was responsible for the decline and loss of other species, for as O'Brien et al. (1982) noted:

> "It is tempting to speculate that similar circumstances [to cheetah] which have produced monomorphism followed by niche perturbation might explain extinction of successful species in the past."

It seems (to us) less likely, given the apparent low density human populations in the days of early man, that subsistence hunting could have eliminated many species as suggested by Martin (1971, 1973), but undoubtedly local populations of highly gregarious, large mammals and a few entire species were severely reduced, contributing to eventual extinctions. By far the greatest number of local, or total, species extinctions in historical times have been of animals existing in relatively low numbers in restricted range areas. For example, the giant sea mink (*Mustela macrodon*) occurred in low density, only along the New England and Bay of Fundy coasts. In the case of the localized subspecific population of Newfoundland wolf (*Canis lupus beothucus*), the animal was restricted to, and occurred in low numbers only on, the island of Newfoundland. But in other cases, species existing in large numbers over extensive range areas also have declined and disappeared in historic times. The best-known North American example is that of the gregarious, migratory passenger pigeon (*Ectopistes migratorius*), which apparently achieved incredibly high populations; 136 million birds breeding within 2200 square kilometers (Schorger, 1937 in Welty, 1962) and 2000 million birds in a flock (quoted from Alexander Wilson in Welty, 1962). Passenger pigeons were highly vulnerable to commercial harvesting in their roosting trees and were slaughtered by the thousands. Their roosting and nesting habitats, which were traditional and critical for the species, were also destroyed!

In an attempt to clarify recent mammalian extinction, MacPhee and Flemming (1997) indicate ninety full species lost worldwide since 1500 or about 2 percent of the 5000 known mammals with 71 percent of the extinctions occurring on islands. They also note that with the exception of Australia, continental land masses have not suffered high rates of extinction in the past 10,000 years. Australia, however, has lost around sixty mammals, reflecting a rate more aligned to those of large islands than to continents. In the western hemisphere an extinction wave, such as that of 11,000–10,000 B.C. when at least 135 species were lost, or the decimation of some 80 percent of land mammals of the West Indies and Galapagos that occurred thousands of years later, is predicted by some scientists in the future where high species diversity and severe habitat loss are present.

Today humans and wildlife continue to vie for food and space, for land and water areas. These conflicts and the action resulting from them, in addition to the natural factors we have alluded to, form the bases of our concern for animals we consider to be endangered. There is no doubt that increasing human populations are pressuring many species and heightening the intensity of change far beyond what would be the natural evolutionary trends occurring without man's presence. Human population size is clearly the common denominator at the core of our wildlife competitive problems. Our food needs allow for elimination of wild habitat through clearing and cultivation. Already several species, including Attwater's prairie chicken (*Tympanuchus cupido attwateri*), the kit fox (*Vulpes macrotis*), and the black-footed ferret, have been threatened or lost in North America largely because we took their lands for grazing our domestic stock and raising crops or we destroyed the prey base for the same reasons.

Commercial harvesting is still a factor as well. The blue whale (*Balaenopterus musculus*), for example, is a mammal with a low reproductive rate which has declined over years of poorly regulated commercial harvesting. Beddington and May (1982) have considered some of the ecological complexities we must cope with in relation to krill harvests and harvest of both toothed carnivores (Odontoceti) and krill feeders (Mysticeti). Since we, as well as the other mammals, are ultimately dependent upon the lower trophic levels of terrestrial and marine ecosystems, the implications of commercial harvest at lower trophic levels are those of which all wildlife managers should be aware. Animals must not be managed as individual species apart from the remainder of their habitat. Competitive release and elimination of dependent species are two possible ecological consequences of such action.

The influence of biocides is another factor that must be considered when we reflect on the effects of human action upon wild species. The peregrine falcon, a bird of prey in generally low numbers under the best of conditions, suffered considerably after World War II due to our use of pesticides. Many other species of birds of prey have also been harmed, including the bald eagle and the osprey, but fortunately these particular birds are rebounding vigorously today.

Oil pollution in the ocean has brought death to thousands of sea birds and has the potential to create a disaster, as was evidenced in Prince William Sound, which could drop populations of some species to dangerously low levels because of their highly social breeding behavior. So far the losses have been generally sporadic in relation to major spills, but the presence of oil is a continuous problem and causes considerable annual mortality in ducks and alcids. It has become a major cause of death that is clearly additive in the total mortality picture.

The separation of populations into isolated units with low numbers of animals may result in reduced breeding success and gradual declines, as suggested by O'Brien et al. (1982). In relation to the low levels of genetic variation reported from two geographically isolated populations of cheetah (*Acinonyx jubatus jubatus*), O'Brien notes: "The extreme monomorphism may be a consequence of a demographic contraction of the cheetah (a population bottleneck) in association with a reduced rate of increase in the recent natural history of this endangered species."

All of these situations, as well as the rapid decline in local habitats because of development, have as their root cause the increase in human population. As long as humans with their demands for energy, food, and comfort increase, the struggle to prevent the losses of populations and species will be a growing problem with which wildlife professionals will become increasingly involved. It seems inevitable that the rate of decline for some species will accelerate in future years in spite of the combined efforts of individuals, agencies, and governments as we increasingly compete with wildlife for food and space.

With all of the changes wrought by man and his activities impacting hundreds of species in dozens of ecosystems, the problem of determining just which animals are threatened or endangered and which are not, by any standards, is difficult indeed. The tolerance levels of each species relate to its social behavior, feeding, spatial requirements, and a host of other factors. With some wide ranging species such as the tiger (*Panthera tigris*), known space, cover, and food requirements clearly dictate the survival abilities of certain of its subspecies in the face of human population expansion and development. With other animals, the survival abilities relative to human disturbance are not so clearly defined. Still it is better for managers to err on the side of conservatism than to continue to exploit populations in the absence of adequate data bases and risk mistakes that are detrimental to the species. It is also true in today's rapidly changing world that what is threatened or endangered at one time could become a pest at another time, in another place. The ability of some predators such as wolves to respond to protection and an adequate food supply is just such an example. On the other hand, an animal that is seemingly doing well might, under pressure of human harvest or some other human induced factor, begin to slide downward rapidly as was shown with the Atlantic cod fish. Thus, definitions must be broad enough and management efficient enough to encompass the changes resulting from an animal's response to range or population disturbance.

Perhaps some of the more pragmatic among us might ask "Why bother?" Why, indeed, bother trying to save what in the long run may compete with us more directly? And why try to save an animal when we know it will cost us a great deal of money, time, and effort often with little success? Others among us liken the endangered species problem to motherhood and apple pie; everyone should want to save the endangered from extinction. There should be no question! Hornaday (1914) held to an animal rights line in his view of this question: "The murder of wild-animal species consists in taking from it that which man with all his cunning can never give back—its God-given place in the ranks of living things. Where is man's boasted intelligence, or his sense of proportion, that every man does not see the monstrous moral obliquity involved in the destruction of a species?"

Others in past years have used economics to justify their positions, as Williams (1917) wrote in defense of birdlife in South Carolina:

> "It is an undisputed fact that the prosperity of the State and nation depends on successful agriculture. Therefore, whatever assists in the production of crops has a money value in proportion to the degree of assistance rendered. The result of the study of the relation of birds to agriculture made by government experts shows that birds are among the farmer's best friends. Mr. Henry W. Henshaw, Chief of the Bureau of Biological Survey, is authority for this statement: 'So great is their value from a practical standpoint as to lead to the belief that were it not for birds successful agriculture would be impossible' . . . If the birds' work in nature be of so much importance, bird conservation should become a part of the constructive work of the State, and any agency or condition which tends to reduce the bird population below the limits necessary to hold in check the countless hordes of injurious insects, should be considered inimical to the best interests of the whole people."

Still others have felt a strong moral obligation and tried to develop a conservation ethic of sorts to help us do the best we can by other creatures. Leopold (1949), for instance, wrote:

> "We all strive for safety, prosperity, comfort, long life, and dullness. The deer strives with his supple legs, the cowman with traps and poison, the statesman with pen, the most of us with machines, votes and dollars, but it all comes to the same thing: peace in our time, a measure of success in this is all well enough, and perhaps is a requisite to objective thinking, but too much safety seems to yield only danger in the long run. Perhaps this is behind Thoreau's dictum: In wildness is the salvation of the world. Perhaps this is the hidden meaning in the howl of the wolf, long known among mountains, but seldom perceived among men."

We have come a long way since these people wrote in defense of their beliefs. Now we have a dozen organizations to protect wildlife and save the endangered when in the past there was one; a hundred books where there were a dozen. Now we have devised more precise ways of assessing values of wild things, and there are also more professionals to work on their behalf. But we have less space for the animals, too, just as there are more polluted areas, more examples of human competition within natural systems, and millions more people. We (the authors) believe, however, that all of us do have a responsibility to help save the threatened and endangered. For some species, it will be possible in our time, while for others we may only be able to delay the inevitable. For a few, perhaps, society unfortunately may deem the costs too great even now.

The present movement to save wildlife has, among its supporters, groups from the entire spectrum of natural history interests. The Species Survival Commission of the International Union for the Conservation of Nature, the

Humane Society of the United States, the World Wildlife Fund, the National Audubon Society, Greenpeace Inc., government agencies, naturalists' and sportsmen's groups, and many industries all show interest in preventing extinction. The organizations, agencies, and individuals concerned have different justifications and motives; their degree of emotional involvement varies and their approaches and techniques differ markedly. It is also true that it is the nonprofessional who is gradually becoming the one influencing the public and public policy (through government) to the greatest degree. This is true partly because the greatest number of professional ecologists and their kin are employees of either governments or international agencies and in positions where they are not entirely free to express their opinions or discuss their study results. Professional ecologists and wildlifers also tend to report their findings in professional publications to solidify and enhance their positions within the organization they work for and to establish themselves among their peers. It is probably also true that most ecologists and wildlifers are not as adept at communicating as are writers, artists, showpeople, and advertising or other media people involved in the struggle to prevent sport or commercial harvesting of a wild animal. Many nonprofessionals now hold positions of influence with one or more of the agencies or organizations dealing with threatened or endangered species, and their views are not always the soundest to enhance the future of a species.

The status of the endangered and threatened is probably less important than the rate of species loss. Each time we lose a race, a subspecies or a species locally, regionally or world wide, the supporting system becomes less diverse. Still, numbers are important because they say something about the magnitude of the problems facing us. For the United States the *Endangered Species Bulletin* of the U.S. Fish and Wildlife Service, which is included in the University of Michigan's *Endangered Species Update,* is one of the best sources of data. As of March 31, 1997, the *Bulletin* listed 335 animals and 523 plants endangered and 111 animals and 111 plants, threatened. In Canada, *Recovery—An Endangered Species Newsletter* produced by the Canadian Wildlife Service and the Committee on the Status of Endangered Wildlife in Canada (COSEWIC) provide up-to-date status information. As of April 1997, COSEWIC listed thirty-seven animals and thirty plants endangered in Canada as well as thirty-two animals and thirty-eight plants threatened. These figures (United States vis à vis Canada) may not be comparable, however, because the U.S. figures can include subspecies or even distinct populations for the purposes of the Endangered Species Act while the Canadian lists are for full species only. Even when species alone are compared, however, there may be some variation between published listings depending upon the acceptance of an organism as a full species rather than a subspecies. Such differences are even more pronounced when measuring extinctions and MacPhee and Flemming (1997) found their list "differed by as much as 50 percent from those cited by others" largely because of disagreement over classification. Regardless of differences in listings, we may expect numbers on all lists to rise as knowledge increases on the status of invertebrates, especially insects, and as human induced environmental perturbations increase worldwide, affecting all systems and all living things.

So let us ask once again. What is an endangered species? The criteria used to determine species status are similar among the major organizations involved, including CITES, the IUCN and both the United States and Canadian Nature Conservancies. They are population size and trend, species distribution and trend, the number of sites occupied by the species, specific population threats, and habitat threats. Agencies may use different ranking processes and different rating scales but a rapidly declining species with a rapidly declining distribution that is faced with serious threats to its populations and habitat can be expected to be a candidate for threatened or endangered status.

The Convention on International Trade in Endangered Species

Perhaps the most effective international approach to date in aid of endangered species has been CITES, aimed at the market place. The first draft of a treaty was produced under the aegis of the IUCN in the mid-1960s. Between 1970 and 1972 a number of governments, including the United States and Kenya, and NGOs including the IUCN, the World Wildlife Fund, the National Audubon Society, and the New York Zoological Society produced several draft alterations. Then in 1973, the Plenipotentiary Conference (a meeting of delegates imbued with full authority to conclude a treaty controlling international trade in wild animals and plants) was held in Washington, D.C. The conference produced a treaty known as the Convention on International Trade in Endangered Species of Wild Fauna and Flora (CITES). Among the more significant provisions of this treaty are the following:

1. A presumption against trade in and a presumption in favor of protecting animals unless it could be shown that trade would not harm animal populations.
2. Protection for certain species, subspecies, and populations.
3. A requirement that species be maintained throughout their range at a level consistent with their role in the ecosystems in which they occur.
4. A requirement that before trade is allowed, a scientific authority in the country of export make a finding that the export will not be detrimental to the survival of the species.

5. Provision that the burden of proof for findings resides with those who would allow export.

In addition to these provisions, appendices listing endangered and potentially endangered species considered at the convention were adopted by consensus of the participating nations. Appendix I species are those considered to be rare or endangered for which trade will not be permitted for primarily commercial purposes. Import permits from the receiving country are required as well as permits from the exporting country when one of these animals is involved. Appendix II species could become rare or endangered if trade is not regulated. Permits are required from the country of export for such animals when importing to signatory nations. Appendix III species are not endangered but may be considered in regulatory management where they exist. For several countries, export permits must be obtained prior to importation (CITES, Control List No. 11, 1995).

For participating nations, the convention (treaty) came into force in 1975. In effect, the agreement was imposed on many nonparticipants as well, because nations that did not issue export permits were prevented from shipping animals or their parts into member nations requiring such permits. In that the convention essentially eliminated markets through control of trade, the treaty was also imposed upon non-participating countries that sought to profit. In meetings held since 1973, the original treaty has been amended to strengthen the effect of the agreement. In particular, the Berne agreement added species to Appendix II that might be harvested more heavily because of trade restrictions on other species listed in Appendix I. This allowed more intensive monitoring of wildcat and lynx as a result of restrictions on the trading of leopard and other tropical cats.

Of course changes are occurring with CITES implementation procedures as governments themselves change, at least within less developed countries. For instance, the Danish government is sponsoring a CITES implementation program to fit new jurisdictional boundaries of the Provinces in the Republic of South Africa. Infrastructure relative to scientific authorities, identification of appendix species, enforcement, administration and all other aspects are being developed to ensure continuity among all the jurisdictions within the Republic. In addition, the project is seeking to establish common CITES policies for all of southern Africa from Zambia to Cape Town, a process of unbelievable complexity but, if successful, highly worthwhile.

In Canada and the United States, jurisdictional responsibility for nonmigratory birds and all mammals, reptiles, amphibians, flora, and fish within their boundaries lies with provinces and states. The control of export and import out of, and into, these countries, however, is a federal responsibility. An international treaty is the supreme law of the land in the United States. Since export is initiated in the jurisdictions of the animal's origin, considerable cooperation has developed between federal and state or provincial jurisdictions in implementing CITES regulations in North America. Scientific authorities, which include representation from both federal and state or provincial agencies, keep abreast of all aspects of the status of endangered species and advise the relevant CITES administrators. Similarly, management authorities advise on minimum requirements for acceptable species management, while others administer the regulation of trade through permit.

The existence of CITES should offer an avenue for increased cooperation between federal and state or provincial agencies. It has been responsible for improved monitoring of populations and has stimulated research throughout North America. In the future, with the interagency and the "bicameral" approaches required for administration along with the thousands of species rated, listed and delisted, problems will arise in North America. In addition, there are new and growing bureaucracies in each participating country involved with CITES. In Canada, the patriation of the Constitution in 1982 has reopened rather extensive considerations of present federal-provincial relations and responsibilities. With the exception of migratory birds, as far as wildlife is concerned it seems likely that provincial jurisdiction and responsibility may be strengthened. If the effects of federal agreements or legislation reduce the ability of a responsible jurisdiction to manage a species, conflict can result regardless of supremacy clauses.

Since 1975, the increasing effects of CITES have been to slow to a trickle the trade in most species considered to be endangered or threatened. Among the animals most affected and presumably benefitted by CITES are elephant (*Elaphus maximus* and *Loxodonta africana*), African rhinoceros (*Rhinocerptidae,* both black and white), leopard, tiger, and cheetah (CITES, Control list No, 11, 1995). Johnson (1979), however, cautions that there is no proof that involvement with CITES has improved the status of any species. He feels that more time is needed before judgments of its effectiveness can be made. In 1989, CITES imposed a total ban on ivory trade in an effort to stem the rapid decline in African elephants. In doing so, countries, such as Zimbabwe that were increasing their elephant population while controlling illegal harvesting and realizing benefits from managed legal harvests, were penalized. However, in recognition of this, limited marketing of ivory stores from certain countries was permitted in 1997. The overall effects of the ban and subsequent limited trade will have to be continually monitored to be certain that neither is counterproductive.

In the United States, the secretary of the Department of the Interior is the management authority for CITES with operational responsibilities delegated to the USFWS. The Endangered Species Scientific Authority advises the management authority on scientific matters of the convention. Other important United States legislation, including the Endangered Species Act (1973) and its amendments, are listed in CITES Appendix 1. The Endangered Species Committee is the listing body for all wildlife under United States legislation. The Endangered Species Act was the subject of intense debate at the time of publication of this book as it underwent the reauthorization process with concerted efforts to weaken its provisions especially as related to private land holdings.

The Canadian Wildlife Service is the management authority and operational body for CITES in Canada. Within Canada's boundaries, COSEWIC has a primary responsibility through its committees of assigning status (vulnerable, threatened, endangered, extirpated, extinct or undetermined) to all wildlife, including plants. The Canadian government was debating the Species at Risk Act that would provide limited protection to threatened and endangered species at the time this book went to press but the bill was highly controversial because of the probability that it would afford little real federal protection for habitats outside federal jurisdiction.

Strategies for Wildlife Managers

There have been management success stories in the past and there are some now. Among the species once threatened in North America, the wood bison (*Bison bison athabascae*) and the whooping crane (*Grus americana*) have been brought back from the edge of extinction, and dozens of other species have responded positively to management practices. It will take all the knowledge and ingenuity wildlife professionals can muster, plus strong political support, to stave off extinction for many animals from now on. In North America, it seems likely that we will have reasonable short-term success. We are not yet extremely pressured by demands for lands inhabited by much of our wildlife and we even have a good chance of protecting some habitat within intensive agricultural areas—at least for a while. We also have the benefit of an aroused public; we have legislation in place with additional enabling legislation ready; we have a body of professional people in established government agencies; and we have a number of well organized citizens' groups lobbying on behalf of wildlife. Nonetheless, in the United States, there have been efforts to weaken the Endangered Species Act when economic and private interests have been threatened and the situation is much more precarious elsewhere particularly in Africa, South America, and Southeast Asia.

"Recovery" has become the key word for wildlife managers dealing with endangered species. In the United States there are, as of this writing, about 500 approved recovery plans, each covering one or more species, subspecies or populations. In Canada, about forty plans relate to vertebrate species. Many earlier plans were little more than lengthy literature reviews with scant direction but today's plans provide detailed biological, ecological and taxonomic backgrounds, together with the all important socioeconomic and fiscal concerns as well as a consideration of the potential for recovery. In addition, a good plan will state goals and objectives clearly, delineate priorities and detail the activities relating to stated objectives. The primary objective of a recovery plan should be to provide a blueprint to follow that will allow for improvement in the status of the plant, animal, population or system the plan considers. A lead agency may be responsible for guiding implementation of a plan or there may be lead agencies cooperating in more than one jurisdiction. In addition to a lead agency, NGOs, industries, various government departments, native people's groups, universities, zoos, and other participants are often involved in research, monitoring, protection, and the innumerable field, laboratory and office activities required. Often, a recovery plan will allow for participation of local naturalist or wildlife groups and individuals thereby increasing interest and providing social benefits not present before. Although recovery plan approaches doubtless will be important in many countries in future years, the following strategies may also be of importance for particular areas and/or species.

Protection Through Area Preservation:
Parks, Preserves, and Buffer Zones

For some nonmigratory species that do not require an extensive home range and even for low numbers of more wide-ranging species, the concept of area protection is practical. In North America, the bison of Wood Buffalo Park illustrate a relatively wide-ranging, gregarious species succeeding in a protected area. Unfortunately, plains bison transplanted to the area in the 1920s brought brucellosis with them so that today the entire park population may have to be destroyed if Canada is to retain its status as a brucellosis free cattle trading nation. If the bison are destroyed, disease free stock will replace them eventually. But in this case, the land is available. The critical conditions exist elsewhere where more than 75 percent of all humans live and where human numbers are increasing at rates of 3 percent a year or more.

In 1980, the IUCN published its World Conservation Strategy (WCS). Two of the three objectives of the WCS were (1) the maintenance of essential ecological processes and (2) the preservation of genetic diversity. While we admit that the publication of a strategy by an international body is, by itself, no solution, adoption by or even agreement of more than 500 member states, government agencies, international and national NGOs and affiliates signals a considerable degree of world support. Both of the objectives also relate directly to wildlife preservation because the absence of any species, or a reduction in a population we are concerned with, signals a reduction in genetic diversity and failure to maintain portions of ecological processes.

It may be, however, that undeveloped or underdeveloped nations do not wish to lose land to a national park or a legislated protected area of another type. The competition for land from various resource bases is strong, and development programs are not usually concerned with wildlife. Nor in many cases will the people using the land want to see a protected area established if it restricts their own use. In such circumstances, we must recognize that the protected area or park concept is specifically a Western one not always relevant to people of cultures different from ours. Indeed, the notion of protecting animals of specific value simply does not fit with the utilitarian concepts of many indigenous people. Furthermore, it makes still less sense to them to give costly protection to an animal that has little value, even as food. If 4 percent or more of the total land area of African countries is a goal for those interested in setting aside wildlife lands, perhaps the costs of establishing and maintaining them should be borne by the industrial nations of the world. It may even be necessary to consider compensation in land, goods, or currency for those people who are deprived through the establishment of a protected area.

Where international concerns are involved, the entire process of establishment, park protection, and maintenance in underdeveloped areas seems to fit best within the IUCN framework. Fragmenting programs and responsibilities among international agencies has long been a problem. Perhaps it is time to channel categorized approaches toward a single agency.

It would also be ideal if most parks had buffer zones around them. The notion that the boundary stops suddenly and any animal stepping across the line is dead is not always true, but in some areas there is real cause for concern. It is particularly serious where extensive migration of animals may take them outside of protected boundaries on a seasonal basis. Such movements occur from some of North America's western parks in the winter and from African parks during the rains. Nairobi National Park, for instance, is a very small area of high quality grass which draws savannah land species in large numbers during the dry season. Their numbers decrease through dispersal when forage is more readily available elsewhere at other times. In this case, buffer zones would certainly be inadequate, and even protected movement corridors would probably be less than adequate, even if possible. Parks such as the Luangwa in Zambia, where movements of many species are less extensive, lend themselves well to the buffer concept. In fact, "game management areas" (GMAs) around the portions of the Luangwa where most animal movement occurs serve as excellent buffers. It is not that animals are not harvested in the GMAs. They are, but the number taken by hunting parties is restricted. Illegal hunting also occurs, but ideally, regulations and enforcement keep the poaching at a low level. Dasmann (1981) discussed Olympic National Park as an example of a park with a buffer zone—the surrounding multiple-use national forest. The biosphere reserve concept of UNESCO's Man and the Biosphere program reflects this same concept, a protected area surrounded by a region in which the land and the ecosystem are essentially protected but where controlled economic development is allowed. In all cases where establishment of protected areas in underdeveloped countries is considered, wildlife managers should seek to implement measures to improve the living conditions of local people through extending services such as potable water, or electricity, or through labor intensive development. Regardless of who pays, the park must be seen to benefit people living near it if we expect them to respect what it stands for.

Protection Through Use: Utilization and the Marketplace

Conservation implies efficient and continuing use of existing supplies for the benefit of present and future generations. To Gifford Pinchot conservation meant "wise use." This connotation has been widely accepted in North America where preservation is but one of the management options included in conservation. Elsewhere in the world, and most particularly in underdeveloped countries, conservation is still often synonymous with preservation. The dictionary definition of conserving—preserving, guarding, or protecting—is the only one understood and accepted. Thus the approach of international agencies in Africa largely has been one of preservation rather than one of use. Conservation had been the international aid catchword, complete with its connotation of locking up the resource. More recently, natural resource conservation has started to assume its broader definition, even in the international aid arena.

We have noted that wildlife protection within parks was a Western concept. Wherever possible, however, the African has coexisted with animals and used them since the origin of humans in the heart of the continent. This was one of the areas where utilitarianism began and it is still present. The preservationist influence has been so strong, however, that in recent years the ability of African governments to use wildlife has often been paralyzed by fear of world censure, particularly because so much help was required from the Western world. Better go along than risk losing a million dollars from the World Bank!

Without use, much wildlife is eventually doomed. A park of today, may on occasion, become a management area tomorrow. Remember that some 80 percent of the world's people live in South America, Africa, and southern Asia and are increasing at about 3 percent a year. It seems unlikely that all protected areas can remain completely inviolate for long in the twenty-first century.

It is easy for us to say that use will help prevent wildlife species from becoming endangered or extinct. How will it though? Preservationists and hunters alike have argued against the concept which they have often simplified to one of "kill in order to save." The heart of the problem relates to values. How does a society perceive a species from which it benefits? This question must be addressed to each of the several societies and publics within them, but it seems quite safe to say that whether the perceived value is utilitarian or esthetics, the end result is similar; that is, each group wants to conserve the species. The difference lies in approach. The esthetics block would protect the species by any means possible, including the elimination of commercial exploitation (as with the harp seal), and the utilitarian block would conserve the species by harvesting under specific regulations while otherwise protecting both the animal and its habitat. Since we can no longer depend upon natural animal population controls because humans are altering the ecosystems which provide those controls, it may be that in an increasing number of situations a pragmatic approach involving use combined with an element of protection will become necessary. Indeed, the controlled-use approach may often be necessary to prevent uncontrolled use from creating further imbalances in the systems we are dealing with. For instance, if an illegal market for ivory persists, elephants will be harvested illegally, in an uncontrolled manner, with the largest "tuskers" being removed. If the elephant continues to be a problem as a garden marauder in areas of settlement, the guilty animals and others will be killed in self defense. A population of harp seals allowed to increase in the absence of major predation from man or other animals will eventually be controlled illegally by those elements of human society directly or indirectly competing with them for a marine food resource. Grey seals in the western North Atlantic have no significant natural predator affecting their numbers today and from a low population twenty years ago have increased to become a pest species in the eyes of fishermen around the Gulf of St. Lawrence. They are now shot and wasted with impunity.

The failure of African governments and international agencies to provide the means whereby villages can benefit from wildlife has often led to a situation where the poacher fills the control niche left vacant by those in authority. It may be that the poacher provides the service the regulatory management should have provided. In the Luangwa Valley of Zambia, Dodds and Patton (1968) recommended the "removal of several thousand elephant" while Naylor et al. (1973), acting on more complete population and range data, recommended a two-step reduction of 5000 elephants. No official action was taken, however, and poachers approximately achieved the recommended capital reduction over the next several years. From the standpoint of the elephants and their range, this illegal removal was probably not terribly detrimental; but the specific animals chosen would probably not have been the same ones taken under controlled removal. The prime bulls, reflecting elements of the best genetic stock were harvested most heavily by the poachers.

Because CITES should continue to operate as an effective control over much commercial poaching, illegal harvesting should be less detrimental to many wildlife populations, nonetheless, the opportunity remains to consider innovative approaches to controlled use. There are several methods the manager might consider. We consider some of them, but each situation has its own set of conditions that will dictate, to a great extent, the direction taken.

1. Perhaps the most logical approach is to allow controlled harvest of animals, within cultural contexts, under a system that would provide direct benefits to the people in food, cash, or their equivalent. The equivalent values must be seen to be directly related to the process if food and/or cash are not provided. There have been schemes using this type of procedure but they have not been generally successful. One such was the Kenya Wildlife Management project conducted by the Food and Agricultural Organization of the United Nations in the early 1970s. The scheme did not involve the people; it was conducted by a new and different body essentially outside the existing agencies of government, and the connection between harvesting and benefits was not clear. The benefits did not exist as far as most people of the district were concerned. There were other problems that led to the abandonment of the project, but we emphasize that the concept itself was good.

To be successful, operations should be small and restricted in area. Whenever possible, they also should be implemented through organizations already in place, and should involve the people who are expected to benefit. In some instances, fresh meat could be a benefit, in others biltong, or currency might be provided and managed for a group, an association, or a village. The approach can be altered to relate best to certain rare or endangered species and it will always vary with species problems. It has the potential to be one of the more workable approaches to managing all wildlife, including the threatened and endangered in many underdeveloped areas—if it is done properly. Such community based management has been viewed as instrumental in elephant population increases in southern Africa while there have been decreases elsewhere in that continent (Getz et al., 1999). Despite this, Inamdar et al. (1999) suggest that community based management has done little to sustain biodiversity. Nonetheless, the strategy has much to offer that is positive and counteracts the negative image of protected areas.

2. The use of subsistence poaching might be considered in certain specific situations. It is unlikely that existing regulations would be rescinded to allow this, but if enforcement was relaxed in an area for periods of time the results might not differ greatly. A similar concept might be considered in areas of

underdeveloped countries where imposed regulations make traditional practices illegal and where regulatory management conflicts with use patterns firmly established in culture. When one of the authors was in Maine, a similar approach was used in rural areas, where it was recognized that white-tailed deer provided winter subsistence for many poor families. Wardens, who knew what was happening, did not intervene when there was no obvious abuse of the resource. This won them respect and support for dealing with other more important illegal activities. Subsistence poaching, of course, differs from commercial poaching. The latter has usually been induced by user societies providing ready markets for what has often been in the past a noncommodity for indigenous people.

3. Not only do the rural people need to benefit from the animals they coexist with, but so also do the governments of underdeveloped countries. If the use of impala by villagers reduces the burden of the government in caring for its people, a larger public benefit is present. Governments, however, need more tangible benefits than this. Several underdeveloped countries are desperate for foreign currency. With a devalued currency of their own and inflation rising as productivity remains stagnant, foreign monies from the industrial nations are highly coveted.

To provide a source of export income, countries may develop a limited tourist industry as Zambia has. In Kenya's case, tourism is a highly developed industry based, in part, on wildlife resources. There are two general categories of wildlife-oriented tourism; one based on viewing and one on hunting. In the first case, development costs may be considerable and capital programs are required for lodges, transportation, viewing facilities, and maintenance. Costs may be relatively low in the case of the "walking safari" where cameras replace guns carried by visitors and the participants stay under canvas. Both types of tourism generate foreign currency; both increase the value of rare and endangered species and other wildlife in the eyes of the tourists and local people; and both depend upon legal protection of the animals. From sport or safari hunting, a great deal of foreign currency is made available to the host country in exchange for the lives of a few animals. Under adequate supervision, this use of animals can elevate the value and encourage protection of the species harvested.

4. The rare and endangered may benefit from game ranching of one type or another. Although North American managers are gradually moving into this arena where lands are available (as in Texas and Alberta), extensive and intensive ranching (farming) has developed most rapidly in Africa. In Canada, the question of whether, and/or how to develop game ranching is being considered and cautious policies are gradually being evolved. The Canadian Wildlife Federation's position, however, is that game ranching is inappropriate.

Nevertheless, the farming of once wild animals is being practiced in some provinces under highly restrictive regulations aimed at preventing escapes, preventing disease, and preventing the marketing of native wildlife. Land owners in the Republic of South Africa often harvest wild herbivores as well as their cattle, an example of extensive ranching. The wild animals are not usually rare or endangered, and as long as they are providing economic return they are not likely to be. Blesbuck (*Damalisus dorcus*), oryx (*Oryx beisa*), and sometimes eland (*Taurotragus oryx*) are killed and marketed within the country and elsewhere. In Texas, both indigenous deer and exotic ungulates are offered by landowners to sportsmen in various sport hunting schemes. Although the ranching of wild ungulates on an extensive basis (conducted within state laws) is available on only a few western U.S. sites, there are several ranches in the Republic of South Africa where the harvest of wild ungulates occurs. In subarctic areas, reindeer have been an exploited captive resource for centuries, just as camels (*Camelus bactrianus* and *Camelus dromedarius*) have been elsewhere. The degree to which once wild animals have come under the control of man has been detailed by Clutton-Brock (1982). For certain species, this control may become the means by which they are perpetuated in future years. The eland has been herded in the Ukraine since 1968 and along with oryx and buffalo (*Syncerus caffer*) is managed today just as cattle are at the Galana River Estate in Kenya. Crocodiles are farmed in the southern United States and Southeast Asia. The list is long but the important things to consider are that both domestication and extensive ranching may allow an animal to continue to exist when others of its kind are eliminated, and that the presence of a variety of wild ungulates under semicontrolled conditions on open range can allow for the continuation of ecosystem elements, such as leopards and coyotes, at the next trophic level. The ranching and farming of animals can benefit endangered predators, but more importantly, the direct role of the domesticated, or exploited captive, in the market place may help prevent those species from ever becoming threatened or endangered.

Protection of Captives: Species Survival Plans

Clutton-Brock (1982) has traced the origins of our current domesticated animals. For some groups the domestic animals we have are the only reflections left of the original source stocks. Such is the case with the cow (*Bos taurus*); its last progenitor, the aurochs or giant ox (*Bos primigenius*), was apparently killed in Poland about 1627. In other instances, the domestic progenitors are still with us. Dogs presumably descended, one and all, from wolves some 12,000 or more years ago. Our own progenitors, watching a competitor for certain prey, apparently struck up a liaison with the wolf to aid them in the hunt.

Endangered Species • 275

It was a successful liaison. Perhaps other successful ones will develop from experiments we have alluded to previously, but there is another, obvious approach—to keep an animal as a wild captive, and this is what zoos are all about. Only one of the purposes of the zoo is to make representative wild animals available for public viewing. In relation to rare and endangered species, zoos are performing a much more important role today.

According to the World Conservation Strategy of the IUCN, captive populations and propagation should be integral components of the global program to preserve endangered species. The American Zoo and Aquarium Association (AZA) recognizes four ways in which zoos can contribute to the WCS objective of "preservation of genetic diversity." They are:

1. To serve as refugia for species that are destined for extinction in the wild.
2. To provide sources of propagules for repopulation of natural habitat.
3. To reinforce natural populations which may be so small and fragmented that they are not viable genetically or demographically.
4. To maintain repositories of germ plasm in addition to, or as an alternative to, populations of animals. (Foose, 1983).

This role has evolved as a result of the Species Survival Plans (SSP) of the AZA. The original intent, however, was probably not completely altruistic in that for years zoos have been experiencing increasing difficulty in obtaining individuals of certain species from the wild. If not available, then they try to raise them in captivity. Foose also notes that the zoos can perform other significant roles "by conducting research that will improve wildlife management and by educating the public to support conservation."

The central theme of both the SSP and Population Management Plan (PMP) is to manage captive, widely dispersed populations through controlled breeding to maintain genetic diversity. The PMP involves only captives in zoos while the SSP also promotes conservation approaches on sites where the species exists. Member organizations of AZA alone house 270,000 vertebrates in addition to invertebrates. As a result, the selection of species to be managed through SSPs or PMPs and the establishment of special requirements for the minimum requisite number are significant problems. For example, Mellen and Wildt (1997) indicate that "optimally" 1600 zoo "spaces" should be allocated to the large cats (eight species). The basic criteria used for species selection are: (1) a breeding nucleus of the species must be available for captive management; (2) continued existence of the species in the wild must be imperiled as defined by IUCN, UCBP, USFWS, or reliable field reports; (3) there must be available an organized group of capture propagation professionals with sufficient support to ensure the program will reach captive preservation status.

When too many species meet the basic criteria, the following are used to select candidate species: (1) the probability of successful captive management is high; (2) the degree of endangerment is relatively high; (3) the degree of rarity is relatively high. These latter criteria are not dissimilar to those suggested by Sparrowe and Wight (1975) for a system of assigning action priorities in the Endangered Species Program of USFWS. The problems of logistics and timing are considerable, of course, particularly if zoos from around the world cooperate. The breeding of each species is directed by a species coordinator and a small propagation group chosen from the institutions housing members of the species in question. A typical master plan for a species such as the snow leopard (*Panthera uncia*) considers general strategies. For example:

1. What the size of the population is presently, potentially, and optimally, in terms of numbers, ages, and sexes.
2. How many institutions should accommodate the species.
3. Which animals should reproduce, how often, and with which mate.
4. Which animals should be maintained or removed from the population.
5. What basic standards of husbandry and considerations of sociobiology should be emphasized. (Foose, 1982).

The master plan also will consider all aspects of demography, genetics, genetic diversity, carrying capacity, subpopulations, spatial and behavioral requirements, and other technical aspects. Students are advised to study examples of SSPs and PMPs, to better understand the captive approach and how it complements other aspects of endangered species management.

A primary concern is the maintenance of adequate genetic diversity in the small remaining captive stocks. Although future genetic engineering developments may ease this problem, current management is designed "to preserve as much as possible of the heritable diversity that has evolved and exists in the wild gene pools" (Foose, 1983). To achieve this, a strategy has developed which includes requiring an adequate number of founders; for example, according to Mellen and Wildt (1997), a population of 200 breeding individuals is needed for large cats to retain 90 percent of the genetic diversity. The strategy then calls for expanding the population as rapidly as possible from the founders to whatever carrying capacity has been determined, and then possibly subdividing the population. The founder question can be a significant problem for extremely rare species. Although as few as ten pairs may provide the needed diversity if selected judiciously, the requirements are that the founders be unrelated, noninbred, and interfertile. In the case of the black-footed ferret, where only one extant population was known to exist, it may be impossible to meet these criteria. The prime objective then becomes to maximize the effective population size to minimize the loss of genetic diversity. The practicability of meeting this objective will vary dependent on captive holding capacity, founder sources, and opportunities for reintroduction, among other factors. Suffice it to say that the more individuals of a species with divergent backgrounds that can be incorporated into captive breeding programs, the greater the chance of ultimate success in species retention.

All animals that are threatened and/or endangered cannot be held in captivity in adequate numbers to allow this kind of population management. In an attempt to address part of this problem, zoo managers are trying to increase zoo capacities to provide more habitat. The overall approach should, however, guarantee the presence of wild representatives of several species for a considerable time period, perhaps several hundred years. A conflict exists between the zoo manager who wishes to show viewers a wide variety of species, including representatives from the endangered list, and the pragmatic advocate of species survival plans. This is because the total number of zoos with representatives of any one species may well have to be restricted in future years.

Reintroductions

Where habitats remain but the animals are gone, reintroduction may be feasible. This depends on the absence, or control of, the original causative factors that resulted in the demise of the species in question. For example, if animals are trapped to a level at which they cannot maintain their numbers, trapping must be controlled. If prey species are poisoned or otherwise eliminated, these activities must cease. It should also be obvious that a knowledge of social behavior, spatial requirements, and feeding habits assists the manager in deciding whether or not to reintroduce the species.

For peregrine falcons, the primary cause of their original decline (DDT use in North America) is no longer present; the habitat is, however, and captive breeding and release of stock from Cornell University and elsewhere have been effective. Other reintroductions depend upon wild-caught stock. Such a program is in effect in Newfoundland, where pine or American marten (*Martes americana*) are being taken from the only remaining population center at Little Grand Lake and moved into the two Newfoundland national parks. This is a reintroduction attempt, of course, but it may also turn out to be a salvage operation in that much of the presently populated forest habitat may be cut eventually.

The peregrine and the marten are unique examples in that there is still available habitat and the animals are either free of the more important mortality factors, or totally protected, or both. Such situations are not likely to occur for many rare or endangered species, especially in Africa, Southeast Asia, and South America where habitats are rapidly disappearing. Even if habitats are present, reintroductions can be failures. Woodland caribou were reintroduced into Maine and Nova Scotia in the 1960s and into Maine in the 1980s and all efforts failed, presumably because a mortality factor (the nematode *Parelaphostrongylus tenuis,*) possibly not present during previous occupation by caribou, was present when reintroduction attempts took place.

Whatever strategies are employed in the wild, two conditions must prevail. One is that effective legislation must be in place to give adequate protection to the species in question, and an enforcement body must be present to effect that protection. A second is that for any endangered species measures, whether involved with protection or use, the wildlife resource base must be incorporated into economic and land-use planning.

Wildlife law is only now being effectively revised and enacted in underdeveloped countries, and many such actions now in effect were instituted by virtue of external (international) pressures. An example of a new body of legislation which may prove to be effective in protecting wildlife from overexploitation is present in Ethiopia. In 1981, the Food and Agriculture Organization of the United Nations assisted in drafting new laws for that country which protect endangered endemics while allowing controlled use of other species.

Recognition of wildlife in economic and land-use planning is necessary if any wildlife is to be effectively managed in underdeveloped countries in years to come. Without such acknowledgement, wildlife will continue to exist as a noncompetitive entity, eventually succumbing to the inevitable losses of habitat that will drive many species toward extinction. Wildlife cannot compete in the marketplace without controls, however, for that would soon decimate both populations and species.

Wildlife managers should understand that there are different ways to approach the endangered species problems and that certain strategies may work for certain species in some places but not for others. A balanced approach must be employed to consider the many alternatives, from use in the wild to captivity in zoos. What works for one species may not work for another because of the animals' varied biological requirements. Protection, it seems, may not be all that is necessary regardless of pressures from animal rights adherents, but it is just as true that use would do more harm than good for other animals, notwithstanding the strong utilitarian views held by some. Perhaps one of the most important changes that needs to occur is for professional biologists-managers to assume influential positions in decision-making processes for endangered species. More than a strong emotional attachment is needed today to manage wild things, especially the endangered ones. Pragmatic decisions must be made daily on wildlife's behalf, and these decisions can only be made by knowledgeable professionals who exhibit a degree of realism. Perhaps a fence is required, as in Kruger or Etosha Park. Or perhaps people and wildlife, including an endangered species, are competing for land, as in Kenya. To the greatest extent possible, bureaucratic and political impediments must be surmounted or circumvented; emotionalism must take a back seat. It will take cool heads and professional skills to cope successfully with the frustrations ahead.

Bibliography

Beddington, J. R. and R. M. May. 1982. The harvesting of interacting species in a natural ecosystem. Sci. Am. 247(5):62–69.

Benson, D. A. and D. G. Dodds. 1980. *Deer of Nova Scotia*. Department of Lands and Forests, Nova Scotia.

Berglund, B. E., Hakansson, and E. Lagerlund. 1976. Radiocarbon dated mammoth (*Mammuthus primigenius* Blamenbach) finds in South Sweden. Boreas 5:177–191.

Canada Communications Group. 1991. The state of Canada's environment. Ottawa, Ontario.

Canadian Wildlife Service. 1994. Recovery planning guidelines for endangered and threatened species. Ottawa, Ontario.

Clutton-Brock, J. 1982. *Domesticated animals from early times*. Univ. Texas Press, Austin, TX.

Dasmann, R. F. 1981. *Wildife biology*. 2nd ed. John Wiley and Sons. New York, NY.

Dodds, D. G. 1976. Evolution of wildlife harvesting systems in Africa. Trans. Fed.-Prov. Wildl. Conf. 40:106–113.

Dodds, D. G. 1983. Terrestrial mammals. In: G. R. South (ed.), *Biogeography and Ecology of the Island of Newfoundland,* pp 509–550. Junk, The Hague, The Netherlands.

Dodds, D. G. and D. R. Patton. 1968. Wildlife and land-use survey of the Luangwa Valley. Report to the Government of Zambia. FAO No. Ta 2591. Rome, Italy.

Environment Canada. 1995. CITES Control List No. 11. Ottawa, Ontario.

Foose, T. J. 1982. *Panthera uncia:* genetic and demographic analysis and management. Intl. Ped. Book of Snow Leopards 3:81–102.

Foose, T. J. 1983. The relevance of captive populations to the conservation of biotic diversity. In: C. M. Schonewald-Cox, S. M. Chambers, B. MacBryde, and H. Thomas (eds.), *Genetics and Conservation*. Benjamin/Cummings Publ. Co., Menlo Park, CA.

Getz, W. M., L. Fortmann, D. Cumming, J. DuToit, J. Hilty, R. Martin, M. Murphee, N. Owen-Smith, A. M. Starfield, and M. I. Westphal. 1999.

Sustaining natural and human capital: villagers and scientists. Science 283:1855–1856.
Grandy, J. W. 1982. Trial, error, and politics in international cat protection. Intl. Cat Symp., Kingsville, TX.
Gustafson, A. F., C. H. Geis, W. J. Hamilton Jr., and H. Ries. 1949. *Conservation in the United States.* 3rd ed., Comstock, Ithaca, NY.
Hamilton, P. H. 1982. Status of the leopard in sub-Saharan Africa with particular reference to Kenya. Intl. Cat Symp., Kingsville, TX.
Harpur, W., A. Harcombe, R. Halladay, G. Court, S. Brechtel, W. Hall and R. Andrews. 1996. Ranking: a proposal. Recovery (Fall). Can. Wildl. Serv., Ottawa, Ontario.
Harrington, C. R. 1977. Wildlife in British Columbia during the ice age. B.C. Outdoors (Dec).
Hornaday, W. T. 1914. *Wildlife Conservation in Theory and Practice.* Yale Univ. Press, New Haven, CT.
Inamdar, A., H. DeJode, K. Lindsay, and S. Cobb. 1999. Capitalizing on nature: protected area management. Science 283:1856–1857.
Johnson, M. 1979. Review of endangered species: policies and legislation. Wildl. Soc. Bull. 7 (2):79–93.
Kaiser, G. W., H. J. Barclay, A. E. Burger, D. Kangasniemi, D. J. Lindsay, W. T. Munro, W. R. Pollard, R. Redhead, J. Rice, and D. Seip. 1994. National recovery plan for the Marbled Murrelet. Renew. Rep. No. 8, Recovery of nationally endangered wildlife. Can. Wildl. Fed., Ottawa, Ontario.
Leopold, A. 1949. *A Sand County Almanac: And Sketches Here and There.* Oxford Univ. Press. NY.
MacPhee, R. and C. Flemming. 1997. Brown-eyed, milk-giving. Losing mammals since A.D. 1500. Nat. Hist. 106:84–88.
Martin, P. S. 1971. Prehistoric overkill. In: T. R. Detwyler (ed.), *Man's Impact on Environment.* pp. 612–614. McGraw-Hill, New York, NY.
Martin, P. S. 1973. The discovery of America. Science 179:969–974.
McMahan, L. 1986. The international cat trade. In: S. D. Miller and D. D. Everett (eds.), *Cats of the World. Proceedings, Second International Symposium, 1982,* pp.161–188. Caeser Kleberg Wildlife Research Institute, Kingsville, TX and National Wildlife Federation, Washington, D.C.
Mellen, H. and D. Wildt. 1997. AZA Taxon Advisory Group profile: Felids. Endangered Species Update 14:15–16.
Millsap, B. A., J. A. Gore, D. E. Runde, and S. J. Cerulean. 1990. Setting priorities for the conservation of fish and wildlife species in Florida. Wildl. Monogr. 111.

Naylor, J. N., G. C. Caughley, N. D. J. Abel, and O. Liberg. 1973. Game management habitat manipulation, Luangwa Valley Conservation and Development Project. FAO, FO: DP/ZAM/68/510: Working Document No.1.
Newell, N. D. 1963. Crises in the history of life. Sci. Am. 208(2):76–92.
Newton, I. 1979. *Population Ecology of Raptors.* Buteo Books. Vermillion, SD.
O'Brien, S. J., D. E. Wildt, D. Goldman, G. R. Merril, and M. Bush. 1982. The cheetah is depauperate in genetic variation. Intl. Cat Symp., Kingsville, TX.
Ratcliffe. 1973. Studies of the recent breeding success of the peregrine, *Falco peregrinus.* J. Reprod. Fert. Suppl. 19: 377–389.
Raup, D. M. and J. J. Sepkoski. 1982. Mass extinctions in the marine fossil records. Science 215:1501–1503.
Rodriguez, J. P., M. W. Roberts and A. Dobson. 1997. Where endangered species are found in the United States. Endangered Species Update. 14:1–4.
Schorger, A. W. 1937. The great Wisconsin passenger pigeon nesting of 1871. Proc. Linnaean Soc. 48:1–26.
Seidensticker, J. 1982. The conservation of tigers in Indonesia and Bangladesh. Intl. Cat Symp., Kingsville, TX.
Seidensticker, J. 1986. Large carnivores and the consequence of habitat insularization: eology and conservation of tigers in Indonesia and Bangladesh. In: S. D. Miller and D. D. Everett (eds.), *Cats of the World. Proc. 2nd Int'l. Symp.,1982,* pp.1–41. Caeser Kleberg Wildl. Res. Instit., Kingsville, TX and National Wildlife Federation, Washington, D.C.
Sparrowe, R. D. and H. M. Wight. 1975. Setting priorities for the endangered species program. Trans. N. Am. Wildl. Nat. Resour. Conf. 40:142–156.
Teer, J. G. 1968. Evolution of wildlife harvesting systems in Texas. Trans. Fed.-Prov. Wildl. Conf. 40:114–121.
U.S. Fish and Wildlife Service. 1978. Endangered and threatened species recovery planning guidelines. U.S. Dept. Int., Washington. D.C.
Welty, J. C. 1962. *The Life of Birds.* W.B. Saunders, Philadelphia, PA.
Wetmore, A. 1952. Recent additions to our knowledge of prehistoric birds. Proc. Xth Intl. Ornith. Congr., Uppsala. Almqvist and Wiksells, Uppsala, Sweden.
Williams, B. 1917. The decrease of birds in South Carolina. In: W.T. Hornaday (Trustee). *The Statement of the Permanent Wild Life Protection Fund. Vol. II.* Published by the Fund, New York, NY.

Recommended Readings

Endangered Species Update. School of Natural Resources and Environment, Univ. Michigan, Ann Arbor. Includes the *Endangered Species Bulletin* of the U.S. Fish and Wildlife Service. *Update* is published six times a year.

Foose, T. J. 1983. A species survival plan (SSP) for the snow leopard, *Panthera uncia*. Genetics and demographic analysis and management. Intl. Ped. Book of Snow Leopards 3:81–102. An example of an SSP. The zoo approach.

Recovery, an endangered species newsletter. Can. Wildl. Serv. An update on the endangered species situation for the Canadian scene.

Schonewold-Cox, C. M., S. M. Chambers, B. MacBryden, and L. Thomas (eds.). 1983. *Genetics and Conservation*. Benjamin/Cummings Publ. Co., Inc., Menlo Park, CA. A valuable reference updating the literature on the genetics of extinction. Highly recommended for students in wildlife.

11 ENVIRONMENTAL IMPACT ASSESSMENT (EIA) — The New Dimension

Following the publication of the *World Conservation Strategy* in 1980, governments at all levels began developing conservation or environmental strategies as well as single resource strategies for forestry, wildlife, clean air, and marine and freshwater fish. In 1987, the World Commission on Environment and Develolpment reenergized these government processes with *Our Common Future* which was followed by a second international cffort by the IUCN, UNEP and World Wildlife Fund in 1991 called *Caring for the Earth*. These were followed by the first U.N. Conference on Environment and Development in Rio de Janeiro (1992) and a second such "earth summit" in New York in 1997. Today, no one with access to a radio, television or the internet is unaware of the need to foster sound environmental policies or the need for individuals to implement sound environmental practices. In the past thirty years, massive government bureaucracies have developed the world over to deal with environmental matters. Citizens' advisory bodies have been established by hundreds of governments and thousands of environmental, conservation, and single resource citizens' bodies have evolved. People from environmentalists to industrialists are discussing the meaning of new concepts such as sustainable management, diversity, and environmental impact. In the United States, forty-one states have specific environmental departments or agencies and the remainder have environmental responsibilities divided among two or more departments. Many state governments have "arm's length" advisory groups and most have from one to several citizens' groups involved in environmental education, resource management, or simply, political lobbying. If you leave an environmental group and move from your present abode to another community today, you will most certainly be able to find an environmental "special interest" group that will allow you to continue your activities. In Canada, all provinces have environment departments and all have citizens' advisory groups in addition to hundreds of Canada-wide, provincial, county, municipal, and community NGOs representing special interests from the broad environmental spectrum to specific resource matters and even single species protection. The "environment" age has arrived.

This new era has come about because of all life forms, only man has totally eliminated natural systems while impacting others to the extent that plant and animal species have been reduced or destroyed. Goudie (1981) traced the effect of man on nature from the time of hunter-gatherer societies and concluded that: "Primary impacts give rise to a myriad of successive repercussions throughout ecosystems which may be impracticable to trace and monitor. Quantitative cause-and-effect relationships can seldom be established." Still, it is these primary impacts, their repercussions and their cause-and-effect relationships that new breeds of wildlife biologists, ecologists, environmental engineers, and resource managers are trying to measure today with the environmental impact assessment (EIA). It is only too obvious that building an airport or highway will destroy, perhaps permanently, that portion of a natural system that is replaced by concrete and buildings. It is not so obvious what effects the same development may have on the flight patterns or resting behavior of gulls in the vicinity, or what effects there may be on the movements of caribou if a highway intersects their migration routes. With many development projects, and often with existing industries, changes in the chemical nature of soils, water, or air are measurable long before the biota show signs of being affected. Ecologists should be able to quantify both the early and later changes and determine with reasonable accuracy what their influences may be.

In addition to the ecological portions of the EIA, there are also social and economic aspects that must be studied. Possibly the wildlife manager only will be involved on one aspect of an EIA, but since the assessment, the resulting statement, the evaluation, and the review must consider the entire spectrum from the original project guidelines, today's wildlifers require a familiarity with all processes involved, including the socioeconomic ones.

The EIA in the United States—Background

We could say that the roots of the U.S. environmental conscience leading to the government's intervention in environmental matters began with Thoreau, whose communion with nature at Walden led him to champion the cause of nature and absolute personal freedom. We might also trace these roots to Theodore Roosevelt or Aldo Leopold, but to try to do so is only to acknowledge the existence of a slumbering environmental conscience that was really not awakened until Rachel Carson wrote *Silent Spring.* Carson's work is of broad significance because it was written to be read, and it was, by millions of North Americans. With momentum from *Silent Spring,* the sixties became a decade of ecology in the United States, with names like McHarg, Hardin,

Ehrlich, and Commoner becoming familiar to laymen and politicians alike. In 1969, the federal government passed the National Environmental Policy Act (NEPA), and by 1970 the Environmental Revolution was in full swing. Odell (1980) has said of this movement: "So sweeping was this revolution that its adherents concerned themselves with almost all of society's and nature's activities—from the fate of the desert pupfish to that of an interstate highway, from soil erosion to sonic boom." Odell also presented the movement's goals as follows:

1. The safety and good health of individuals, including their psychological and physical well-being as affected by the natural environment.
2. The long-range survival and welfare of society, including the life-supporting environment on which these depend.
3. The achievement of a richer and fuller life, including desirable environmental characteristics.

There is little doubt that both public sentiment for the preservation of nature and public apprehensions and fear concerning the future of human life were somewhat responsible for the interest exhibited in the 1970s by federal and state governments. The Environmental Revolution was primarily a grass roots movement spawned by the people. It was a broader approach to conservation than the great conservation movement for wildlife of the 1930s. Perhaps this was so because the problems of environmental degradation accelerated so swiftly following World War II and thus affected each citizen so much more directly. It was a movement that had a niche somewhere for almost every North American citizen.

The response of most governments was initially similar: pass legislation to protect the environment, but do not box government or private enterprise in to the extent that development would be unnecessarily impeded. Exceptions were continued expansion of the national parks system and the classification of wilderness areas. A significant aspect of most legislation passed was the environmental impact assessment. The NEPA, for instance, required all federal agencies to consult with each other and to employ systematic and interdisciplinary techniques in planning and decision making. For every recommendation or report on proposals for legislation or other federal actions significantly affecting the quality of the human environment, the act required a detailed statement (environmental impact statement or EIS) concerning:

1. The environmental impact of the proposed action.
2. Adverse environmental effects which cannot be avoided should the proposal be implemented.
3. Alternatives to the proposed action.

4. The relationship between local short-term uses and ... enhancement of long-term productivity.
5. Any irreversible and irretrievable commitments of resources ... involved ... in the proposed action (U.S. Fish and Wildlife Service, 1976).

In addition, the relevant Section 102 (C) read:

"Prior to making any detailed statement, the responsible Federal official shall consult with and obtain the comments of any agency which has jurisdiction by law or special expertise with respect to any environmental impact involved. Copies of such statement and the comments and views of the appropriate Federal, State, and local agencies, which are authorized to develop and enforce environmental standards, shall be made available to the President, the Council on Environmental Quality and the public as provided by section 552 of title 5 United States code, and shall accompany the proposal through the existing agency review process." (Nelson, 1973)

According to Zigman (1978), this federal legislation served as a foundation for similar legislation adopted by the states such as the California Environmental Quality Act of 1970 (amended 1972 and 1976) and the State Environmental Policy Act of 1972 in Washington. Some state legislation, however, serves a broader function in overall planning than the several state environmental acts. Florida's Environmental Land and Water Management Act passed in 1972 provides for state, regional, and local involvement in areas of critical state concern or in any development producing regional impacts. All states soon had environmental legislation as separate acts and/or as sections incorporated within one or more legislative instruments. In addition, cities, towns and countries across the land passed bylaws and regulations, established commissions and committees, and amended industrial development and residential planning to help guarantee healthier environments.

With legislation in place, the EIA and its resulting statement had become, by 1975, a normal part of the development process accompanying the planning, construction, and operational phases of projects. Sometimes, the environmental assessment was carried out by in-house government agencies or combinations of agencies or departments. Often private consulting firms were called in by government either to project an image of impartiality or because civil service staff did not have the time to devote to such matters in light of their previously planned (and budgeted) responsibilities. For the most part, governments have now established specific agencies with responsibilities to monitor, coordinate, review, and sometimes direct the EIA. Today, professional biologists, economists, and engineers in the government employ are closely involved with people undertaking the EIA, from beginning to end.

The EIA in Canada—Background

The shock waves from *Silent Spring* were also felt in Canada, and the defensive attitude of both industries and governments quickly became apparent. Carson and others prominent in the decade of ecology were often castigated in the press and within government circles in the early 1960s as being against progress, and the "environmentalist" label took on an almost antisocial connotation. Environmentalists were cast not as people concerned with environmental matters but as people out to stop development at any cost and consequently, out to take dollars from the pockets of adults and bread from the mouths of children! Nevertheless, in Canada the 1960s did become a decade when the word "ecology" became a part of the average citizen's vocabulary. The Environmental Revolution smoldered first in the universities. By the early seventies many an environmental, ecological, and antipollution organization was active and primarily involved young people. Unfortunately, perhaps, many such youth-led groups projected an image in common with some other "rights" movements of narrow minority interests or political ideologies. This made their positive environmental actions vulnerable to attack from conservative and centrist elements in society. These activist groups obtained little sympathy and little general public support. In the beginning, ecological "facts" were sometimes misleading or even dead wrong, and the environmentalists' credibility was often questioned. An early attempt to form a nationwide environmental lobby by amalgamating ecological action groups with resource institutes and associations across Canada was short lived and the Canadian Association on the Human Environment, led by D.H. Pimlott, Don Chant, and others died over a period of two years. It eventually fell to a few individuals projecting a moderate image to counsel the elected representatives toward reasonable, necessary action. As a result, the movement in Canada lagged somewhat behind the United States, although the federal government had reorganized a number of departments to create the Department of the Environment in 1970. Still, by 1972, the pace had quickened and a federal cabinet decision on 8 June 1972 directed that all proposed federal projects be screened to identify potential pollution effects. Then, on 20 December 1973, the Minister of the Environment was directed, again by cabinet decision, to establish in cooperation with other ministers a process to ensure that federal departments and agencies:

1. Take environmental matters into account throughout the planning and implementation of new projects, programs, and activities.
2. Carry out an environmental assessment for all projects which may have adverse effect on the environment before commitments or irrevocable decisions are made; projects which have significant effects have to be

288 • *The Philosophy and Practice of Wildlife Management*

submitted to the Federal Environmental assessments in planning, decision making and implementation (FEARO, 1978).

The decision further established definitions, responsibilities, and procedures which had recently undergone a policy review, and on 11 July 1984 new guidelines for environmental assessment and review were published by the federal government (Parks Canada, 1984) as an order-in-council:

1. Federal projects requiring assessment are those projects initiated by federal departments or agencies, those for which federal funds are committed, and those involving federal property. Thus, under complicated general development agreements between the federal government and provinces and within cost-shared programs such as those considered under DREE (the Department of Regional Economic Expansion) and other federal agencies, most large development projects in Canada are subject to review and assessment.
2. Proprietary corporations and regulatory agencies are directed to participate in the process as long as it is corporate policy to do so and there is no legal impediment or no duplication of effort involved. The Canadian National Railway and the Canadian National Museum, as examples, would undertake a review only if it were corporate policy to apply the FEARO process. Although exemption seems to be a possibility, it is unlikely that these organizations would avoid an environmental review.
3. It is the responsibility of the initiating department in a public review to ensure that the proponent and its own staff meet the responsibilities of the review process.
4. The public review is conducted by an environmental assessment panel appointed by the minister. Panel members are outside the political realm and generally experts on the anticipated technical, environmental, and social areas to be affected by the proposal under review. Departments and agencies are responsible for ensuring "in co-operation with other bodies concerned with the proposal, that any decision made by the appropriate Ministers as a result of the conclusions and recommendations ruled by Panel" are incorporated into the proposal.

The initiating department assesses each proposal to determine whether the review process outlined above is actually carried out. Among the possibilities are the following:

1. No adverse environmental impacts are expected and the proposal automatically proceeds.

2. It is a type of proposal that would produce significant adverse environmental effects and the proposal is automatically referred to the minister for public review by a panel.
3. The potentially adverse environmental effects that may be caused by the proposal are insignificant or mitigable with known technology, in which case the proposal may proceed or proceed with the mitigation.
4. The potentially adverse environmental effects that may be caused by the proposal are unknown, in which case the proposal shall either require further study and subsequent rescreening or reassessment or be referred to the minister for public review by a panel.
5. The potentially adverse environmental effects that may be caused by the proposal are significant, as determined in accordance with criteria developed by FEARO in cooperation with the initiating department. in which case the proposal shall be referred to the minister for public review by a panel.
6. The potentially adverse environmental effects that may be caused by the proposal are unacceptable, in which case the proposal shall either be modified and subsequently rescreened, reassessed, or abandoned.
7. If public concern about the proposal is such that a public review is desirable, the initiating department shall refer the proposal to the minister for public review by a panel.

The order-in-council complemented regulatory requirements of several federal acts such as the Government Organization Act (1979) of which it is a part; Fisheries Act; Clean Air Act; Canada Water Act; Migratory Birds Convention Act; the Northern Inland Waters Act, and others.

In June 1992, a new bill entitled the Canadian Environmental Assessment Act was assented to by Parliament. It was amended in 1994 and finally implemented in January 1995. The Canadian Environmental Assessment Agency (CEAA) was established to replace FEARO and charged with putting the new act into practice. Although participants and responsibilities in the environmental assessment (EA) process are not changed greatly from before, the new act does establish a more rigorous, detailed and complex review process. It also calls for a review of the act and the very process it details at the end of five years. The CEAA has four specific roles:

1. Administer the EA process
2. Provide advice to the responsible minister relative to his responsibilities under the Act
3. Provide opportunities for public participation in the EA process
4. Promote sound EA practices

290 • *The Philosophy and Practice of Wildlife Management*

Environmental legislation and EA review processes present in all Canadian provinces also are being updated as knowledge of process pitfalls develops. Most such legislation and the regulations thereunder provide the responsible minister with considerable authority, however in practice this authority lies primarily with civil servants. In some instances, bureaucratic responsibility is specifically stated within legislation, as in Nova Scotia's new environment act where the relevant Section 36 reads, "Where the Minister decides that an environmental assessment report is required, an administrator shall (a) prepare proposed terms of reference for the environmental assessment . . . etc." Regardless of the detail present in both legislation and regulations and regardless of the specific assignment of responsibilities, processes will vary. The complexities involved preclude the development of generic processes that fit all situations. For this reason, and others, oversight on both environmental policy and review processes is deemed important. To help provide such oversight and to incorporate citizen input, the federal government and each province have established "Environmental Round Tables." Although such bodies are expected to obtain elements of environmental expertise, they are politically appointed and as such may not be very professional.

Some Approaches to the EIA

Elements of environmental analysis appeared in several early land-use planning exercises across North America. One method developed in Ontario by G.A. Hills is detailed in his 1961 text, *Ecological Basis for Land-Use Planning*. The Hills technique depends upon the classification of land units into physiographical land types, classes, and site types. It eventually results in landscape units with capability, suitability, and feasibility ratings for wildlife, recreation, and other resources (Figure 11.1). The approach was helpful to those involved in planning the nationwide Canada Land Inventory that classified Canadian lands with a one (highest) to five (lowest) capability rating for agriculture, forest production, wildlife, and recreation. The Hills method was one of the first to recognize the importance of ecological factors as determinants in land-use considerations; however, neither use impacts nor long-term effects were considered. Ecological entities were not considered as the dynamic, changing elements they really are in natural systems.

Another planning device used considerably in the United States was the technique of determining "environmental corridors" based upon the occurrence of identifiable landscape patterns (Figure 11.2). This method was worked out by the architect, Philip H. Lewis, and its primary objective was to identify, preserve, protect, and enhance the most outstanding "natural values

Environmental Impact Assessment • 291

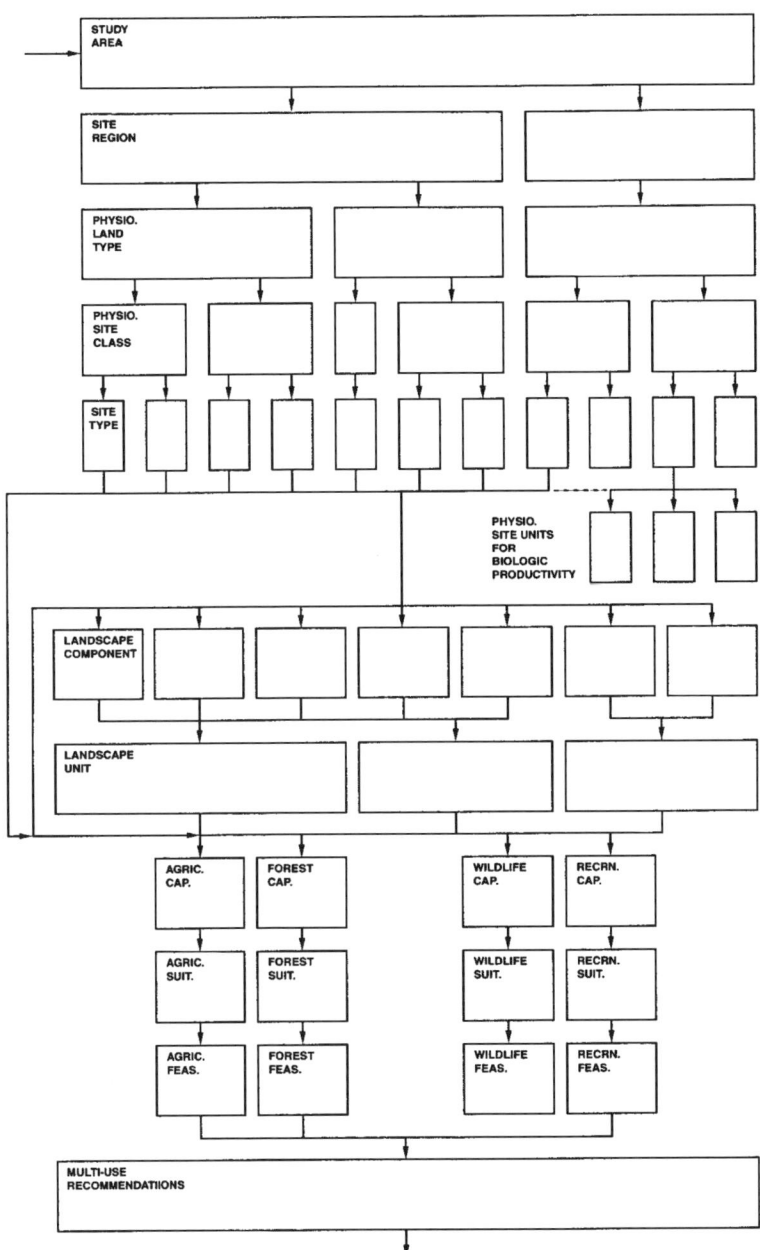

Figure 11.1—Diagram of method, G. Angus Hills. (Parks Canada, 1973).

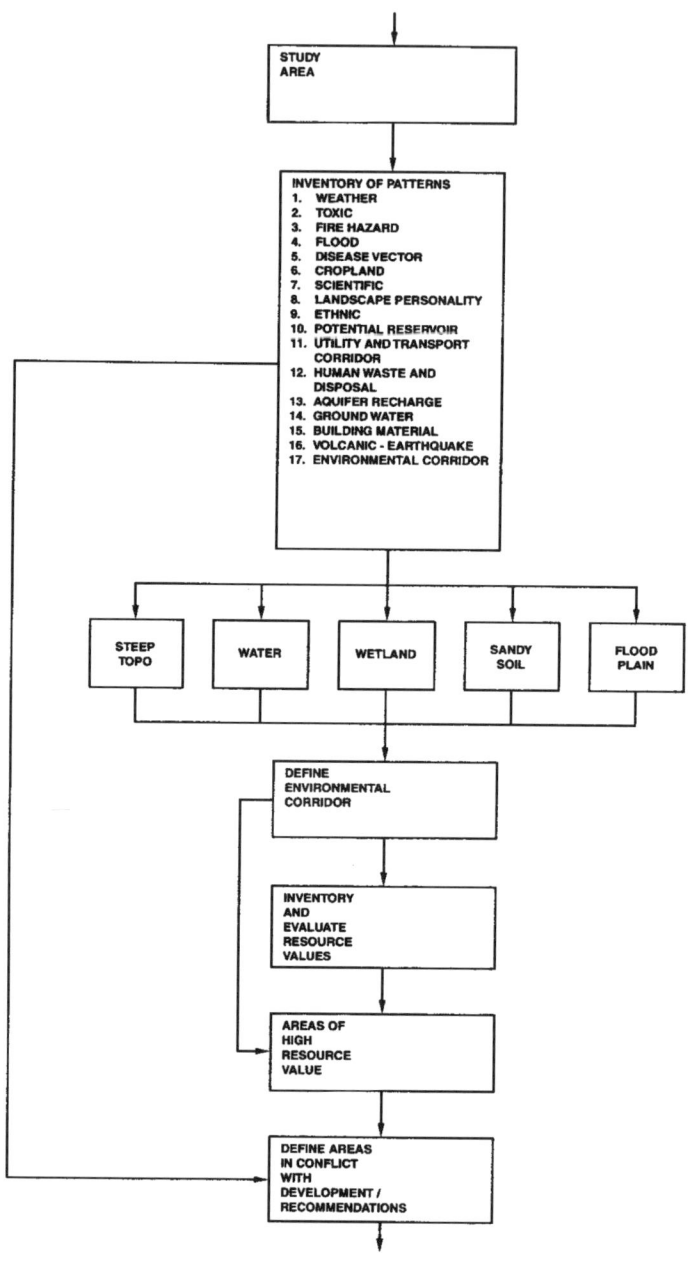

Figure 11.2—Diagram of method, Philip H. Lewis, Jr. (Parks Canada, 1973).

of land units" and see that man-made values were developed in harmony with the quality resources recognized.

Like Lewis, Ian McHarg brought a landscape architect's vision to land use planning with an eight step method which could be subdivided. Primary elements were:

1. An environmental inventory.
2. Determination of the dominant prospective uses for each land unit.
3. An assessment of land-use values.
4. A determination of the suitability of land units for dominant uses.
5. A determination of the compatibility of various land uses.
6. An economic inventory.
7. Establishment of criteria for visibility.
8. Establishment of criteria for form and design. (Abbreviated from Parks Canada, 1973)

The method was based upon the concept that any land area is valued as the sum of its historical, physical, and biological processes which are all dynamic and constitute social values. The method recognized that each value has an intrinsic suitability for a particular land use and that certain areas lend themselves to multiple, coexisting uses. Figure 11.3 diagrams the process thoroughly detailed and articulated in McHarg's *Design with Nature,* a book that influenced North American planning greatly following its 1969 publication.

Beginning with the seventies decade, the need to develop procedures actually to measure the impacts or effects of implementing a particular change in land use, or of projects which would change entire systems, became imperative; legislation was being implemented throughout the continent requiring that environmental assessments be made. The earlier land-use inventories and planning approaches were valuable, but they did not meet the specific demands required by the biological impact analysis, nor did they always involve socioeconomic factors. The problems now were how clearly and accurately to measure the biological, social, and economic bases, determine their values, assess the effects of the known measured changes to take place with proposed developments, and determine mitigation procedures, costs, and benefits. All of these problems have not been completely solved. New methodologies are regularly being developed as the need for more precise measurements increases. The entire process of environmental assessment is one of change, and because of the dynamic aspects of natural systems, changing social values and rapidly altering economies, the processes are likely to remain fluid for many years. Public attitudes and political priorities are also dynamic and will continue to evolve.

One of the earliest attempts to quantify and compare impact data was published by Luna Leopold et al. (1971). The authors indicated that since "there is

294 • *The Philosophy and Practice of Wildlife Management*

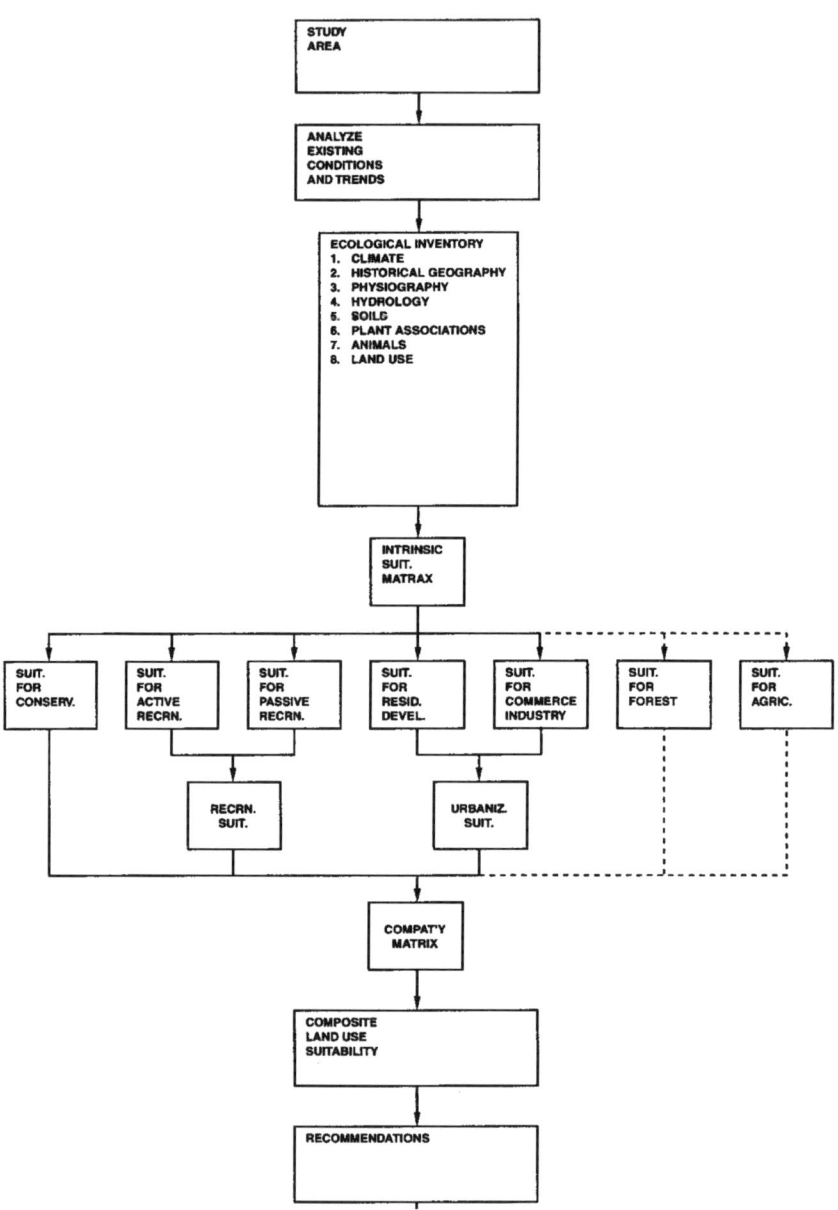

Figure 11.3—Diagram of method, Ian McHarg. (Parks Canada, 1973).

no uniformity in approach or agreement upon objectives in an impact analysis ... this generalized matrix is a step in that direction." The authors further noted that the EIA should consist of three basic elements:

1. A listing of the effects on the environment which would be caused by the proposed development, and an estimate of the magnitude of each.
2. An evaluation of the importance of each of these effects.
3. The combining of magnitude and importance estimates in terms of a summary evaluation.

The terms *magnitude* and *importance* are explained as follows. Magnitude is used in the sense of degree, extensiveness, or scale. For example highway development will alter or affect the existing drainage pattern and may thus have a large magnitude of impact on the drainage. Importance is the significance of the particular action on the environmental factor in the specific instance under analysis. Thus the overall importance of impact of a highway on a particular drainage pattern may be small because the highway is very short or because it will not interfere significantly with the drainage.

The matrix approach, in general, has proven to be a useful tool employed in many EIAs. Environmental characteristics present in an area are listed on one axis and the construction and operational phase activities on the other. A matrix example is shown in Figure 11.4, where the scale is 1 to 5 with 1 being the lowest. The magnitude of the action is given at the upper left and the importance on the lower right. Looking at the horizontal axis, for instance, it is noted that for ungulates (deer and moose in this case) inundation of the Cheticamp Lake area is of maximum magnitude in that the area would be eliminated as a terrestrial, or usable aquatic habitat, for moose or deer. It is also noted to be of maximum importance since the area to be flooded is prime summer moose range. Because a much smaller area of less important range inhabited by fewer animals is flooded in the other cases, the magnitude is rated lower and the importance almost nil. If an area is totally destroyed for ungulates, it is possible to have a very high magnitude while having a low importance if few of them are affected. For the Cheticamp inundation, Figure 11.4 illustrates this type of importance—magnitude relationship including amphibians and reptiles. For most developments today, matrices will be far more complex than the example shown in Figure 11.4. Socioeconomic, aesthetic, ecological, behavioral, physical, chemical, and toxicological effects may be considered in relation to activities throughout stages of project development, operation, and maintenance. No potential impact is left out and impacts may be remeasured by a number of means over months or even years before final decisions are made.

Others have considered impact assessments at the ecosystem level (Auerbach, 1978; Odum and Cooley, 1980) and have proposed specific procedural

296 • *The Philosophy and Practice of Wildlife Management*

PROJECT ACTIVITIES	PHYSICAL										BIOLOGICAL								
	INUNDATION			IMPOUND-MENT		STREAMS					TERRESTRIAL						AQUATIC		
	LAND	MARSHES	LAKES	FREE FLOWING STREAMS	WATER LEVEL FLUCTUATIONS	GROUND WATER	WATER QUALITY	MEAN FLOWS	MINIMUM FLOWS	WATER QUALITY	SEDIMENT & EROSION	LENGTH AFFECTED	FOREST	WETLANDS	UNGULATES	BIRDS – UPLANDS & PASSERINE WATERFOWL & SHORE BIRDS	OTHER MAMMALS & BIRDS	UNIQUE SPECIES	SALMONIDS	AQUATIC VEGETATION	AMPHIBIANS & REPTILES	EELS
–CHETICAMP:																						
CONSTRUCTION													3/-1	1/-1		5/-4						
INUNDATION TO MAX. LEVEL	5/-2	5/-5	5/-2	4/-2		4/-2	5/-5	5/-1		5/-3			4/-1	5/-4	5/-3	4/-2	2/-2		4/-4	4/-5	4/-2	5/-4
DIVERSION OPERATIONS				4/-5			5/-1	5/-2	5/-2		5/-4								4/-4	5/-5	5/-3	
–INGONISH:																						
CONSTRUCTION																			2/-2			
INUNDATION TO MAX. LEVEL	1/-1	3/-3	1/-1	1/-1		3/-1	1/-1	3/-1		1/-2			1/-1	1/-1	1/0	1/-1	1/-1		1/-1	1/-5	1/0	
DIVERSION OPERATIONS				1/-2			2/-1	5/-2	3/-4		3/-4								1/-1	1/0	1/0	
–EAST INDIAN:																						
CONSTRUCTION													2/-1	1/-1		4/-4						
INUNDATION TO MAX. LEVEL	1/-1	1/-1	3/-3	3/-4		3/-1	5/-4	3/-3		5/-3			5/-1	2/-2	2/-2	2/-1			2/-2	3/-5	3/-1	
DIVERSION OPERATIONS				5/-3			4/-2	5/-3	5/-4		3/-3								4/-4	4/-3	3/-2	

Figure 11.4—Matrix developed to show environmental impacts of proposed hydroelectric development at Cheticamp, Nova Scotia.

approaches. Barske (1978) and Truett (1979) were concerned with mitigation of impacts on natural resource entities, primarily through improved planning and design. A large number of authors have been concerned with various other aspects of the EIA including Hirsch (baseline studies, 1980), Lucas (statistical aspects, 1976), and Stover (general procedure, 1972).

Although precise determination of the effects of development is seldom possible, changes in physical factors such as flow, depth, and often temperature as well as chemical and biological characteristics can sometimes be accurately calculated for aquatic systems. In some situations, modelling results may closely reflect reality when the input data are adequate or when field environments can be replicated under controlled conditions. This may sometimes be possible in the laboratory. Accuracy is severely reduced for terrestrial populations and habitats where more variables present complex control problems and where input data for models may be far less precise. The wildlife student must become familiar with as many approaches to obtain measures of change or impact as possible. The significance of any change is really a measure of change in productivity within the ecosystem. If an ecosystem is irreversibly damaged or destroyed to the point that it will not self-correct, there is no question—the impact is significant. At this point, it becomes a matter of values, social losses, and trade-offs. Society may make the ultimate value judgement by what it considers to be important. For example, loss of the supporting ecosystem of the whooping crane may have greater value than a similar loss of California condor habitat.

The Process—From the Beginnings

The steps in the process from the first project proposal to the submission of an EIS with recommendations and subsequent reviews will vary depending upon whether the project is initiated by a federal agency within the United States or Canada or whether it is of provincial, state, or other origin. The legislation within which it is considered may also affect the steps in the process. Regardless of the steps involved, you, the biologist, will become involved if the proposed project has, what may be considered, a significant impact on some environmental entity affecting wildlife. Leopold et al. (1971) have described the possible sequence of events as follows:

1. A statement of the major objective sought from the proposed project is made.
2. The technologic possibilities of achieving the objective are analyzed.
3. One or more actions are proposed for achieving the stated objective. The alternative plans which were considered as practicable ways of reaching the objective are spelled out in the proposal.
4. A report which details the characteristics and conditions of the existing environment prior to the proposed action is prepared. In some cases, this report may be incorporated as part of the engineering proposal.
5. The principal engineering proposals are finalized as a report or series of separate reports, one for each plan. The plans ordinarily have analyses of monetary benefits and costs.
6. The proposed plan of action, usually the engineering report, together with the report characterizing the present environment, sets the stage for evaluating the environmental impact of the proposal. If alternative ways of reaching the objective are proposed in 3, and if alternative engineering plans are detailed in the engineering report separate environmental impact analyses must deal with each alternative. If only one proposal is made in the engineering report, it is still necessary to evaluate environmental impacts.

298 • *The Philosophy and Practice of Wildlife Management*

7. The text of the environmental impact report should be an assessment of the impacts of the separate actions that comprise the project upon various factors of the environment and thus provide justification for the determinations presented in 5. Each plan of action should be analyzed independently.
8. The Environmental Impact Statement should conclude with a summation and recommendations. This section should discuss the relative merits of the various proposed actions and alternative engineering plans and explain the rationale behind the final choice of action and the plan for achieving the stated objective.

Projects in Canada that are considered by the CEAA may require several dozen steps depending upon the complexity of the development, the stakeholder groups involved, the projected benefits of the project, and the political reaction to project approval, disapproval, or approval with mitigation. The relevant act suggests the following general procedures:

1. Screening or comprehensive study and preparation of report.
2. Determination of EIA scope and factors to be considered.
3. Public review of screening or study report.
4. Establishment of EIA review panel.
5. Conduct of EIA (government, consultants etc with panel oversight).
6. Preparation of EIS.
7. Conduct of public hearings and EIS alterations.
8. Panel recommendations and submission to responsible authority.
9. Decisions concerning approval, disapproval, conditional approval (further study), and mitigation.

Screening is initiated by the responsible agencies as soon as possible after the project has been proposed. If the screening process or initial comprehensive study indicates there will be no significant impacts the entire process may be short cut. If significant impacts are indicated the process continues and will probably resemble the example shown in Figure 11.5 taken from the Canadian federal approach. Any of three decisions may be reached following screening, two of which can lead to the provision of guidelines, an assessment (EIA) and its resulting statement (EIS).

Nowadays, both public (government sponsored) and private enterprise developments must wait for a process to be carried out that is similar to those noted here; it is still true, though, that because political commitments or investment decisions have already been made, the project contemplated may be a fait accompli. Nonetheless, those involved with an EIA should take their work seriously, for the possibilities of mitigation through alternative design or development still may be present. In past years, there have been examples such as the hydroelectric development at Wreck Cove in Nova Scotia when project construction was well underway before the environmental process was even started. In these cases, the proponent of the project may have spent very little on the environmental assessment processes, or if pressured by the public, spent a great deal more than necessary! Mitigation recommendations were seldom implemented to any extent unless regulations required them. Then the regulations would have been recognized and some action taken even without the EIA. Updated legislation and/or regulations now help prevent such occurrences.

The Assessment Proposal

Usually consulting companies are invited to submit a proposal for an EIA that includes a general budget and breakdown of costs. The invitation comes from the proponent-the company or political jurisdiction undertaking the project having potential environmental effects. An invitation may be very detailed or it may only indicate that a project of a given magnitude is being considered in a particular location and provide spare details. Detailed invitations are the norm as everyone in any particular state, province, or country in North America now operates within the same or similar regulations. If the invitation is not detailed, it is possible that a decision has already been made as to what consulting firm will be conducting the EIA.

Usually the invitation will include information on the proposed project design, its construction, and its operation (plus alternatives, if any), the time frame involved, the kinds of environmental baseline data required, an outline of environmental impacts relating to each environmental data base, outlines for mitigation, and recommendations for both construction and operational phases of the project. The nature of the project will dictate the content and to some extent the approach used. Guidelines for the EIS in Canada are usually available to the proponent at an early date after a panel review has been initiated and can be used by those bidding on the EIA study or portions of it. These guidelines are very complete and will probably cover all necessary environmental entities as they are prepared by professionals serving on an assessment panel for the specific project.

If you have been supplied with detailed information, your proposal should generally follow the outline and approach provided. If you know that the information you have is incomplete, you are responsible for securing whatever clarification your professional judgement deems necessary.

The proposal you may be asked to draft should reflect activities related to each of the environment data bases noted on the invitation or guidelines as well

300 • *The Philosophy and Practice of Wildlife Management*

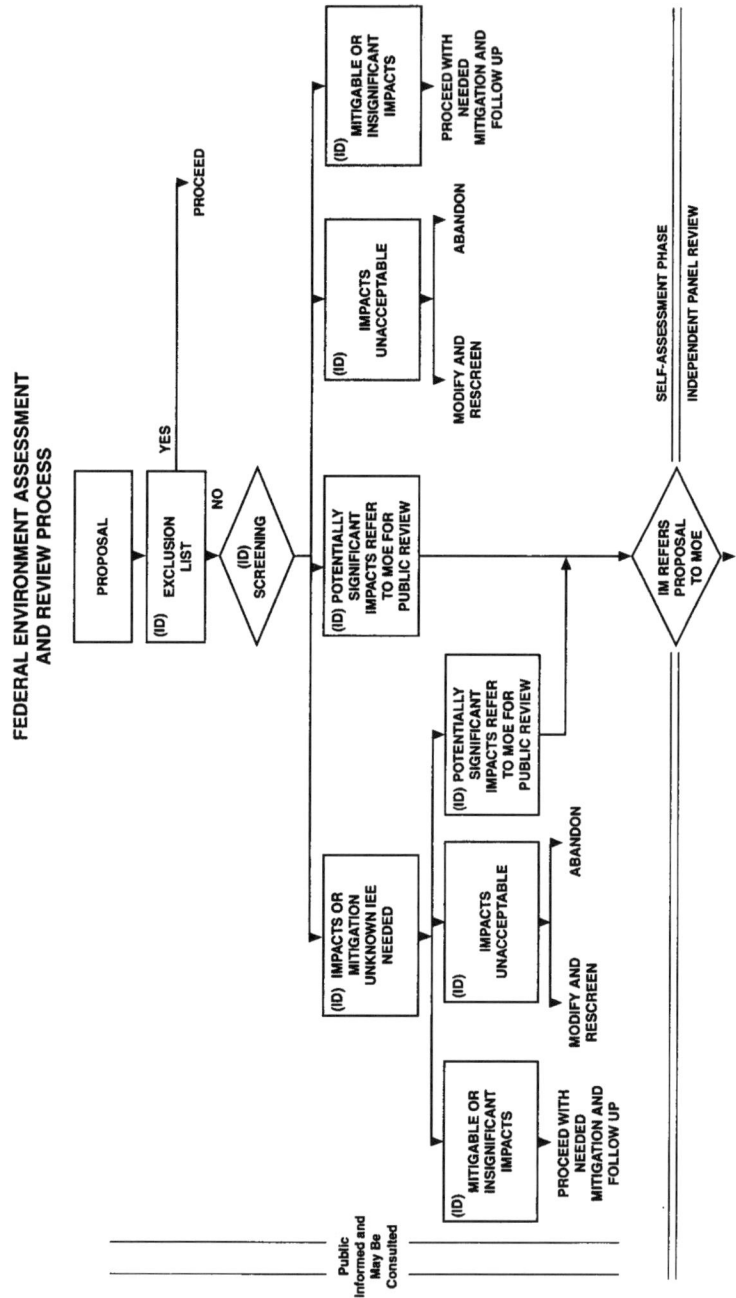

Environmental Impact Assessment • 301

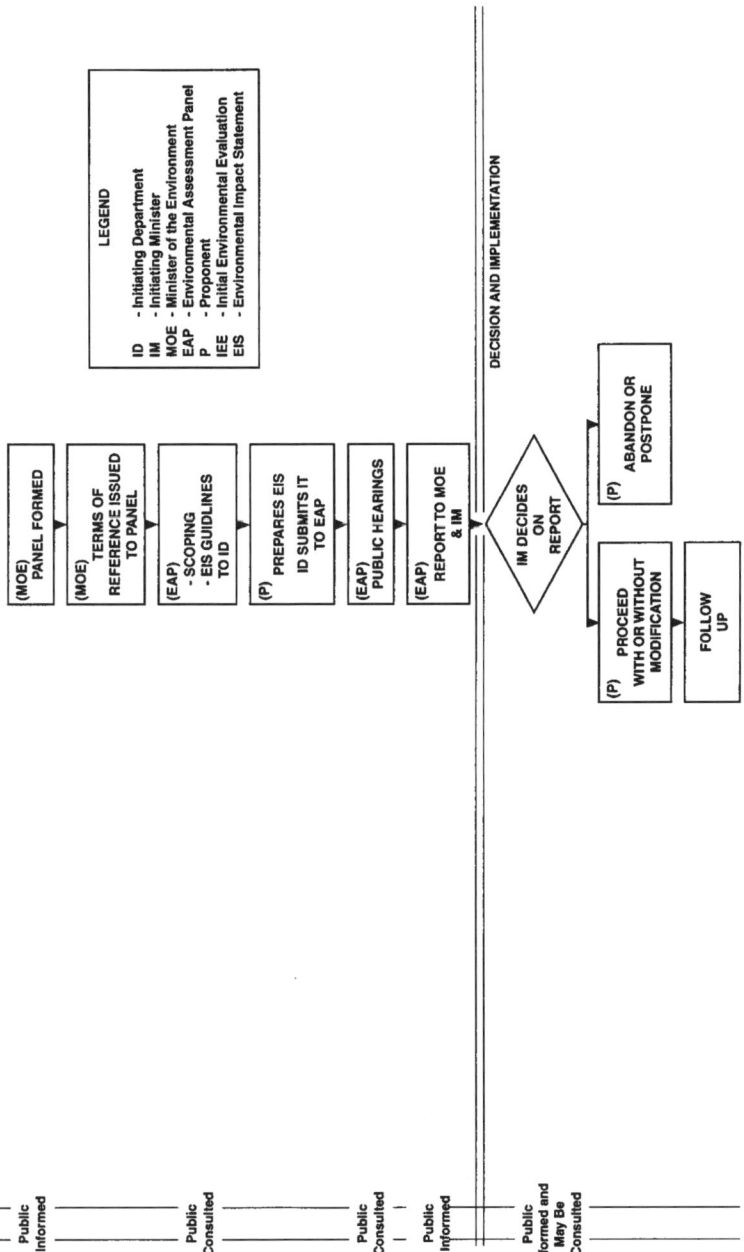

Figure 11.5—Schematic diagram of the Canadian federal environmental assessment and review process.

302 • *The Philosophy and Practice of Wildlife Management*

as activities pertaining to impact determination for each ecosystem or data base under each construction and operational alternative. The consequences (to ecosystem and/or species) of the impact with, and without, mitigation are also significant and must be recognized in the proposal. Note that you should provide mitigation recommendations and cost estimates for implementing them. Also include community effects, benefits, and costs. These are in the realm of socioeconomic concerns important in all projects.

In addition, the proposal should indicate the time frame you expect to follow, the approach you will take in completing the work (the management, administration, and coordination), a budget for each phase (and professional entity), methodologics to be employed, and who makes up the team (what expertise they have and how their special abilities relate specifically to the project concerned). If the proposal is well conceived and if the design of the EIA study remains essentially unchanged, the proposal can serve as a reasonable outline for the EIS. This is the way it should work out unless the unexpected occurs on a grand scale!

The Environmental Impact Statement

Few reports purportedly of a scientific or professional nature have been criticized and even ridiculed as much as many of the environmental impact statements prior to the late 1970s. But, until the EIS came around, few reports were available for criticism. In-house government reports were discussed in the involved offices, and usually problems were ironed out before senior people saw them, or perhaps the seniors were not even able to digest the contents fully enough for a critical analysis. Government reports that were prepared for others were tidied up before release. Some poor ones escaped, but by and large they were capable of withstanding the ordinary critical review. Of the millions of other reports reaching (or within the reach of) the public, most were of limited interest and used only by a few special interest groups. With the EIS, we had a different situation. Here was a report prepared for a proponent who might not always be capable of dissecting it thoroughly, but the public and a large body of professionals from both opponent's and proponent's sides of an issue were able to read it and pick it apart. This happened often in the past and still does on occasion in spite of critical reviews and rewrites of several drafts. Sometimes impact statements are too big, too complex, and too descriptive, generalizing on environmental aspects and failing to pinpoint important impacts. Many times the EIS is stuffed with irrelevant data and generalized statements of no use to anybody. Do not pad your EIS. Remember that you have a proposal which can serve as an outline. If you find items in the proposal to be inapplicable, then say so. Remember to justify any changes you make, and put general data packages in an appendix or make them available on request. Do not put your EIS in between expensive covers. Here is what the FEARO (now CEAA) had to say about the EIS in 1979.

> "It describes the project, its location, the need for it and any alternative methods of achieving the project other than the one proposed. The IS also describes the area's existing environment and current patterns of resource use, social factors such as population characteristics, community life style and the economic base of the area. It provides a detailed description of the potential effect of the proposal on the area's environment and identifies the measures the proponent intends to take to reduce those impacts. Any impacts that might remain after these mitigating measures have been taken must also be identified."

If the proponent has not yet indicated plans for mitigation, such measures should appear in the proposal as recommendations. Similarly, remaining impacts should be identified and their suggested mitigation procedures written in as well. In the Leopold et al. (1971) approach to the EIS there are four basic items. The authors listed these items, in part, as follows:

1. A complete analysis of the proposed action.
2. An informative description of the environment to be involved, including a careful consideration of the boundaries of a project.
3. A discussion of the pertinent details of the proposed action.
4. An assessment of the probable impacts of the variety of specific aspects of the proposed action upon the variety of existing environmental elements and factors.

In enlarging on these items, the authors note that point 1 (the analysis) should also be a justification "which considers the full range of values to be derived, not simply the usual cost-benefit analysis." Concerning point 2, the authors indicate that special emphasis should be placed "on those rare or unique aspects, both good and bad, that might not be common to other similar areas . . . The description should include all the factors which together make up the ecosystem of the area . . ." And in considering point 3 the authors signal that details of the proposed action ". . . should include discussion of possible alternative engineering methods or approaches to accomplish the proposed development." Their point 4 is an explanation of the Leopold matrix and its use.

So follow the outline of the proposal. Add and subtract the newly relevant and irrelevant, justify your statements, and use an acceptable guide such as those normally available from appropriate state, provincial, or federal agencies. If you write clearly and simply, reducing jargon and complex theory as much as possible, and if your statistical handling of data is appropriate and thorough, your EIS should be acceptable.

Assessment Problems

Perhaps the most difficult is one that may not become apparent until after you enter professional ranks. That is the overall management of the assessment project. Management is less difficult if the successful bidder put together a proposal with no need to contract for expertise outside the company whose employees probably worked as a team in agreeing on time frames, the minimum data bases required, and the methodologies needed for baseline and experimental study. Costs, meetings, and deadlines also were agreed to. In such cases, the modus operandi falls into place with time and experience and it becomes a simpler matter to prevent overexpenditure on any aspect or phase of the assessment. All budget procedures must also include a margin for increasing costs, editing, design, and printing as well as those extra few meetings of team members with company engineers or proponent representatives. But the smoothly functioning unit assessing ecosystem entities in the field is the exception. There are always unexpected problems in field studies even when the personnel are familiar with the terrain, the waters, and the ecosystems involved. The biologist never knows in advance all of the kinds of problems that may be found or all of the biological organisms that may be affected (positively or negatively)—all factors that may require added time in the field, in the lab, or on the computer. Good management is able to account for such situations within the parameters of time, money, and people yet still meet deadlines with a sound product and without losing money.

Cooperation is part of the management picture, something each team member should be concerned with. Much of the background and most of the data base are often obtained from existing information already in government agency files, in the files of university researchers, or in publications or files elsewhere. Publications and data are available, but getting data from a wildlife biologist working for the state of Washington, or the province of Nova Scotia, is often a problem. Freedom of information legislation, if present, may be relevant in such cases. Nevertheless, a biologist seeking out information from another person in government usually depends on that source to provide all the necessary data—although the person might not be willing, or able, to do so.

Neither is it always easy to extract data from university research people. Sometimes, it is impossible and the assessment team may have to duplicate a great deal of work. We have been in both positions—having to ask others for data, and sometimes data interpretation, as well as having consultants request unpublished information. In the first instance, familiarity with an agency and personally knowing the people involved have helped. In the second instance, we admit that we often are unwilling to cooperate fully with a stranger who is unfamiliar with the area, sometimes unfamiliar with the subject, and works for a company from another jurisdiction 2000 miles away. Like that of others in these circumstances, our attitude is often one of protecting data for personal use.

Schindler (1978) has argued from a government employee's point of view as follows:

> "I am often interviewed by someone from a university or consulting firm who is hoping to obtain insight into a particular impact. Almost without exception these are people with bachelor's degrees. Also almost without exception, they were not totally capable of assimilating all of the complex information which I could put at their disposal. Upon questioning them more closely I found that they (or their firm) had been contracted by the government (my employer) to investigate a problem in which our department had several international experts. Even more ludicrous, the salaries paid to these undereducated individuals to interview government experts were higher than those paid to the experts themselves."

Even though all assessments subject to the federal review process in Canada are now carried out by an independent panel of experts, which reduces such possibilities as those raised by Schindler, it is still possible in many states and provinces to have such problems repeated. Surely it is not the fault of the interviewing biologist that others become upset at his presence, but if an appropriate protocol is not in operation, the consulting firm should make the initial approaches and establish a cooperative climate for its people before the interviewer calls or makes personal contact. As has been recognized by the government of Canada, the problem is one of establishing a level of expertise in the overall process and determining a protocol to be followed. The project team is almost always short of time for portions of work deemed important to them. We have seldom found field biologists to be really satisfied with their efforts, even though they themselves may have helped establish their own budgets and time allotments. Something always occurs to make more time and work desirable. Good planning reduces the extent of problems but they are always present. That one last sampling season you really want is not available because deadlines must be met! The unexpected notwithstanding, field biologists must often cut the cloth to fit the available time and money. In doing so, they have to indicate to the project manager and (directly or indirectly as the process requires) to the proponent, the limitations of their data pack and the resultant data analyses. Except in rare circumstances, the EIS will not include original research and will seldom provide new findings in ecology. The short-term nature of the EIA usually demands that scientific precision be reduced in favor of good scientific judgment. This fault in the entire process has been one which goes against the grain of all of us, at one time or another. It is the basis for many attempts on the part of ecologists to find ways of improving EIS ecosystem data packages and their interpretation and analysis.

Different Viewpoints

Depending upon the roles they play by choice or necessity, people obviously view the EIA and the EIS in very different manners. We are indebted to Beanlands and Duinker (1983) for an examination of many attitudes and an excellent digest of process data that end on a relatively positive note. We here present extracts portraying viewpoints of the government administrator, the proponent, the consultant, and the research scientist.

The Administrator's Perspective

Government administrators tend to view EIA as the fulfillment of required procedures as set by policy or legislation. For these people, it often becomes a matter of whether the assessment guidelines have been met. In most cases, the first priority is on running the administrative machinery of assessment with less regard for the details of the resulting studies. Although the agencies may retain outside experts for the preparation of guidelines, such terms of reference usually amount to lists of "things to do" rather than providing any form of scientific direction or performance standards. It is only at the review stage that the administrators are faced with determining the scientific or technical substance of the assessment studies undertaken. At this time, outside experts may be brought in to give their opinion. In doing so, such experts almost invariably adopt a fairly rigorous interpretation of the guidelines-a perspective which may have helped at the beginning of the assessment but which can be very disruptive at the end.

The Proponent's Perspective

In industry, the objective of EIA is tied directly to the project approvals and licenses. Because of the high public profile that is often adopted in review procedures, impact assessment is also important to industry from a public relations perspective. With project approval in mind, the proponent's main objective is to develop an acceptable EIS. They will "do what has to be done" to get that document approved, but are understandably reluctant to consider anything beyond that as part of the impact assessment process. This EIS focus may present problems when it comes to implementing impact assessment in a much broader time frame as implied by the inclusion of operation-phase monitoring. It seems, that not all industrial proponents believe it is in their best interest to have the scientific quality of impact assessment studies improved. A certain degree of flexibility in study results can sometimes be used to advantage when debating potential impacts. On the other hand, there is ample evidence to indicate that industrial proponents in Canada have generally adopted a positive attitude toward EIA. As stated on a number of occasions during the workshops by various industrial representatives, "Any reasonable study will be funded."

The Consultant's Perspective

In Canada, the task of conducting environment impact assessment studies and preparing an EIS most often falls to consultants in the employ of the proponents. They find themselves caught between the differing perspectives on the assessment process held by government agencies and proponents. The consultants' main role is to translate assessment guidelines, which are often generalized and vaguely worded, into a number of field or laboratory studies, or both. Basically, they try to establish a short-term applied research program. In so doing, they are normally directed by their clients to limit their efforts to a level that is necessary to get the project approved. However, they must also consider the possibility of project delay or refusal if the studies are found unacceptable to the reviewers. In effect, the consultants are expected to practice good science in a politically motivated system. In many respects, the role of the consultants in EIA is the most difficult of all. They do not have the luxury of working according to their own fundamental objectives for the assessment process. They must develop a compromise between the approval required by the client and the scientific and technical standards which they would like to adopt to ensure acceptance within a process that is essentially a peer review.

The Research Scientist's Perspective

Research scientists in government and universities have not generally been attracted to EIA. From their perspective, the overriding constraints of time and politics usually preclude the conduct of acceptable science in assessment studies. They are, however, often called upon to assist in the preparation of assessment guidelines. Since the guidelines are seldom written in a contractual format which would guarantee the conduct of acceptable work, their basic suspicion of impact assessment tends to be confirmed.

As well, government and university researchers and staff of resource management agencies are often called upon to review the results of impact assessment studies. In so doing, they wear their scientific hats and evaluate the studies according to standards of excellence which are rarely established at the outset. In effect, they undertake a peer review of the work in much the same way as they would evaluate an article submitted for journal publication. This amounts to implementing a quality control program at the end of an assembly line with no feedback loop. It is frustrating to both the reviewers and authors of the documents.

308 • *The Philosophy and Practice of Wildlife Management*

We believe these statements include most of the concerns except those the general public may have. Because the EIS is often provided to interested groups within the public and since public hearings frequently occur in the process, we should also be aware of the public's concerns. There are a variety of public opinions, but in most cases, the portion of the public that is against the proposed project will be the one dealing most actively with the EIS. These people, or groups, may even provide their own experts to find whatever weaknesses they can in an EIS. Should the EIS suggest little in mitigation, or even in serious impact, the public may try to denigrate the work, if possible, as unscientific and erroneous. It is not uncommon in public hearings for people to attack the EIS, the consulting firm, and the proponent, together. Here is where good public relations at the outset on the part of the consulting group and the proponent can help lessen the conflicts in the public hearing processes. Keeping various elements of the public aware and knowledgeable can often be more helpful to everyone than keeping them in the dark. The public will not disappear!

Areas of Concern to the Wildlife Biologist

A good way to familiarize yourself with the EIA process is to study maps and land-use patterns of an area before a development project begins and then visit it when completed. The before-and-after look can be quite striking in the case of projects resulting in large scale impoundments or super highways and less so in specific restricted-site, industrial developments. It is difficult to assess the before-and-after when air pollution is involved or for large-scale offshore arctic developments. In these instances and some others, a review of the process steps, complete with the engineering, ecological, and socioeconomic data packages is helpful. Detailed study of several case histories will help give you a feel for the EIA.

For most environmental projects, the wildlife manager should be knowledgeable about eight information sets, which we have expanded on here:

1. Most states and provinces have legislation (guidelines) which indicates when an EIA is required and often outlines the general processes. The wildlife employee is advised to be just as familiar with this type of environmental legislation as with wildlife or other land-use and pollution legislation. Most particularly, government employees will need to understand the legislation involved, for they will be conducting their work and representing their agency under its regulations. Some jurisdictions also publish guidelines which expand on the process and indi-

cate the general format required for their IS. Copies of this material should be within reach of all professional wildlife workers.
2. The reasons for the proposed project and the proponent's position must be clearly understood. Justifications for the proposed development may not satisfy you as an individual, but whether they do or not, a clear understanding of why the project is being undertaken is necessary because you may be balancing certain wildlife habitat losses with the benefits and needs indicated by the proponent, the initiator, or both. Often the proponent's position is a defensive one. The more clearly you understand why a defensive posture exists, the easier it will be for you to state the positions dictated by your results. You are not likely to change a proponent's attitude, but you can help avoid confrontation by the approach you take, both orally and in writing.
3. An understanding of the project is necessary. This is where your ability to absorb engineering, construction, and operational details comes in. You must learn to understand the language of civil and hydrological engineers, petroleum geologists, resource economists, foresters, and others. The wise wildlifer has already prepared himself for this role by carefully planning the university courses taken. If this wasn't the case, then he must learn on his own. If you do not understand the portrayal of various sets of data, ask questions. Don't stay in the dark or your inability to interpret will surely show in your writing of the IS.
4. The ecological aspects must be clear enough to you so that you can articulate them for others, ecologists or nonecologists. Remember that your work is to influence engineers, politicians, and the public. Still, you must work with enough detail, accuracy, and statistical thoroughness to please yourself and other professionals. If you are satisfied, your fellow biologists will usually be reasonably well pleased. Ghiselin (1980) provides considerable guidance for the wildlifer here and for impact assessments generally.
5. Social and economic concerns should also be understood by the wildlife biologist. It is important to know how many families will have their lifestyles altered detrimentally in the development area and how many will have their's improved, both on site and away. Is the project labor intensive? Is the operational phase after construction labor intensive? What are the attitudes of the people and what are the costs to the taxpayer? You should have some grasp of these and other socioeconomic considerations but need not become involved in a people versus wildlife issue. If you know what the public pulse is, and why, you can deal more effectively with conflicts as they arise.

6. Clearly understand the project's impacts for the present and for the future. The impacts from alternative construction and operating models have also to be thoroughly understood. Your report may influence the selection of alternatives; you need a thorough understanding of the ecological impact aspects of all present and potential phases. Your data will not usually allow you to be exact, particularly if you are dealing with a terrestrial system—all the more reason for you to know and understand the differences among alternatives, even if you are inexact as to the degree. Subjective opinions based on guesses are not good but subjective opinions from a professional knowledgeable in the field are respected.
7. The wildlife professional is the one who can recommend and defend actions to alleviate or reduce the magnitude of impacts. Mitigation is sometimes not possible, but where it is (within cost parameters) it may be the only positive note in an otherwise dreary tale for the environment and for wildlife. This means that you may have to think beyond the project itself. If a wetland or stream is to be dredged, maybe a tributary can be dammed to continue the area's waterfowl and muskrat production. Or mitigation may mean setting out food plants, stocking fish, providing nest sites, altering maximum drawdowns, realigning a pipeline route slightly, providing road or pipeline crossings for animals, or any of a host of other activities appropriate to the type of project you are dealing with.
8. Finally, understand both the costs and benefits. There are apt to be some cost-benefit analyses conducted using different inputs relative to revenue and expenditure. Be sure to understand these statements, for they are probably crucial in decision making prior to and during construction. Your own cost-benefit concern about ecosystem impact might be enough to deny continuation of the project when added to the specific economic analyses available. Decisions will usually be made on a political basis with economic justification, of course, but if there is resistance to the project from a segment of the public, your recommendations can help substantiate one or the other position and may be the strongest factor in a decision.

Wildlife professionals are more likely to hear of and understand the sentiments of the nonprofessional public. The opinions of the opponents to the project must not influence you, but knowing them may help create bias on your part if you are not careful. Working alone with your fellow professionals, completely apart from the pressures of all interested parties, you would be much happier than you are likely to be in a realistic situation where both the press and the public try to give you direction. The professional wildlife worker plays it straight and keeps his opinions to himself even in private. A high proportion of graduates in conservation biology and wildlife will find themselves employed by consulting firms. Many will form their own limited companies or partnerships with other ecologists, foresters, or engineers. Although much of their workload may be involved in environmental assessment, they likely will be called on to carry out specialized work for government agencies or private enterprises.

The Future of the EIA in North America

The EIA is here to stay; that seems certain. It should be clear that environmental legislation probably cannot protect wildlife or natural systems from the pace of development. What has occurred since the late 1960s is the evolution of a general process that admits that environmental concerns exist. To be sure, in many cases, the worst possible impacts have been avoided and in others alternative procedures or mitigation activities have partly replaced what was lost. Even in these latter instances, the ecological changes have usually been considerable. There are a few places where development has been stopped or delayed because of impact studies, but there are probably many more cases in which a vocal public has caused the government to put a moratorium on an activity that it deemed either to violate the integrity of natural systems or to threaten human health, or both. Nevertheless, the EIA will endure. It is sometimes a valuable tool to politicians. The process also creates employment. It has added to the short-term costs of government and industry. It may be responsible for some increases in taxes as well as higher costs on consumer items. Still, as decisions to conduct impact assessments are gradually moved from political decision making processes to legislation making them mandatory, the benefits to wildlife and habitats will increase. The EIA will become a major tool in building the road toward sustainable development.

Whereas impact assessments have the potential to provide environmental data for decision making in developments where risk factors can be determined with some confidence, they may be less helpful in situations where human error or noncompliance may occur, as exemplified by the disastrous oil spill of 1989 in Prince William Sound, Alaska or coastal Japan in 1997. One example of a potential development where risks are well defined and where mitigation efforts should be understood and applied at the time development occurs exists in the upper Bay of Fundy. Someday a Bay of Fundy site may become a giant project destined to elicit energy from the world's highest tides. Consideration of the use of Fundy's tides began soon after French tidal power projects were successful and continues today. A growing data bank of biological and physical information is being developed as a result of studies from several universities and government agencies, primarily in the Atlantic region. One data set includes the approximately 1.4 million shorebirds which use the extensive tidal mudflats as staging areas prior to their nonstop, transoceanic flights to South America. Two areas, Shepody Bay (Mary's Point) and Minas Basin (Evangeline Beach) have already been included as Ramsar Convention sites and are a part of the Western Hemisphere Shorebird Reserve Network (WHSRN). The WHSRN is a private support group with participation from several federal and more than sixty state or provincial governments in the Western Hemisphere. The presence of the shorebird network and the Ramsar sites in the Fundy region, along with a myriad of fishery and marine mammal interests, should help ensure a sound EIA through the Canadian federal process, should the project approach development reality.

Environmental information and sources are available today to everyone on the internet and no professional wildlifer should neglect to search for what is required. For example, in the fall of 1997, over twenty years of Canadian environmental assessment publication titles could be accessed and searched. Thousands of other natural resource titles, information web pages and even professional biologists' viewpoints can be incorporated by consultants from their computer screens and e-mail. It is definitely easier to access information than in the past but it is also easier because of the mass of information to miss that unique data set that could be of most significance to your work.

The EIA Elsewhere

Attempts to improve the lot of indigenous people through large-scale developments have often been unsuccessful. The condition of people after the project became operational was no better and sometimes worse than it was before. Until recently, these development projects have proceeded without environmental assessments. In some cases, special studies were conducted by sociologists, demographers, or anthropologists. The wildlife people usually ended up on a search and rescue type of salvage operation as in the Kariba hydro impoundment in Zambia. Today the EIA is sometimes, but not always present. It seems unlikely, however, that a development which is touted as a major source of employment or which might help thousands to irrigate their lands will be held up because of possible impacts or ecological boomerangs like disease or even starvation. This is especially true if these conditions are not envisioned or are predicted but not articulated.

Consideration of the great land improvement programs of the Third World are beyond the scope of this text. We note only that initial impact statements and some intensive studies have been produced for some of them. Their effectiveness will probably depend as much on the existing wildlife lobby as on the accuracy and completeness of the statements themselves. If serious impacts are detrimental to human populations and also the wildlife, then wildlife may occasionally obtain a reprieve. Policy and the resultant management of wildlife are still unsettled throughout much of the world today. Until the utilization-preservation issue is settled and until wildlife resources are a part of planning data, we can expect a variety of responses from governments to impact assessments. Baseline studies such as those of Manning and Moss (1976) may be considered in development plans for the Bangweulu area of Zambia or they may be buried. It seems unlikely that further impact studies will ensue.

As one might expect, the situation is considerably different in the Republic of South Africa, where the Environmental Conservation Act was passed in 1982 (Government Gazette, 1982). According to J. L. Benade, director general of the Department of Environment Affairs (personal communication, 27 January, 1983), "this Act provides, among other things, for the establishment of a Council for the Environment. A function of this Council is to advise the Ministry of Environment Affairs and Fisheries on the co-ordination of all actions directed at, or liable to have an influence on, any matter affecting the conservation and utilization of the environment. It is anticipated that the Council will examine the feasibility of mandatory EIA through its technical committees." Here, as in North America, we see the beginnings of a process. It may not provide panaceas for wildlife/development conflicts, but at the very least it will reflect an awareness that conflicts exist. The Republic of South Africa is well advanced in environmental affairs and many EIAs have been conducted even without legislation. There was an Environmental Planning Professions Interdisciplinary Committee as early as 1947 which drafted a set of guidelines to assist developers in effectively taking environmental aspects into account. Within government itself, policy has been articulated by the 1980 *White Paper; National Policy Regarding* Environmental Conservation. As in other matters relating to wildlife resources, there is much the Republic of South Africa has in place that should benefit other African countries in future years.

In Spain, the minister of Development Planning is responsible for economic and social development plans prepared by sectoral commissions which he appoints. The commissions have representatives from the ministries, public and private organizations and include experts from specific disciplines. The plans are reviewed by the minister and then submitted to the cabinet, which forwards a coordinated plan to Parliament (Cortez). The Parliament examines and approves or amends the national plan government must follow but which is only advisory for the private sector. A project may be initiated by public or private interests. The possible alternative, cost and benefit estimates, and impact on the environment are all detailed by the proponent. The environmental assessment process is not yet guided by statutory regulations.

The proponent's design is channelled to the appropriate ministry, and if a review process shows the assessment is adequate and the impacts on the environment are acceptable, the plan proceeds to the public review process. This review period lasts a minimum of thirty days, after which the ministry determines the conditions to be imposed on the proponent and grants a license to proceed. One of the conditions may be a requirement for monitoring, either during construction or operation of the project.

Gradually we are seeing a worldwide acceptance of the principle that we can no longer afford to insult the plant's air, water, and soil as we have in the past. For instance, the United Nations World Commission on Environment and Development (1983) produced a final report, *Our Common Future,* more popularly known as the Bruntland Report after its Norwegian chairperson in 1987. The recommendations there have stimulated action to establish environmental/economy structures and processes in several countries, worldwide. We are only beginning to see the effects of environmental assessment, and while the process may be slowing the rate and the degree of environmental insults, we have a heritage of abuse dating from the early years of the Industrial Revolution. Considerable remedial response still is necessary before substantive improvement in environmental conditions will be seen.

Bibliography

Auerbach, S. I. 1978. Current perceptions and applicability of ecosystem analysis to impact assessment. Ohio J. Sci. 78:163–174.

Barske, P. 1978. Environmental concerns in planning and development at the local level. In: S. Bendix and H. R. Graham (eds.), *Environmental Assessment, Approaching Maturity.* Ann Arbor Science Publishers Inc., Ann Arbor, MI.

Beanlands, G. and P. Duinker. 1983. *An Ecological Framework for Environmental Impact Assessment in Canada.* Inst. Resour. Environ. Stud., Dalhousie Univ., Halifax, Nova Scotia.

Burchell, R. W. and D. Listokin. 1975. *The Environmental Impact Handbook.* Center for Urban Policy Research, Rutgers University, New Brunswick, NJ.

Carson, R. 1962. *Silent Spring.* Houghton-Mifflin, Boston, MA.

Department of Water Affairs, Forestry and Environmental Conservation. 1980. National policy regarding environmental conservation. W.P.O. 1980. Republic of South Africa.

FEARO. 1978. Federal environmental assessment and review process. Federal Activities Branch, Environmental Protection Service and Federal Environmental Assessment Review Office, Ottawa, Ontario.

FEARO. 1979. Revised guide to the federal environmental assessment and review process. Supply and Services, Canada, Ottawa, Ontario.

FEARO. 1988. 1988 Summary of Current Practice. William. J. Couch (ed.), Canadian Council of Resource and Environment Ministers. Federal Environmental Assessment Review Office. Ottawa, Ontario.

Ghiselin, J. 1980. Preparing and evaluating environmental assessments and related documents. In: S. D. Schemnitz (ed.), *Wildlife Management Techniques Manual*. The Wildlife Society, Washington, D.C.

Goudie, A. 1981. *The Human Impact. Man's Role in Environmental Change*. The MIT Press. Cambridge.

Government Gazette 205 (8291). 1982. Environment Conservation Act. Cape Town, Republic of South Africa.

Heiser, D. W. 1978. Implementing the environmental policy—Where we've been and where we're going. In: S. Bendix, and H. R. Graham (eds.), *Environmental Assessment, Approaching Maturity*. Ann Arbor Science Publishers Inc., Ann Arbor, MI.

Hills, G. A. 1961. The ecological basis for land-use planning. Ont. Min. Nat. Resour., Res. Rep. No. 46, In: Parks Canada, 1973. *Environmental Analysis. A Review of Selected Techniques*. Planning Studies Section, Planning Division. National Parks Branch, Parks Canada, Ottawa, Ontario.

Hirsch, A. 1980. The baseline study as a toll in environmental impact assessment. In: *A Compendium of Selected Papers on Ecology and Environmental Impact Assessment*. Inst. Resour. Envir. Stud., Dalhousie Univ., Halifax, Nova Scotia.

Leopold, L. B., F. E. Clarke, B. B. Hanshaw, and J. R. Balsley. 1971. A procedure for evaluating environmental impact. U.S. Dept. Int. Geol. Surv. Circ. 645.

Lucas, H. L. 1976. Some statistical aspects of assessing environmental impact. In: R. K. Sharma, J. D. Buffington, and J. T. McFadden (eds.), *Proceedings: Workshop on the Biological Significance of Environmental Impacts*. Nuclear Regulatory Commission, Washington, D.C.

Manning, I. P. A. and P. F. N. de V. Moss. 1976. Appendix C. Environmental studies. Report on ecology. Report on feasibility of producing hydro-electric power on the Luapula river. Watermeyer, Legge, Piesold, and Uhlman in assoc. with Merz and McLellan, Lusaka, Zambia.

McHarg. I. 1969. Design with nature. The Falcon Press, Philadelphia, PA., In: Parks Canada, 1973. Environmental analysis. *A Review of Selected Techniques*. Planning Studies Section, Planning Division. National Parks Branch, Parks Canada, Ottawa, Ontario.

BIB:Milledge, A. and E. G. Gallop. 1978. Environmental assessment: the Florida perspective. In: S. Bendix and H. G. Graham (eds.), *Environmental assessment, approaching maturity.* Ann Arbor Science Publishers Inc. Ann Arbor, MI.

Munn, R. E. (ed.). 1975. Environmental impact assessment: Principles and procedures. SCOPE Rep. No. 5, Toronto, Ontario.

Nelson, J. G. 1973. Some background thoughts on environmental impact statements. Park News 9(2):34–44.

Odell, R. 1980. *Environmental Awakening.* Ballinger, Cambridge, MA.

Odum, E. P. and J. L. Cooley. 1980. Ecosystem profile analysis and performance curves as tools for assessing environmental impact. In: *Symposium Proceedings, Biological Evaluation of Environmental Impacts.* pp. 94–102. FWS/OBS-80–26, Council on Environmental Quality, Washington, D.C. Parks Canada. 1973. Environmental analysis. A review of selected techniques. Planning Studies Section, Planning Division. National Parks Branch, Parks Canada. Ottawa, Ontario.

Schindler, D. W. 1978. Presentation at Can. Soc. Zool. Annu. Mtg., Univ. Western Ontario, London, Ontario. (unpubl.).

Stover, L. V. 1972. Environmental impact assessment: a procedure. In: *Environmental Analysis, a Review of Selected Techniques.* Parks Canada, Document 1042OUC (June 1973).

Truett, J. C. 1979. Pre-impact process analysis: design for mitigation. In: *Mitigation Symposium: A National Workshop on Mitigating Losses of Fish and Wildlife Habitats.* pp. 355–360. Gen. Tech. Rep. RU-65, Rocky Mountain For. Range Exper. Sta., Fort Collins, CO.

Truett, J. C., H. L. Short, and S. C. Williamson. 1996. Ecological impact assessment. In: T. A. Bookhout (ed.), *Research and Management Techniques for Wildlife and Habitat.* pp. 607–622. The Wildlife Society, Bethesda, MD.

U.S. Fish and Wildlife Service. 1976. Final environmental statement. Operation of the national wildlife refuge system. U.S. Fish Wildl. Serv. Dep. Int., Washington, DC.

World Commission on Environment and Development. 1987. *Our Common Future.* Oxford Univ. Press, Oxford, England and New York, NY.

Zigman, P. E. 1978. The California Environmental Quality Act and its implementation. In: S. Bendix and H. G. Graham (eds.), *Environmental Assessment, Approaching Maturity.* Ann Arbor Science Publ. Inc., Ann Arbor, MI.

Recommended Readings

Beanlands, G. and P. Duinker. 1983. *An Ecological Framework for Environmental Impact Assessment in Canada.* Inst. Resour. Envir. Stud., Dalhousie Univ., Halifax, Nova Scotia. An invaluable review of ecological problems and processes applicable to wildlife workers everywhere. The examples are Canadian.

Bendix, S. and H. Graham (eds.). 1978. *Environmental assessment, approaching maturity.* Ann Arbor Science Publisher Inc., Ann Arbor, MI.

Truett, J. C., H. L. Short, and S. C. Williamson. 1996. Ecological impact assessment. In: T. A. Bookhout (ed.). *Research and Management Techniques for Wildlife and Habitats.* pp. 607–622. The Wildlife Society, Bethesda, MD. A balanced consideration of most wildlife aspects in relation to environmental impact with five case histories.

12 WILDLIFE INTERNATIONAL AID POLICIES, PROBLEMS, AND MANAGEMENT

Most people visiting the United Nations plaza in New York City are unaware of the massive complexity of the United Nations Development Program (UNDP) office (Figure 12.1). The political arm of the UN dwarfs the development arm, both in budget and in the number of people involved, yet the programs approved and monitored through the UNDP reach every facet of progress in our underdeveloped world. There are three UN "family" agencies that may be involved with wildlife or related concerns: the Food and Agriculture Organization of the United Nations (FAO) in Rome; the United Nations Educational Scientific and Cultural organization (UNESCO) in Paris; and the United Nations Environmental Program (UNEP) in Nairobi.

The FAO wildlife involvement may include legislation, administration, research, management, and policy as well as overall country programming. Projects in progress may also be reviewed by an FAO or UNDP representative, and short studies leading to development programs are often carried out over a period of several weeks. Larger projects established through FAO are a part of the UNDP development strategy for the jurisdiction in which the project is located. Such projects are usually management oriented and are seldom specific to the wildlife resource alone. They most often involve tourism, agriculture, or even several resource bases. In other instances, FAO provides wildlife or park advisors who work very closely with the governments of underdeveloped countries and who may work directly with wildlife and local staff in the field. FAO also provides wildlife expertise for universities and technical schools in a number of less developed countries. A major interest for UNESCO has been the biosphere reserve approach for the preservation of ecosystems in which resident and seasonal wildlife populations are protected at the same time economic development is encouraged, as noted by Dasmann (1981). In other instances, UNESCO scientific advisors may become involved in wildlife or parks projects through their advice to relevant ministries, or

320 • *The Philosophy and Practice of Wildlife Management*

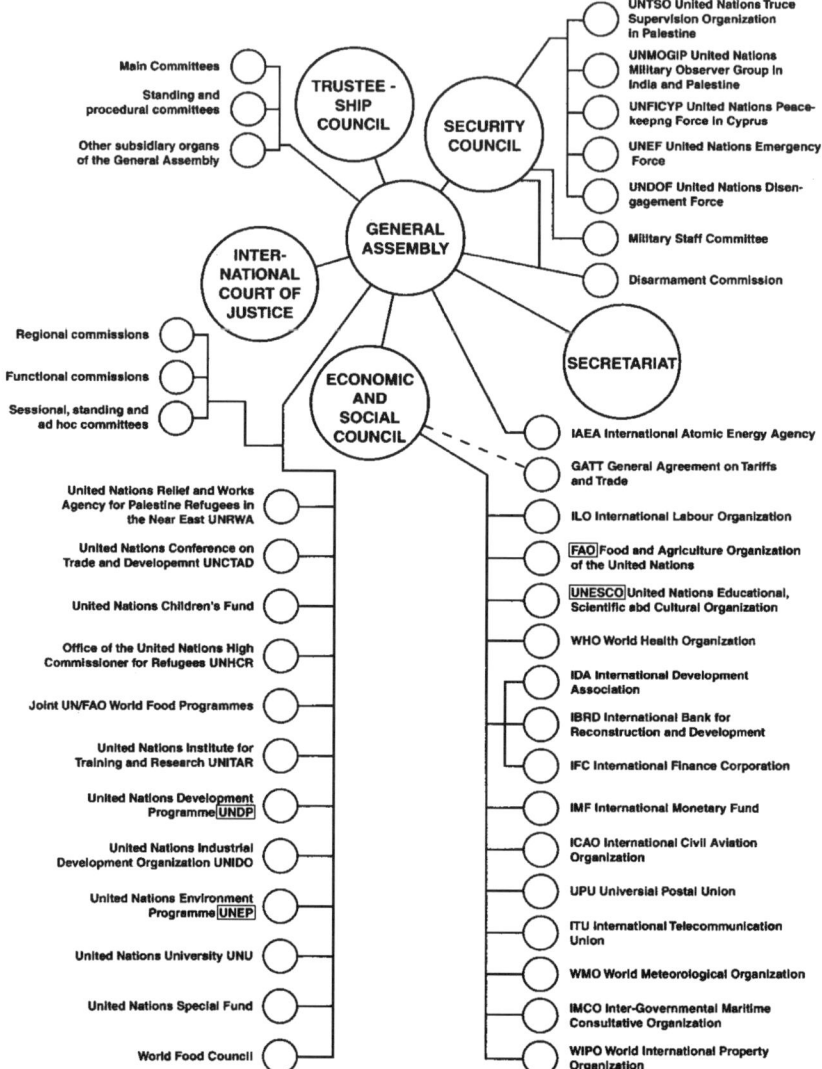

Figure 12.1—An abbreviated organizational chart of the United Nations system. The units which may be involved with wildlife are boxed. Overall administration of wildlife oriented development programs is a responsibility. The interlocking nature of funding and ancillary support and cooperation tends to make the autonomous agencies such as FAO, responsible organizations not only to UNDP but to other countries both developed and underdeveloped. (Courtesy, C.J. Lankester, UNDP, New York, Chart dated December 1979).

through personal contact with UNDP or FAO project personnel. UNESCO also may respond to requests for short-term consultants in relation to land use, tourism, national parks, and wildlife associated with World Heritage sites.

The youngest of the UN family groups with wildlife conservation interests, UNEP, was initiated in 1972 and is based in Nairobi, Kenya. UNEP concerns are specific to humans and their natural environment. UNEP cooperates closely with other UN agencies and assists the cause of wildlife conservation through the International Union for the Conservation of Nature and Natural Resources (IUCN) in Gland, Switzerland. Representatives from UNEP liase with agencies and NGOs worldwide on environmental matters relating to both terrestrial and aquatic ecosystems and may on occasion be closely involved with wildlife.

The IUCN, also known as the World Conservation Union, was established through efforts of UNESCO in 1942. IUCN oversees a network of several thousand scientists and other professionals around the world who promote and engage in scientifically based conservation action directed toward sustainable resource use. IUCN's six commissions cover broad areas of interest relating to ecosystem survival and the maintenance of biodiversity. To many people this agency is best known because of its red data books' listings of threatened species and other efforts of its Species Survival Commission. The IUCN is not a UN family member and depends upon public support for much of its funding.

Some developed countries provide "associate experts" to international aid projects. These people are usually well trained but may lack the longer-term professional experience of those normally employed as experts by a UN family organization. They are obtained through bilateral aid and assigned, on agreement of the host country, to a particular project or special development problem. Many countries also support volunteer aid organizations such as the U.S. Peace Corps, the VSO (Voluntary Service Overseas) in the United Kingdom, CUSO (Canadian University Service Overseas), and the UN volunteers from UNDP. People working for such agencies are usually recent university graduates or are professional, but comparatively untrained or inexperienced adults who offer their services to developing countries through their own countries' volunteer agencies. They may also be retired professionals who volunteer their considerable experience. On occasion, such people may be engaged in a park, wildlife or fish management project, or a special conservation program for two years or more.

Bilateral aid to help maintain natural systems and for wildlife conservation in less developed countries is in 1999 by far the most important source of expertise and funding. Gone are most of the grandiose multilateral or UN sponsored programs which have been replaced by more precise, targeted con-

322 • *The Philosophy and Practice of Wildlife Management*

sultancies and developments from individual countries. Canada has the Canadian International Development Agency (CIDA), the United States has the Agency for International Development, and other developed countries have similar organizations. Much expertise in wildlife and parks also comes from Great Britain, Germany, the Netherlands, France, the Scandinavian countries, New Zealand, and Australia. Not all less developed countries are comfortable accepting aid from all developed jurisdictions because foreign influence through economic control may be a concern. A less developed country might accept a particular foreign national expert and yet not accept certain types of aid from the expert's country. It also should be understood that bilateral aid is not completely altruistic. One of the reasons that donor countries are willing to help the underdeveloped countries is that trade ties may be established allowing the donor both future markets and the import of goods, including raw materials. The "north-south dialogue" is in part, a reflection of the desire to increase trade opportunities that represent potential future sales to the large and increasing populations of underdeveloped countries. The image donor countries have in the Third World is often so important to their governments that they tend to seek bilateral aid avenues with underdeveloped nations and shun multilateral approaches. Sharing the development limelight with other countries or with an international agency is not as desirable as taking full credit! Far more money is available and spent through bilateral aid than through UNDP. The development picture is complicated in its politics, funding arrangements, management, and number and variety of projects. The number of projects and experts only tells a part of the story, for projects have counterpart experts, associate experts, technical and clerical help, and one or more guiding management committees. Usually a wildlife or park development operation is initiated by UNDP after a host country understands it has a specific problem or wishes to provide assistance to one of its endangered or undeveloped resources. International aid offices are contacted locally, and a consultant or a group of consultants eventually reviews the situation. Recommendations favorable to both the host country and the counterpart agency (usually a wildlife, parks, tourism, or game department, authority, or ministry) go to the UN family agency sponsoring the review and to the regional UNDP office. Perhaps further reviews will take place, but if the host country and the UN agency approve, a project document (called a plan of operation) is eventually drafted and considered by the regional resident representative of UNDP (an ambassadorial status position). Others in the regional office, senior people in the UN agency, interagency groups, and the UNDP office in New York also review the plan. If the wildlife project is implemented, it may call for certain experts for specific periods, counterparts, associate experts, buildings, equipment, cooperative research, and management assistance. A number of designated inputs from the

UN, other organizations, and agencies in the host country, itself, may also be specified in the plan of operation.

Competition is sometimes keen between UN agencies and bilateral aid offices in underdeveloped countries. On occasion both a UN agency and a bilateral aid representative will compete to establish the same project. This may well confuse the host country officials who represent a potential counterpart agency or a planning ministry in government. In 1974, a senior UNDP official asked why the Canadians (CIDA) believed they could do a better job than FAO with a range-wildlife program. Both were involved in the same (potential) project but each had its own approach. The same query was repeated by people in the Ministry of Finance and Planning, including visiting expert advisors to the ministry. The Canadians won out and the KREMU (Kenya Rangeland Ecological Monitoring Unit) project replaced the FAO project that was in progress and under review. In an earlier situation (1967), a Canadian CIDA representative ceased to be interested in a project in Zambia when it appeared that the operation was heading toward a multilateral approach. Sometimes competition even shows up among UN agencies themselves, and often one agency representative in the field will openly criticize the work of another, even in reports to senior officers. Such infighting and disagreement is as discouraging to host country officials as is the lack of clear intent expressed by representatives of some donor agencies, or as discouraging as the attempts to sort out the ideas expressed by a myriad of donor experts who all too often give conflicting advice on identical issues.

Wildlife problems are often bypassed in Third World countries even if they seem crucial to wildlife professionals from the United States or Canada. In underdeveloped areas they are considered to be less important than agriculture, education, communication, trade, energy, and a host of other development needs. When a stagnant economy, poverty, and starvation are serious socioeconomic and political concerns, wildlife must be understood to be related to them if it is to be included in planning and development.

Interagency meetings of the UNDP and host country plus program planning help reduce conflict in international aid but with hundreds of donor agencies, NGOs, universities, and the UN all involved, problems remain. To complicate matters further, a UN/FAO project may have experts from six or more different countries communicating in several languages. Counterpart personnel may be well educated or only semiliterate, clerical staff may or may not be computer literate, and support staff may be well trained or untrained. In addition, most of the 142 member nations of FAO want a share of the development business and each wants representation at the head office. A great many member nations are underdeveloped and each hopes to influence the manner in which aid is directed and who receives it! The wildlife agencies of these countries also may have problems. They are invariably understaffed, undertrained, and underfunded. Politically, they have less influence than agriculture, education, finance, and many other government agencies. This often results in land-use planning proceeding without wildlife agency input. Thus, wildlife professionals from Canada and the United States will find the same problems, though usually magnified in the underdeveloped country, that they find at home. Wildlife agencies everywhere are no better, or worse, than their personnel and their progress no better than the political will of their elected or appointed representatives allows. Whether it is permanent secretaries, directors, and wardens in Zambia, or commissioners, directors, and biologists in a state, those responsible for wildlife policy face similar problems. There is a feeling of familiarity in the offices and field stations of wildlife agencies almost anywhere in the world.

In addition to the UN family, the IUCN, and bilateral aid, universities and NGOs may also be involved in cooperative wildlife research and management in underdeveloped countries. Among the many important organizations contributing primarily to research, protection, and training are the Wildlife Conservation Society with projects in some fifty countries, the African Wildlife Foundation, the Serengeti Research Institute, and the World Wildlife Fund.

Wildlife Policy and Management: Some Concerns in Africa

Hundreds of ecologists and professional wildlife people from developed nations have worked in Africa yet in many instances conditions are becoming worse. From the press and popular writings we tend to get a picture of a magnificent wildlife resource being rapidly decimated through mismanagement, trophy hunting, and poaching. The word *endangered* is usually present somewhere in the narrative, and this is what we tend to key on. African wildlife is endangered: but is it?

Perhaps our starting point should be an understanding of what is happening to the land and the people of many African countries, for wildlife populations are responding naturally to both human population increase and the resultant land-use changes. Although there are thousands of local and regional complexes, a general view of many African countries south of the Sahara is as follows.

Soon after World War II human populations began to increase rapidly because of improved medical and health facilities, disease control, and limited industrial and agricultural development. Population distribution also changed, and a rural-to-urban migration surfaced with large concentrations of people gathering around cities and towns to obtain employment and reap the "benefits" of Western life—for through a new communications explosion, most could now learn of the wonderful gadgetry available. Transistors in the bush, new paths or roads for bicycles and four-wheel drives, and increased contact with others outside the village or tribal unit made millions more aware and caused hundreds of thousands to have aspirations new to the people and their cultures. Population increases soon outstripped the results of development programs and altered lifestyles weakened cultural ties. Divided loyalties were not in favor with newly independent governments. The influence of the Christian church, tribal chiefs, and elders was often undercut to direct people toward a single, nationalistic objective: to build the economic resource base and infrastructure of the country, everyone's country. The old ways were often lost and among them, sometimes, was a successful subsistence means of living. Dependency on governments increased and development, good or bad, was slow in part because of a culturally based inefficiency in production and management within the people and sometimes because development programs were inappropriate for the people and/or area. Population growth continues to forge ahead of development and efforts to move people back to the land have not always been successful. But as the land mass has neither increased in area nor improved in fertility, the competition for land also grows daily. As populations explode, land-use changes often occur in an ad hoc manner. In addition, large scale agricultural production schemes have not always been economic successes nor have they always met the needs of the people. Regardless of the successes or failures, land for wildlife is declining necessitating that wild animals and humans compete to a greater extent each day.

Wildlife must become a competitive source of economic benefit to both governments and local people if the resource is to survive. Nationally, countries need sources of foreign currency, so wildlife must provide an economic return through both consumptive and aesthetic (nonconsumptive) uses. The people must benefit locally from the existence and use of the resource. They must understand that the social and/or direct benefits received are obtained because of the wildlife resource. The inability to direct parts of the wildlife resource base into the market or national economy is a primary problem facing those responsible for the wildlife of Africa. Until wildlife pays its way it will not be recognized as a viable renewable resource to manage and will continue to be upstaged in government planning by other natural resource bases. This really means that when it comes to wildlife and people, it is people who must come first. It may be that wildlife can be used to help a people, and if so, wildlife will be maintained. If not, wildlife may be lost. This philosophy, difficult as it may be to implement, is also logical in view of the indigenous African's utilitarian attitude toward wildlife. Wildlife cannot compete in an unregulated market and its uses must be carefully controlled.

The wildlife resource base not being represented adequately in land-use planning and primary resource development schemes is allied to this basic problem. The potential to produce both protein and export income through utilization, whether this be subsistence and safari hunting, game viewing, or cropping, should be measured against capital and recurrent expenditures and the production levels employed by agricultural and other primary resource people. In Kenya, the Range Ecological Monitoring Unit was an attempt to provide a data base that would allow the inclusion of the wildlife resource in economic land-use planning. Another project with a socioeconomic base was established in the Luanga Valley of Zambia in the late 1980s, but it may be that for many wildlife populations the efforts will be too weak and arrive too late. Unless wildlife can be shown to be economically viable and is included in government planning, the resource will be lost. The trend toward the necessary multidisciplinary planning seems slow when we realize that practical models allowing rural incomes from wildlife have been available for nearly a quarter century.

The current endangered species may well continue to exist long after much of Africa's wildlife resource is gone if the world concentrates only on rhinos and leopards! Endangered species do exist in increasing numbers, and it is logical to try and save most of them. But if the attention of the world is focused only on endangered animals, it will become increasingly difficult to provide support to the many species constituting the bulk of the resource base. It is even questionable, for some species, whether the Convention on International Trade in Endangered Species (CITES) and the resultant legislation in several countries or the annual IUCN publicity on endangered species are always entirely positive. Does legislating and publicizing save or destroy? Or both? Making an animal more valuable because of its rarity clearly creates markets, although they may be illegal, and poachers will attempt to fill those markets. Rhino horn is not declining in value and the spotted cat market is bullish although often illegal and usually restricted. Profits are also limited to fewer people because of new legislation and better enforcement, but the animals remain vulnerable and underground distribution continues. Perhaps an approach that included rhinos and leopards in nonconsumptive use and management programs, along with controlled consumptive use in which revenues were directed to the people might have been more successful. If local people realized direct benefits their cooperation in maintaining the resource might have been realized, allowing for continued healthy populations. The value of such approaches can never be measured for many threatened and endangered species already immune to most management except preservation.

There are a great many other problems. One is the combination of staffing, training, and education. After independence, gradual Africanization of the civil service in many countries became a political necessity. Similar staffing by indigenous people occurred in other postcolonial countries in Asia. Obviously it was more desirable from consistency and continuity standpoints to staff wildlife posts in Zambia with Zambians and in Kenya with Kenyans. For thirty years after World War II, American, British, Canadian, German, and other nationals trained in animal ecology drifted into Africa and back out. They started projects in development, management, or research and then disappeared, leaving only reports behind. Until recently, impact studies and development schemes have invariably been directed by non-Africans. Boshe (1990a) has suggested that results of research conducted by visiting non-Africans have not necessarily been useful and that they have not been implemented because the link between outside researchers and local managers has been weak and the research has not addressed the most immediate conservation problems which often are more socioeconomic than biological. Lamenting on the shorter term consultant approach to wildlife employed by many agencies, Ian Manning, long-time cropper, white hunter and consultant, noted that the consultant "gives fleetingly of his service, never tarrying long enough to take responsibility for the long term happiness and well being of the client and often leaving those on whom he has bestowed his services in a diseased state".

Training and fellowship components of projects have not always been successful either. If an important government ministry needed someone with advanced training (beyond high school), a wildlife-trained Fellow might be lost to his own ministry by moving him to a more critical post. If they proved to be capable, the counterparts on wildlife projects would more often than not, be lost at the close of a project. There are many other fields more attractive to young men and women than wildlife management. All professions became new opportunities at independence, and few young people knew what an ecologist or wildlife biologist was. These problems, coupled with the lack of money to provide training—common to most governments in Africa and elsewhere in the Third World—and the difficulties sometimes brought about by major changes in government, as in Ethiopia since 1970, have resulted in wildlife agencies that, until recently, were often not staffed with well-trained professionals. With improved education and training for present and future staff members, the operation of many wildlife agencies will become more efficient in day-to-day work and more competitive in influencing policy within whatever framework may exist. Certainly this is another key to wildlife's future in Africa.

We could examine other problems: the need for an improved data base; better monitoring; improved organization of interagency cooperation (tourism or agriculture and wildlife, for instance); the role of energy and fuel requirements versus forest cover and wildlife; the role of disease and parasites on populations within protected areas; the effects of reduced genetic variation on reproduction and survival in low density populations or the strengths and weaknesses of cropping and game farming or safari hunting. Still the basic problem throughout the world is that there are too many people eliminating or altering ecosystems and their constituent species and the problem perhaps is most acute in southeast Asia, India, and in several African countries. Among underdeveloped countries, however, there are a few problems which, if addressed effectively might make a difference. They are:

1. Wildlife may not be considered to be a major source of economic benefit that might improve the lifes of local people.
2. Wildlife may be inadequately represented in land-use planning and development.
3. Wildlife management is underfunded and wildlife agencies may be understaffed with professionals.

Wildlife Management: The Status in Central Europe and Elsewhere

Wildlife policy and management in Africa have evolved recently from colonial systems, but have been influenced by the veterinary services people because of the importance of tse-tse, rinderpest, and other diseases harbored in wildlife species that are transmitted to humans or domestic livestock. In contrast, the Central European system has been evolving for hundreds of years and reflects the primary right of the landowners to reduce game to possession. Bubenik (1976) enumerated the key points of several systems in Central Europe before and after World War II and we have modified some of the points to reflect current realities:

1. The right to reduce the game into possessions belongs still to the landowner, if his land has the minimum acreage of revier. For using his right he must be an accredited hunter. This means that:
    a. He has to apply to attend a special educational course for accredited hunters, pass the examination and, in some countries, must be a mandatory member of the hunters' organization.
    The applicant must prove that he has no criminal record. The certificate of accredited hunter is also his gun licence and he can buy any hunting gun (in certain countries only a shotgun) without any certificate. In countries such as the Czech Republic, Slovakia, Poland, or Hungary, a special certificate for buying rifles and ammunition is necessary. In Russia it still is very difficult to get such a permit.

b. Education is theoretical and takes about thirty to sixty-five hours. The disciplines are: hunting and conservation laws, biology. basic physiology and anatomy of game, game ecology, hunting and management operations, game diseases, feeding of game, keeping and using of hunting dogs, gun types, mechanisms, and ballistics. Also included is mandatory training in shooting with gun and rifle. In some countries, at intervals of some years, mandatory shooting at a target is ordered. In Fennoscandia, nobody can buy a moose license unless he presents a new certificate from an official shooting station proving that he has hit the necessary minimum score for a moving moose target. Generally, the applicant can pass the examination shortly after the course. In the Czech Republic he must voluntarily serve one year as a helper in a revier, before he can apply for the examination.

2. The revier-owner must have at least three years experience as an accredited hunter. In the Czech Republic, he has to attend a special course for revier-manager and pass a new examination. The revier-owner is liable for damage caused by game on crop or forest.
3. A revier with small game, for example, winged game, furbearers, and roe deer, must be leased for six years, with large game, for nine years. The deposit of a one-year fee must be put into a special account in case the contract is cancelled during the time when feeding of game is necessary.
4. Feeding game is mandatory and the hunters' organizations have the right to check and see if the game is being properly fed.
5. Inventory of game (not only census of numbers) is mandatory. In some countries it is done under official supervision. The "harvest of shooting" plan prescribes the number of game in both sexes and age, or the "social classes." About 50 percent of the planned kill is the class of fawns and calves. With few exceptions, no one sex is protected, but according to the structure of specific species, one or other sex could be more or less hunted.
6. The revier-owner is liable for not fulfilling the plan or for overshooting. In both cases this could be the reason for cancelling the contract: especially he could lose his certificate of accredited hunter.
7. Species belonging to trophy game have a mandatory yearly trophy show, to avoid overhunting of the best breeders.
8. The revier-owner or his deputy, and/or the members of the hunting society, have the right to kill dogs and cats preying, or being, in the revier.

9. The department of game and the hunters' organizations are stratified after the same pattern. On the top is the government department and the presidium of the hunters' organization. Then there are provincial, regional, and district offices. Experienced hunters and accredited game managers can be elected representatives for game management with a rank similar to the stratification pattern of the department of game. These jagermeisters cooperate with the departments and are responsible for the game planning and management.
10. Very large reviers must employ gamekeepers. In some countries they are organized into labor unions.
11. There is a mandatory search for wounded animals.
12. By collective hunts on small game, the number of hunting dogs per distinct number of hunters is also prescribed to avoid losses.
13. The game belongs to the revier-owner. He can utilize or sell it. The trophy and viscera belong to the hunter as his hunting right.
14. The hunters' organizations have their own hunters' court. Any offence to the hunting law or hunting ethics must be handled independently from the judiciary procedure, before the hunters' court. If this court orders that the certificate of an accredited hunter should be removed, nobody can reverse this decree and the offender is expelled from the hunters' fraternity. He automatically loses the right to possess hunting guns. The number of accredited hunters varies from country to country: between 1.0 up to 2.5 percent of the whole population, or about 2.5 to 5.0 percent of all citizens in the age allowed by law for hunting persons.

The revier system and the hunting regulations, as mentioned above, represent surely the most efficient game management. The most important points of this system are the mandatory courses, examinations, and accreditation. The revier system is the most economic for other reasons. The revier-owner has so many duties before the law that he must be year-long in the revier, to feed, make inventory, and to familiarize himself with the ecology and behavior of his game, so that he will be capable of shooting the prescribed number and social class of game. For this reason the open seasons are long enough, that is, for small game about two or three months, for large game six to seven months.

One benefit to wildlife managers should be quite clear under such a system. All necessary data are available to allow for intensive harvest management. We often depend upon the cooperation of interested sportsmen in North America, but hunter, researcher, and manager can operate as a unit under some European systems. In much of Africa, however, almost all biological data are lost from safari, sport, and subsistence hunting operations. The only research materials available come from a few research and management projects that may cover short periods of time. We should recall that for hundreds of years the existence of animals on a European land holding has been tantamount to legal possession. Today, this type of possession continues in relation to both state and private land in Europe. How much simpler this makes regulatory management! The restricted use, disciplined approach in central Europe must be considered a success in relation to several species and their habitats. Both remain after hundreds of years of exploitation and perturbation. The intensive management common to the European system is exemplified further by Nagy and Benaze (1973) who indicate game ranching, winter feeding, habitat management (plantings), game-damage control and payment, and the maintenance of salaried professional game keepers by hunter's associations to be among the approaches used in Hungary. Nagy and Benaze noted harvests by 1100 hunters on 500,000 ha (1,250,000 acres) to be 11,400 hare, 11,120 pheasant, 4,250 partridge, and 187 roe deer. Foreign hunters shooting for paid fees may kill up to 100 pheasants a day or, if they wish to pay more per bird, up to 200 a day! It is also interesting that these authors indicated there was " . . . little if any, antihunting or antihunter sentiment." They attributed this lack of "anti" sentiment to the fact that hunters are disciplined and "devoted to customs and traditions which tend to de-emphasize the killing aspect of hunting." They described ritualization following the kill and emphasized that hunters are "devoted and successful game managers." But that was in 1973; since then there has been growing antihunting sentiment in much of Europe, some generated by animal rights groups and some by those opposed to the elitist nature of the sport.

For the former Czechoslovakia, the intensive management of wildlife and land through hunting associations is reviewed by Newman (1979). He indicates the importance of determining population densities through censuses to aid in establishing quotas. As in Hungary, it is the hunter who is the important game manager, although professional game keepers are employed. The potential changes that political and socioeconomic changes may cause in Hungary are expressed by Csanyi (1997). The primary factor for change is land privatization, a phenomenon becoming common in many formerly communist countries.

In the former West Germany, where most of the land available to hunting was privately owned (versus the former state ownership in Hungary and Czechoslovakia), the revier system was in effect. About 0.4 percent of the population hunts, and game may be marketed. Foresters and hunters have a close relationship and forest managers also serve as hunting managers. Most hunters (86 percent) belong to the German Hunter Society, and until recently were respected by the nonhunting public.

Although no one person and no public authority owns the game in Norway, landowners have an exclusive right to hunt and trap on their own property. On much state owned land, hunting and trapping are open to all citizens on payment of a fee. On other classes of state land, however, the number of hunters may be restricted by the control of special licenses. Game management in Norway is conducted primarily through the approximately 450 Game Boards existing at the community level, as well as the state organized wildlife bodies in each of the eighteen counties. Most research information or management is provided by the privately operated Norwegian Institute for Nature Research which obtains much of its funding from the state and is administered by the directorate for Nature Management under the Ministry of Environment. Hunting associations are prominent and obligatory exams must be taken by first-time hunters both in Norway and in Sweden, as has been the practice in Finland and Denmark for some time. Management activities are related to regulatory measures (harvest management), agricultural and forest damage control, depredation control, habitat management in the context of forestry and agriculture, translocation, and total protection.

> "In Finland matters pertaining to hunting fall under the jurisdiction of the Department of Fishing and Hunting in the Ministry of Agriculture and Forestry. The Department supervises and regulates hunting, game management and the marketing of game, handles any questions arising from damage inflicted by game, and issues special hunting permits and regulations. The Ministry of Environment serves as the highest authority for those species protected under the Nature Protection Act. The authorities responsible for the enforcement of the Hunting Act are the police and game-wardens under the jurisdiction of the county governments, the Frontier Guard Establishment, the Customs, and National Board of Forestry.
>
> The Hunters' Central Organization (Metsastajain Keskusjarjesto) was officially established in 1982 to look after the affairs of hunting. Run by the hunters themselves, its chief tasks are to distribute information, provide advice, and make the practical arrangements for game preservation. Earlier the comparable tasks were the responsibility of the Finnish Hunter's League (Suomen Netsastajaliitto-Finlands Jagarforbund r.y.), which was established in 1931 and which continues to act in a non-obligatory way as a link between hunting organizations in different parts of the country. Alongside it there operates, likewise on a voluntary basis, the Finnish Association of Hunters and Fishermen (Suomen metsastajaja Kalastajalitto) founded in 1946.
>
> The Hunting Act presently in force dates from 1962. The underlying premise in Finland is that the right to hunt in any area is the property of the owner of the land. The right to hunt can be transferred to another for a specified period, 5–25 years, by means of a written rental agreement. In this way those without land are assured the possibility of hunting. The holder of the right to hunt may also surrender his right to another in the form of a permit. Often the species of game that can be hunted and the bag limits are specified in the permit along with the time

period for which it is granted. Separate provision is made for hunting on state lands. There hunting takes place either by special permit or by payment of a rental fee. In addition, in the County of Lapland and in certain municipalities of Oulu County, the local population has the right to hunt in state lands in their own municipality. Every Finnish citizen has the right to hunt on the sea outside of village boundaries and in coastal waters and on islands belonging to the state. Likewise the residents of a municipality have the right to hunt on large open lakes outside of village boundaries and on islands there belonging to the state.

In Finland, game management is entrusted to the hunters and is supported by the state out of funds accumulated from game fees. The national Board of Forestry participates in and promotes the work of game management, by setting aside special areas of forest as game reserves, for example, where the attempt is make to combine the different interests of game and forest management. Nowadays, game management means modifying the environment so that it better serves the needs of game-by feeding in winter when the natural food supply is insufficient, by restocking for the vitalization of a declining population or establishing of a new one, and regulating hunting in such a way that the structure of the game population becomes more favorable and that unintentional overharvesting does not occur.

In 1942, the Finnish Game Foundation established an institute for the purpose of game research. This institute was taken over by the state in 1964, becoming the Finnish Game Research Institute. Today its activities are continued by the Game and Fisheries Research Institute, founded in 1971 under the Ministry of Agriculture and Forestry." (Ermala, 1989).

In 1990, Svein Myrbeget outlined European wildlife management systems, game populations, special problems, management authority and faunal protection. The following quotes come from his summary statements.

"Game management is practiced at many different levels by: central and local authorities, landowners, hunters and other persons interested in nature. To a varying extent, hunting rights are associated with landownership rights. Excluding the former Soviet Union, there are about 8 million hunters in Europe, or approximately 1.6 hunters per km. Membership in hunters' associations is either voluntary or obligatory. In most countries, an obligatory hunter's examination has been introduced. The examination involves theoretical knowledge as well as the practical use of firearms.

In contrast with common practice in North America, there is no system in Europe to control the hunting pressure and yield of migratory birds along flyways. The management of migratory birds in Europe is highly dependent on international agreement, in particular the Berne convention of 1979, the EEC Bird Directive of 1979 and the Ramsar Convention of 1971.

Game management practices used for non-migratory small game species are largely left up to landowners. In some countries in central and eastern Europe, certain management measures are obligatory. Among practices often employed are predator control, introduction of game, and feeding. Measures for conserving or increasing the production capacity of natural habitat or diminishing the negative effects of human activities are increasingly important.

Wildlife management increasingly emphasizes the importance of research. However, at the present time there is no international organization to coordinate wildlife research in Europe. Much modern game management is a continuation of age old traditions and biased attitudes.

Although the threat to many game species and their natural habitats increases, the desire to solve problems has also become stronger. Solutions are dependent upon considerable local and national effort and international cooperation."

Elsewhere in the world, wildlife management is either increasing or is primarily relegated to the preservation of areas and species (several South American countries and India). In China, the recognition of wildlife conservation as being socially, economically, and ecologically important to the country and the people is resulting in a slow increase in the training of professionals and an effort to upgrade the quality of wildlife education. For many countries, the recognition of wild animals as valuable cultural, social, and economic entities has come too late. In some, there may still be time for species like the giant panda (*Ailuropoda melanoleuca*) or the snow leopard, which may occur in sufficient numbers to ensure their continued existence, given proper management attention. In summary, wildlife policy and management around the globe tends to be grouped into intensive, extensive, and developing categories. In Europe, where the system has evolved over a longer period than anywhere else, management is most intensive, but policy often tends to be more elitist than democratic. In North America, management is primarily extensive, but policy with few exceptions, tends to be democratic, although the trend toward intensive private land management and control is growing. Some countries, such as the Republic of South Africa, fall somewhere in between the European and North American situations. In the underdeveloped countries, both management systems and policy are evolving. Whether human population pressures, land-use changes, and industrialization will restrict the effectiveness of improvements in wildlife conservation now in progress is a question only time will answer. It seems likely that what now is a beginning in many jurisdictions has arrived too late.

A Problem to Ponder

Management systems that are implemented result from policy and the directions policies take are reflections of complexes; economic, social, and ecological. To help us understand why things are as they are we need to explore beyond what may seem obvious to the wildlife manager. One of the better known examples of a wildlife management complex in underdeveloped countries is the African elephant. Thousands of such examples of wild species and their systems and of wildlife policy in nations and within regions can be found throughout much of the tropical and subtropical world.

Ivory art is almost as old as art itself. Ivory was an early medium for artisans carving for the religious and the wealthy. For centuries, ivory (primarily elephant) has been known as "white gold." Collecting was also linked to the slave trade in Africa, for what better means was there of getting it to coastal ports? Once there, the ivory and the slave could both be sold! Elephants themselves have played a major role in the tradition and ritual of indigenous people in Africa and Asia. They were domesticated early in Asia, perhaps 4000 years ago, and are still used as beasts of burden. Elephant and man have coexisted for thousands of years as man has exported and imported the animal and its parts almost worldwide. The ivory trade is now well documented. Between 1976 and 1980 it could have accounted for between 37,000 and 55,000 elephant fatalities, although as many as 20 percent may have been natural deaths.

The furor over the decline of elephants began in the early sixties, increased in volume, and reached a crescendo in the 1970s. Elephants were declared threatened, were listed in Appendix II of CITES, were the subject of detailed and complex regulations under legislation in the United States (The African Elephant Conservation Act) and other countries, and became a great fundraising symbol for WWF and other agencies. According to the media and their suppliers, elephants were being systematically eliminated by legal hunting, illegal hunting, and by commercial harvesting. Conservationists, mostly nonprofessionals, in the West misunderstood the problems of the elephant-man-land complex and were largely to blame for the failure of what might have been responsible management—including cropping by some African game agencies. The momentum was toward preservation: save and do not kill.

There was truth in the belief that elephants were declining. Slaughter was common. Poaching was heavy in Kenya, Sudan, Zaire, Central African Republic, Zambia, and elsewhere. Sometimes it occurred when legal commercial harvesting stopped, but wherever it took place, it did so in the absence of sound policy and enforcement. Social upheavals and military struggles also combined with illegal activities to decimate elephant numbers. This was the case in Uganda, Ethiopia, and elsewhere. It is true that elephants were killed in an uncontrolled manner almost anywhere they occurred and some still are being taken illegally. What some preservationists did not understand was that elephants existed in many areas that were beyond the means of their range to support them. They often concentrated in protected areas and increased, destroyed their range, and sometimes died. This happened in Tsavo, where conservationist pressure had effectively prevented culling.

"Guesstimates" on elephant numbers were apparently low for several years. Figures given in 1968 for the Luangwa (Dodds and Patton 1968) for instance were only 20 percent of those presented by Naylor and et al. (1973) five years later after more intensive aerial surveys and statistical testing. Other estimates suggested populations of elephants either higher or lower than actually existed. Douglas-Hamilton (1979) provided a conservative estimate of 1.3 million African elephants, with a number of low density, threatened populations, particularly in west Africa, but also a number of high density, healthy populations. It is now clear and generally accepted that the African elephant is not endangered. There are currently more than 600,000 animals and some countries have witnessed recent population increases. Preservationist pressures from outside Africa prevailed in 1989 when CITES listed the African elephant on Appendix I with any nation requesting a return to an Appendix II listing required to ensure that certain management and trade control criteria would be met. In 1997, CITES permitted trade in elephant ivory with Japan by Botswana, Namibia, and Zimbabwe. This prompted concern by other national bodies and environmentalists that renewed legal trade would jeopardize elephants outside the three African countries involved such as those in India.

Wherever elephants and men exist near one another, man suffers. Darling (1960) wrote of the conflict in eastern Zambia. Gardens of villagers are still being destroyed today. Elephant control has been a part of game department work in Africa since the departments were first established under early colonial regimes. Control exists now as it has in the past, as elephants continue to trample and eat crops, sometimes destroying a season's food in a single night. Wild elephants and people do not usually mix well.

Elephants are valuable. Tusks, tails, ears, and feet are commercial entities and the meat is both nutritious and tasty. In 1989, ivory brought up to US$350/kg (US$160/lb) at auctions in Zimbabwe and a big game hunter may have spent $60,000 for a safari. Today consumptive and nonconsumptive values under a managed system combining controlled harvests and game viewing would amount to hundreds of millions of dollars annually. Programs such as Zimbabwe's Campfire model, which both protect and harvest animals, allow people to benefit both ways. They also hold out hope for at least a portion of some 250 million Africans living on the equivalent of US$1.00/day or less.

In many countries, less than half the elephants are located within park boundaries and some parks are burdens on national economies. If park lands remained generally inviolate and controlled culling was allowed, elephants probably could continue to exist in low numbers even if animals outside the protected areas were lost. But land, for elephants and other wildlife may not continue to be an affordable luxury as human populations increase and more land is required to produce food.

Elephants can be managed. When productivity is known, range-carrying capacities understood, and populations and movements are also known, "excess" can be harvested from the herd(s) at a profit. This allows the range to be kept in reasonable condition, as in Kruger National Park and in Zimbabwe. Within park or reserve boundaries, protection usually is a prime component of management. The cost of protection may well be prohibitive in the future and recently was estimated at U.S$500/square mile/year. In 1990, Leader-Williams estimated costs at only U.S$180/square mile/year. Regardless, such amounts are excessive for any African country today and the cost is obviously increasing.

Lands for elephants will still decline, social (human) uprisings will occur again where elephants exist, and human populations will continue to increase. Management must relate closely to the spatial requirements of people first and elephants second. Where elephants exist, they should be husbanded and used for the benefit of the people. Elephant management has to be a primary responsibility of African governments. All national and international (outside Africa) efforts, including conventions, legislation, and IUCN groups must serve to support and advise on sound internal African management and not controvert or emasculate its efforts. At the very least, the work of all nations concerned must be harmonious.

Bibliography

Ables, E. D., S. Shen, and Q-Z. Xiao. 1982. Wildlife education in China. Wildl. Soc. Bull. 10(3):282–285.

Boshe, J. I. 1990a. Linking wildlife research with conservation and management programmes in Africa. Vol. 2. Trans XIX Intl. Union Game Biol. Congr. NINA, Trondheim, Norway, 621–626.

Boshe, J. I. 1990b. Convenor's report. Wildlife management in the third world. Vol. 2. Trans. XIX Intl. Union Game Biol. Congr. NINA, Trondheim, Norway, 613–614.

Bubenik, A. B. 1976. Evolution of wildlife harvesting systems in Europe. Trans. Fed.-Prov. Wildl. Conf. 40:97–105.

Csanyi, S. 1997. Challenges of wildlife management in a transforming society: examples from Hungary. Wildl. Soc. Bull. 25:33–37.

Clutton-Brock, J. 1981. *Domesticated Animals from Early Times*. Univ. Texas Press, Austin, TX.

Darling, F. F. 1960. *Wildlife in an African Territory*. Oxford Univ. Press, London, U.K.

Dasmann, R. R. 1981. *Wildlife Biology*. 2nd ed. John Wiley and Sons, New York, NY.

Dodds, D. G. 1976. Evolution of wildlife harvesting systems in Africa. Trans. Fed.-Prov. Wildl. Conf. 40:100–113.

Dodds. D. G. and D. R. Patton. 1968. Wildlife and land-use survey of the Luangwa Valley. UNDP-FAO, No. TA2591, Rome, Italy.

Douglas-Hamilton, I. 1979. African elephant ivory trade study. Final Rep., IUCN-WWF-NYZS Elephant Survey and Conservation Program.

Ermala, A . 1989. Hunting Atlas of Finland. Folio 232–233.

Gottschalk, J. S. 1972. The German hunting system, West Germany, 1968. J. Wildl. Manage. 36(1):110–118.

Komarov, B. 1980. *The Destruction of Nature in the Soviet Union*. M.E. Sharpe Inc., White Plains, NY.

Leader-Williams, N. 1990. Allocation of resources for conserving African pachyderms. Vol. 2. Trans. XIX Intl. Union Game Biol. Congr., NINA, Trondheim, Norway. 633–639.

Manning, I. 1995. *With a Gun in a Good Country*. Trophy Room Books, Agoura, CA.

Myrberget, S. 1971. Game management in Norway. Norwegian Game Research Institute. 2. Series, Nond. 35.

Myrberget, S. 1990. *Wildlife Management in Europe Outside the USSR*. NINA, Utredning. Trondheim, Sweden.

Nagy, J. and L. Beneze. 1973. Game and hunting in Hungary. Wildl. Soc. Bull. 1(3):121–127.

Naylor, J. N., G. J. Caughley et al. 1973. UNDP/FAO. FO:DP/ZAM/68/510. Working Document No. 1.

Newman, J. 1979. Hunting and hunter education in Czechoslovakia. Wildl. Soc. Bull. 7(3):155–161.

Parker, I. S. C. 1979. The ivory trade. U.S. Fish and Wildl. Serv. Rep. (Unpubl.).

Parker, I. S. C. and E. B. Martin. 1982. How many elephants are killed for the ivory trade? Oryx 18:235–239.

Petrides, G. A. and W. G. Swank. 1958. Management of the big game resource in Uganda, East Africa. Trans. N. Am. Wildl Conf. 23:461–477.

Ricciuti, E. R. 1980. The ivory wars. Anim. Kingdom 83(1):6–58.

Riney, T. 1964. The economic use of wildlife in terms of its productivity and its development as an agricultural activity. FAO African Regional meeting on animal production and health, Addis Abada. UN-FAO, Rome, Italy.

Salo, L. 1976. History of wildlife management in Finland. Wildl. Soc. Bull. 4(4):167–174.

Sinclair, A. R. E. and M. Norton-Griffiths. 1980. *Serengeti: Dynamics of an Ecosystem*. The Univ. of Chicago Press, Chicago, IL.

Swank, W. G. 1972. Wildlife management in Masailand, East Africa. Trans. N. Am. Wildl. Nat. Resour. Conf. 37:278–287.

Teer, J. G. and W. G. Swank. 1977. Status of the leopard in Africa south of the Sahara. The Office of Endangered Species. U.S. Fish and Wildlife Service, Washington, D.C.

Thresher, P. and W. J. Lusigi. 1994. Rural income from wildlife. A practical African model. UN paper (unpubl. summary).

United Nations Food and Agriculture Organization. 1974. Report of a UNDP/FAO Mission. Project Kenya. 71/526., Rome, Italy.

Recommended Readings

Students can keep in touch, internationally, by consulting publications representing different approaches including those of the Wildlife Conservation Society (Wildlife Conservation), UNEP (Our Planet), FAO (Unasylva), The African Wildlife Foundation (African Wildlife News), World Wildlife Fund (FOCUS) or by visiting websites of these organizations. Information on endangered species is available from CITES national offices and the IUCN Species Survival Commission.

Owen-Smith, R. N. (ed.). 1983. *Management of Large Mammals in African Conservation Areas*. Haum. Pretoria. S. Africa.

Riney, T. 1982. *Study and Management of Large Mammals*. John Wiley and Sons, New York, NY.

Epilogue

As we were putting the finishing touches to this third edition, Don asked whether we were being too negative. Would we turn our readers off? After thinking about it, the conclusion was that it would be misleading to be overly optimistic about the future of wildlife management. The planet is under seige, truly wild places no longer exist, and wildlife species are disappearing. As human populations grow, opportunities for, and understanding of, traditional uses of wildlife like hunting are declining and outright opposition to the taking of wildlife is on the ascent. So-called nonconsumptive uses such as ecotourism and other recreational pursuits that cause their own disruptions of wildlife biology and ecosystems are growing. Consider that the U.S. Forest Service could derive more income from recreational uses of the National Forests than by exploiting the trees and you have a good idea where we are heading. Consider too that recreational pursuits in North America or the developing world bring with them water and air quality degradation, as well as alteration or destruction of vegetation. Also recognize that human development activities are eliminating the habitat for many species, in particular the megafauna. For example, is there really sufficient room left in the lower forty-eight states for long-term viable populations of grizzly bears and wolves? While I have been fortunate to work and live in places where wolves still howl and grizzly bears still forage, how many others will have those opportunities in the future? The reintroduction of wolves to Yellowstone Park and the accounts of wolves being run down by snowmobilers elsewhere show that many still harbor strong resentments to the species.

We live in a world where the dangers we face increasingly derive from fellow human beings. Yet despite this the deaths of a few recreationists at the paws or jaws of grizzly bears cause more press and demand more retribution than the human slaughter at the hands of humans by murder or vehicular homicide. The demands of livestock ranchers, many of whom graze their animals on public lands, to rid the range of "vermin" like wolves and the ranchers' preferential treatment as evidenced by low grazing fees, are no different now than when they were responsible for the elimination of wolves to begin with. Vested interests conspire to maintain an unsustainable situation.

Youth are turning away from hunting in large numbers and despite the efforts of management agencies this is a trend likely to continue. The decline is not just in the percentage of the overall population participating but in some jurisdictions is in absolute numbers as well. Partially because of this the funding for many state and provincial wildlife agencies has neither kept pace with inflation nor competed well with the support for other government agencies. Increasingly the roles of wildlife managers has become marginalized. As scientists and managers involved with environmental issues, it is important that we be activists for the sustainable use of ecosystems even when the knowledge base is limited. We need to fight for the basic principles of ecosystem conservation with whatever ammunition is available until the necessary research has been done and more definitive answers are in hand. But too many wildlifers are content to pursue the prosaic practice of species management. What would have happened to the remaining old-growth forests of the Pacific Northwest were it not for the Endangered Species Act, the spotted owl, and some environmental activists? Note that the U.S. Fish and Wildlife Service did not take professional action until literally forced to do so. All around us warning signals are flying. Coral reefs are dying, fisheries are collapsing, increased UV radiation due to ozone depletion in the atmosphere is affecting amphibian populations, El Niños are becoming more powerful, more frequent, and more sustained, global temperatures continue to rise, aquifers are contaminated and being depleted, soil erosion continues, plague and famine are more frequent visitors worldwide, and still many discount the warnings.

So how can we remain optimistic that technology and common sense will save the day when the megalopoli continue to expand and suck the energy from the planet like so many vortices of consumption? How too can we expect the wildlife management profession to adjust to these realities? Fred Wagner (1989) flatly stated that it was unlikely that the profession would in fact make the adjustment although he clearly pointed out that what needed to be done from the revamping of wildlife degree programs to changing the nature of the research that academics pursue to focusing the activities of the management agencies. In all likelihood by clinging to traditional values too long and in too many venues the profession risks becoming irrelevant.

Part of our message has been similar to Wagner's. If the politics and people issues are not understood and wildlife professionals do not become involved in the true arenas of decision making, wildlife management will be but an anachronism in a vastly changed world. Conservation biologists seem to understand the urgency of the situation far better.

Audubon Magazine published an article "Farewell to Africa" in 1990. The farewell was to the last bastion of large, complex, wildlife populations and so it was a farewell to wild Africa which would become, like North America, devoid of the great free-roaming herds of herbivores and the carnivores and scavengers they supported. African wildlife would go the way of the eastern elk herds, the plains bison, the plains grizzly, and the wolf. What would be left would be vestiges of those once great assemblages, captured in reserves and viewed to death, not unlike a San Diego Wild Animal Park on a slightly grander scale.

Heaven forbid that under these circumstances we should be pessimistic. After all the planet has healed itself in the past; the experiment with the big brained mammal may be a passing evolutionary phase; there is the chance that we may come to our collective senses in time. Be that as it may, the two of us, in all probability, will live out the rest of our lives hoping we are wrong and clinging to that residual wilderness that has sustained us personally over our short human life spans.

Don and I wish you and the planet well and thank you for reading what we have had to say over almost two decades in the various editions of this book. We thank Krieger Publishing Company for giving us the chance to do that. The effort to create this third edition has been difficult and both of us are becoming more observer than participant in the wildlife profession. There will not be a fourth edition. What we have done has been a labor borne of our commitment to that profession and the wildlife it serves.

Bibliography

Jones, R. F. 1990. Farewell to Africa. Audubon Mag. (Sept.): 50–104.

Meyer, S. M. 1995. The role of scientists in the "new politics". Chron. Higher Ed. 41:B1–3.

Wagner, F. H. 1989. American wildlife management at the crossroads. Wildl. Soc. Bull. 17:352–360.

Appendix:
Important Canadian and United States Federal Legislation Affecting Wildlife Resources and Management

United States (listed alphabetically)

1996 Farm Bill (FACT Act)
African Elephant Conservation Act
Agriculture Appropriation Act of 1907
Alaska National Interest Lands Conservation Act of 1980
Anadromous Fish Conservation Act of 1965
Antarctic Protection Act
Bald Eagle Protection Act
Classification and Multiple Use Act of 1964
Clean Air Acts of 1970, 1990
Clear Water Restoration Act of 1966
Coastal Zone Management Act
Endangered Species Act of 1973
Endangered Species Conservation Act of 1969
Endangered Species Preservation Act of 1966
Estuarine Areas Act
Federal Aid in Wildlife Restoration Act (Pittman Robertson Act)
Federal Environmental Pesticide Control Act
Federal Farm Act
Federal Insecticide, Fungicide and Rodenticide Act
Federal Land Policy and Management Act of 1976
Federal Water Pollution Control Act
Fish and Wildlife Act of 1956
Fish and Wildlife Conservation Act of 1980
Fish and Wildlife Coordination Act
Food Security Acts of 1985, 1990, 1996
Forest and Rangelands Renewable Resources Planning Act of 1974
Forest Reserve Act of 1981

Forest Reserve Transfer Act of 1905
Fur Seal Act
Global Climate Change Prevention Act
Insecticide Act of 1910
Knutson-Vandenburg Act
Lacey Act
Land and Water Conservation Fund Act
Marine Mammal Protection Act
Migratory Bird Act of 1913
Migratory Bird Conservation Act
Migratory Bird Hunting Stamp Act
Migratory Bird Treaty Act
Multiple Use-Sustained Yield Act
National Environmental Policy Act
National Forest Management Act of 1976
National Forests Organic Act (1897)
National Park Service Act
National Wildlife Refuge System Administration Act of 1966
National Wildlife Refuge System Improvement Act of 1997 (PL 105-57)
North American Wetlands Act (1989)
Organic Act (1897)
Partnerships for Wildlife Act
Refuge Recreation Act of 1962
Refuge Revenue Sharing Act (amended 1974)
Rhinoceros and Tiger Conservation Act of 1994
Sikes Act of 1960
Sikes Act Extension
Taylor Grazing Act
Toxic Substances Control Act of 1976
Water Bank Act of 1970
Wetlands Loan Act
Whaling Convention Act of 1949
Wild Bird Conservation Act of 1992
Wild Free-Roaming Horses and Burros Act
Wilderness Act

Canada (listed alphabetically)

Agriculture Rehabilitation and Development Act
Animal Contagious Disease Act
Canada Water Act

Canada Wildlife Act
Canada Wildlife Week Act
Canadian Environmental Protection Act
Customs Act
Environmental Contaminants Act
Export and Import Permits Act
Farming and Food Protection Act
Fisheries Act
Forest Development and Research Act
Game Export Act
Indian Act
Maritime Marshland Reclamation Act
Migratory Bird Convention Act
National Parks Act
Plant Quarantine Act
Prairie Farmer Rehabilitation Act
Species at Risk Act (pending)
*Wild Animal and Plant Protection and Regulation of International and Interprovincial Trade Act

*Will replace Export and Import Permits Act and Game Export Act.

Index

Abell, D. H., 90
Adams, L. W., 244
Administration and policy, 28–36
Advisory Board on Wildlife Protection, 20, 21
Africa, 270–274, 324–328, 335–337, 342
African Wildlife Foundation, 324
Agency for International Development (AID), 322
Akeley, Carl, 19
Allen, Arthur A., 16, 18
American Game Policy, 15, 16
American Zoo and Aquarium Association (AZA), 275
Andelt, W. F., 256
Anderson, R. C., 127
Anderson, W. L., 89
Anthrax, 120–122
Arboviruses, 116
Atlantic flyway, 210–213
Atlantic Waterfowl Council, 212
Auerbach, S. I., 295

Barske, P., 296
Basket, T. S., 171
Bean, M. J., 77
Beanlands, G., 306
Beaver management, 206–210
Behavior, 91–98
Bennet, Rudolph, 18
Bergerud, A. T., 199
Bible, record, 9

Biological Resource Division, 18
Bishop, J. S., 229
Black duck management, 210–214
Black-footed Ferret Advisory Team, 221, 222
Black-footed ferret management, 220–224
Blake, J. E., 122
Bluetongue, 115–116
Boertije, R. D., 199
Boldt decision, 76
Boone and Crockett Club, 18
Boshe, J., 327
Boyce, M. S., 99
British North America Act (BNA), 22
Brucellosis, 120
Brunetti, O., 123
Bubenik, A. B., 11, 93, 328
Buechner, H. K., 129
Bump, G., 215, 233
Bureau of Biological Survey, 16, 18
Bureau of Indian Affairs, 23
Bureau of Land Management (BLM), 23, 178, 221
Byrne, R., 48

Cain, S. A., 247
Canada Committee on Ecological Land Classification, 172
Canada Wildlife Act, 24
Canadian Arctic Resources Committee, 37

349

350 • Index

Canadian Environmental Assessment Agency (CEAA), 289
Canadian International Development Agency (CIDA), 322, 323
Canadian Nature Federation, 37
Canadian Parks and Wilderness Society, 37
Canadian Society of Environmental Biologists, 37, 184
Canadian Society of Fish and Wildlife Biologists, 37
Canadian Society of Zoologists, 21, 37
Canadian University Service Overseas (CUSO), 321
Canadian Wildlife Federation, 37, 186
Canadian Wildlife Service, 21, 23, 36, 172, 197, 218
Canine distemper, 114–115
Canine parvovirus, 117–118
Caribou management, 197–203
Caring for the Earth, 283
Carson, Rachel, 284
Caughley, G., 84
Chalmers, A. W., 112, 115
Chase, A., 228
Choi, D. Y., 118
Choquette, L. P. E., 120, 121
Chronic wasting disease, 123
Chrysler, W. P., 229
Clutton-Brock, J., 50
Commission of Conservation, 20, 21
Committee on Seals and Sealing, 204, 205
Committee on the Status of Endangered Wildlife in Canada (COSEWIC), 263, 267
Computer models, 83–86
Conomy, J. T., 92
Conservation Foundation, 37
Convention on International Trade in Endangered Species of Flora and Fauna (CITES), 253, 264–267, 272, 326
Convention on the Conservation of Migratory Species of Wild Animals, 241

Cook, J. G., 97
Cornell School of Game Farming, 19
Cornell University, 277
Cowan, I. McT., 92
Csanyi, S., 331

Dahlgren, R. B., 98
Darling, J. N. "Ding," 16, 17, 19
Dasmann, R. F., 319
DeAngelis, D. L., 83
Demarais, S., 134
Depredations, 246–250
Disease, 111–135
Dodds, D. G., 271, 336
Domestication, 274
Douglas-Hamilton, I., 336
Doxiadis, C., 242
Ducks Unlimited, 36, 186
Dwyer, Judge W., 178, 179

Easter-Pilcher, A., 209
Ebedes, H., 121, 122
Ectoparasites, 123–124
Elaeophorosis, 125–126
Elephant, African, 334–337
Endangered species, 257–282
   captives, species survival plans, 274–277
   Convention on International Trade in Endangered Species of Fauna and Flora, 264–267
   preservation, 268–270
   reintroductions, 277–279
   strategies, 267–268
   utilization and the market, 270–274
Endangered Species (Preservation) Act (U.S.), 24, 174, 222, 263
Endangered Species Bulletin, 263
Endangered Species Program, 276
Endangered Species Update, 263
Environmental assessments and wildlife professionals, 308–311
Environmental Impact Assessment (EIA), 283–317
   approaches, 290–297

Canada, 287–290
concerns for wildlife biologists, 308–311
Environmental Impact Statement, 302–303
future, 311–312
international, 312–314
problems, 304–305
process, 297–299
proposal, 299–302
United States, 284–286
viewpoints, 306–308
Environmental Impact Assessment perspectives, 306–308
Environmental Impact Assessment status and future, 311–314
Environmental Impact Statement, 302–303
Environmental Protection Agency, 173
Euler, D., 244
European Economic Council, 205
European Union (EU), 254
Exotic species, 232–237
Export, Import and Interprovincial Transport of Wildlife Act, 24
Extinction, 257–259

Fay, L. D., 115
Federal Environmental Assessment Review Office (FEARO), 288–289
Federal-Provincial Committee for Humane Trapping (FPCHT), 54, 252–253
Federal-Provincial Wildlife Conference, 21
Fenstermacher, R., 127
Fire, 199–200
Fischer, C. A., 215
Fish and Wildlife Service (U.S.), 18, 22, 161, 177, 178, 188, 221, 222, 226, 249, 263, 276
Flueck, W. T., 88
Food and Agriculture Organization of the United Nations (FAO), 278, 319, 323
Foose, T. J., 276
Forest Service, 23, 172, 178, 180, 221

Foreyt, W. J., 118
Forrester, D. J., 132
Francis, C. M., 212
Fraser, D. E., 89
Fuller, T. K., 98
Furbearers, 206–210
Fur Institute of Canada (FIC), 253
Fur Seal Treaty, 16, 24

Gabrielson, Ira, 16, 18
Game management areas (GMAs), 269
Gaston, T., 231
Gates' equation, 94
Geist, V., 242
General Agreement on Tariffs and Trade (GATT), 254
Genetic swamping, 211
Geographic information system (GIS), 172
Gesell, G. G., 98
Gilbert, F. F., 37, 90, 97, 127–128, 181
Giles, R. H., 56, 227
Golet Wetland Classification, 172
Goyal, S. M., 118
Greenpeace, Inc., 263
Griffen, Donald, 47
Grimwood, I. R., 117
Grinnell, George Bird, 18, 226
Gullion, G., 215
Guo, W., 114

Habitat Conservation Areas (HCAs), 178
Habitat conservation plans, 173
Habitat evaluation, 165–173
Habitat loss, old growth coniferous forest, 176–183
Habitat management, 165–195
achievability, 183–186
declining habitat—old growth forest, 176–183
habitat evaluation and land-use planning, 165–173
incentive programs, 175
Lakeshore Capacity Study, 186–187
land-use planning controls, 173–174
purchase of property rights, 174–175

Habitat Suitability Index (HSI), 188
Hanley, T. A., 88
Harp seal management, 204–206
Harrison, D. J., 100
Heinz, G. H., 98
Heller, Edmund, 19
Hemorrhagic disease, 115
Hewitt, C. Gordon, 20–21
Hibler, C. P., 125
Hills, G. A., 290
Hirsch, A., 296
Historical record, 9–21, 153–155
Hornaday, W. T., 261
Houston, D. B., 228
Howell, F. C., 43
Hoxie, F. E., 70, 71
Humane Society of the United States, 263
Humane Trap Development Committee (HTDC), 252
Humane trapping, 251–256
Huot, J., 87
Hutchins, Peter, 73

International Association of Fish and Wildlife Agencies, 36
International Biological Program (IBP), 231
International Commission Northwest Atlantic Fisheries (ICNAF), 204
International Standards Organization, 253, 254
International Union for the Conservation of Nature (IUCN), 230, 283, 321, 324, 326

Jack H. Berryman Institute, 249
Johnsgard, P. A., 215
Johnson, D. R., 200
Johnson, H. N., 114

Karstad, L., 115, 116
Kearney, S. R., 128
Kellert, S. R., 51–53, 55
Kelsall, J. P., 128, 200

Kenya Rangeland Ecological Monitoring Unit (KREMU), 323, 326
Kertzen, M. N., 45
Kim, S. C., 70
King, C., 232
King, Ralph T., 16, 56, 60
Klein, D. R., 200
Kozicky, E. L., 237

Lacey Act, 24
Lakeshore Capacity Study, 186–187
Land and wildlife, 26–28
Land claims—native, 75
Land-use controls, 173–174
Land-use incentives, 175
Land-use planning, 165–173
Lankester, M. W., 127, 128
Laurier, Sir Wilfred, 20
Leader-Williams, N., 337
Legislation, 24–25
Leopold, Aldo, 9, 10, 15, 18, 19, 262
Leopold, A. S., 247
Leopold, Luna, 293
Lewis, Philip H., 290
Longcore, J. R., 211
Lucas, H. L., 296
Lungworm-pneumonia complex, 129–132
Lyme disease, 125

MacPhee, R., 264
Maikawa, E., 200
Maine Audubon Society, 212
Management systems, 153–163
Marchand, P. J., 95
Marine mammals, 204–206
Martin, Calvin, 41
Maximum sustained yield (MSY), 84, 204–205
May, R. M., 114, 133
McAtee, W. L., 16
McCorquodale, S. M., 78
McHarg, Ian, 293
McKeating, G. B., 244
McNicol, J. G., 97

Mellen, H., 275
Merriam, Clinton Hart, 19
Migratory animals, 239–242
Migratory Bird Conservation Act, 18, 226
Migratory Bird Convention Act, 20, 23
Migratory Bird Hunting Permit Act, 18, 24, 226
Migratory Bird Treaty, 16, 20, 21, 24
Migratory Bird Treaty Act (U.S.), 219
Miller, F. L., 201
Miner, Jack, 155, 229
Model Forests, 181–182
Moen, A. N., 86
Myrbeget, S., 333

Nagy, J., 331
National Audubon Society, 37, 263, 264
National Environmental Policy Act (NEPA), 181, 285
National Environmental Protection Act, 171
National Institute for Urban Wildlife, 243
National parks and sanctuaries (first), 227
National Parks Association, 37
National Park Service, 23
National Resource Conservation Service, 23
National Wildlife Federation, 16–17, 36, 221
Native Americans and wildlife, 67–78
 after European contact, 69–70
 before European contact, 67–69
 future, 77–78
 government policies in the U.S. and Canada, 70–72
 Native American–wildlife complex, 72–77
 North American population, 70, 77
Natural Resources Conservation Service, 175
Nature Conservancy, 37, 174
Nature Conservancy of Canada, 37
Naylor, J. N., 336
Newman, J., 331

New York State College of Forestry, 19
New York Zoological Society, 264
Nongovernment organizations (NGOs), 36–37
Nordkvist, M., 119
North American Game Breeders and Shooting Preserve Association, 239
North American Waterfowl Management Plan (NAWMP), 36
North American Wildlife Conference, 16
Northwest Game Act, 20, 21
Noyes, J. H., 244

O'Brien, S. J., 133
Onderka, D. K., 119
Our Common Future, 283, 314

Paleo-Indians, 67–68
Parasites and disease, 111–135
 anthrax, 120–122
 arboviruses, 116
 bluetongue, 115–116
 brucellosis, 120
 canine distemper, 114–115
 canine parvovirus, 117–118
 chronic wasting disease, 123
 ectoparasites, 123–125
 elaeophorosis, 125–126
 hemorrhagic disease, 115
 lungworm-pneumonia, 129–132
 Lyme disease, 125
 parelaphostrongylosis, 126–129
 pasteurellosis, 118–119
 pseudotuberculosis, 122
 rabies, 112–114
 rinderpest, 116–117
 strongyloidosis, 132
 tularemia, 119–120
 yersinosis, 122–123
Parker, G. R., 127, 128
Parliamentary system, 30, 35
Parrish, C. R., 118
Pasteurellosis, 118–119
Peregrine falcon management, 217–220
Peterson, M. J., 120

354 • Index

Pienaar, U. DeV., 121
Pinchot, Gifford, 18, 270
Pirnie, Miles, 16
Pittman-Robertson Act, 16
Poole, D. A., 35
Population Management Plan (PMP), 275
Populations, 98–99
Porter, W. P., 94
Post, G., 119
Potvin, F., 87
Preservation, 268–270
Project-Wild, 244
Protected areas, 225–231
Provost, A., 117
Puri, G. S., 242

Rabies, 112–114
Ralls, K. 83
Ramsar Convention, 241
Raskob, J. J., 229
Rasmussen, G., 95
Rau, M. E., 128
Recovery—An Endangered Species Newsletter, 263
Regulatory management, 158–163
Reintroduction, 277–279
Resource partitioning, 100
Revier system, 328–330
Richards, S. H., 115
Rinderpest, 116–117
Robel, R. J., 209
Roosevelt, Franklin, 16, 71
Roosevelt, Theodore, 15, 18, 226
Roosevelt conservation doctrine, 15
Roosevelt Wildlife Experiment Station, 19
Rosatte, R. C., 113
Royal Canadian Mounted Police, 21
Ruffed grouse management, 215–217

Sagebrush rebellion, 155
Saint-Exupéry, A. de, 49
Sandler, B. E., 98
Saunders, B. P., 128
Schindler, D. W., 305
Schoenwald-Cox, C. M., 230

Schultz, S. R., 134
Seed stock refuges, 229
Seip, D. R., 98
Serengeti Research Institute, 324
Severinghaus, C. W., 127
Sheldon, Major Charles, 18
Shooting preserves, 237–239
Sierra Club, 37
Sifton, Sir Clifford, 20
Skoog, R. O., 200
Slade, W. Jr., 11
Snow, D. R., 67
Society of American Foresters, 18, 181
Soil Conservation Service, 18
Specialized areas of management, 225–256
 depredations, 246–250
 exotic species, 232–236
 humane trapping, 251–256
 migratory animals, 239–242
 protected areas, 225–231
 shooting preserves, 237–239
 urban wildlife, 242–245
Species management, 197–224
 black-footed ferret, 220–224
 furbearers—beaver, 206–210
 marine mammals—harp seals, 204–206
 raptors—peregrine falcon, 217–220
 ungulates—caribou, 197–201
 upland game birds—ruffed grouse, 215–217
 waterfowl—black duck, 210–214
Special Survival Plans (SSPs), 274–277
Species Survival Commission, 262
Spotted owl, 177–180
Starfield, A. M., 85–86
Stein, C. D., 129
Steinhoff, H. W., 56
Stephenson, A. B., 98
Stewart, P. L., 89
Strongyloidosis, 132
Swift, Ernest, 20
Switzer, G. L., 90

Index • 355

Telfer, E. S., 128
Thomas, Jack Ward, 167, 178, 179
Thorne, E. T., 120
Threats to wildlife, 259–261
Tongass National Forest, 180
Trainer, D. O., 127
Trefethen, James B., 15, 18
Trigger, B. C., 69
Trinidad, 61–63
Tuchman, E. T., 180
Tularemia, 119–120

Ungulates—caribou, 197–201
United Nations Conference on Environment and Development, 283
United Nations Development Program (UNDP), 319–323
United Nations Educational Scientific and Cultural Organization (UNESCO), 270, 319
United Nations Environment Program (UNEP), 241, 283, 319
Upland game birds—ruffed grouse, 215–217
Urban wildlife, 242–245
Urban Wildlife Research Center, Inc., 243
U.S. Agency for International Development (AID), 322
U.S. Peace Corps, 321
Utilization and the marketplace, 270–274

Verner, J., 171
Voluntary Service Overseas (VSO), 321

Wagner, F., 342
Walters, C. J., 85, 99
Washburn, W. E., 71
Whisker, J. B., 41, 43, 49
White, D. H., 98
Wilderness Society, 37
Wildlife
  administration and policy, 28–36

behavior, 91–98
computer models, 83–86
culture and society, 41–63
endangered, 257–282
habitat management, 165–195
history, 9–21, 153–155
Indians, North American, 67–78
international, 319–339
jurisdiction, 21–24
land and wildlife, 26–28
legislation, 24, 25
management systems, 153–163
Native Americans, 67–78
nongovernment organizations, 36–37
nutrition and energetics, 86–91
parasites and disease, 111–135
populations, 98–100
resource partitioning, 100
Society, 16, 18, 184
specialized areas, 225–256
species management, 197–224
Wildlife Conservation Society, 324
Wildlife Habitat Canada, 36, 186
Wildlife Management Institute, 17, 36
Wildlife Preservation Trust, 221
Wildlife Refuge System (U.S.), 226–227
Wildt, D., 275
Williams, B., 262
Woolf, A., 127
World Conservation Strategy (WCS), 230, 268, 275, 283
World Wildlife Fund (WWF), 263, 264, 324
Worley, D. E., 125–126

Yersinosis, 122–123
Yuill, T. M., 111

Zarnke, R. L., 118
Zigman, P. E., 286